Pesticide Chemistry and Toxicology

Pesticide Chemistry and Toxicology

Contributors

Davor Zeljezic and Marin Mladinic et al.

www.aurisreference.com

Pesticide Chemistry and Toxicology

Contributors: Davor Zeljezic and Marin Mladinic et al.

Published by Auris Reference Limited
www.aurisreference.com

United Kingdom

Copyright 2016
Printed in 2017 for Sale in the Indian Subcontinent

The information in this book has been obtained from highly regarded resources. The copyrights for individual articles remain with the authors, as indicated. All chapters are distributed under the terms of the Creative Commons Attribution License, which permit unrestricted use, distribution, and reproduction in any medium, provided the original author and source are credited.

Notice

Contributors, whose names have been given on the book cover, are not associated with the Publisher. The editors and the Publisher have attempted to trace the copyright holders of all material reproduced in this publication and apologise to copyright holders if permission has not been obtained. If any copyright holder has not been acknowledged, please write to us so we may rectify.

Reasonable efforts have been made to publish reliable data. The views articulated in the chapters are those of the individual contributors, and not necessarily those of the editors or the Publisher. Editors and/or the Publisher are not responsible for the accuracy of the information in the published chapters or consequences from their use. The Publisher accepts no responsibility for any damage or grievance to individual(s) or property arising out of the use of any material(s), instruction(s), methods or thoughts in the book.

Pesticide Chemistry and Toxicology

ISBN: 978-1-78154-876-9

British Library Cataloguing in Publication Data
A CIP record for this book is available from the British Library

Printed in the United Kingdom

Exclusively distributed by CBS Publishers & Distributors Pvt. Ltd.

Sales & Distribution Rights only for India, Pakistan, Bangladesh, Sri Lanka, Nepal and Bhutan. This book is not to be sold outside these territories.

Contents

List of Abbreviations .... *vii*
List of Contributors .... *ix*
Preface .... *xiii*

Chapter 1   **Novel Approaches in Genetic Toxicology of Pesticides Applying Fluorescent in Situ Hybridization Technique** 1
Davor Zeljezic and Marin Mladinic

Chapter 2   **Toxicology of the Bioinsecticides Used in Agricultural Food Production** 35
Neiva Knaak, Diouneia Lisiane Berlitz, and Lidia Mariana Fiuza

Chapter 3   **Genotoxicity Induced by Ocupational Exposure to Pesticides** 57
Danieli Benedetti, Fernanda Rabaioli Da Silva, Kátia Kvitko, Simone Pereira Fernandes, and Juliana da Silva

Chapter 4   **Biosensors for Pesticide Detection: New Trends** 85
Audrey Sassolas, Beatriz Prieto-Simón, and Jean-Louis Marty

Chapter 5   **Quantitative Estimation of Pesticide-Likeness for Agrochemical Discovery** 133
Sorin Avram, Simona Funar-Timofei, Ana Borota, Sridhar Rao Chennamaneni, Anil Kumar Manchala, and Sorel Muresan

Chapter 6   **Evaluation of a Dry Extract System Involving NIR Spectroscopy (DESIR) for Rapid Assessment of Pesticide Contamination of Fruit Surfaces** 161
Umesh Kumar Acharya, Phul Prasad Subedi, and Kerry Brian Walsh

Chapter 7   **Introduction and Toxicology of Fungicides** 181
Dr. Rachid Rouabhi

Chapter 8   **Genetic Toxicological Profile of Carbofuran and Pirimicarb Carbamic Insecticides** 207
Sonia Soloneski and Marcelo L. Larramendy

Chapter 9   **Pesticides and Agricultural Work Environments in Argentina** 231
M. Butinof, R. Fernández, M.J. Lantieri, M.I. Stimolo, M. Blanco, A.L. Machado, G. Franchini, M. Gieco, M. Portilla, M. Eandi, A. Sastre, and M.P. Diaz

Chapter 10　Photosynthetic Response of Two Rice Field Cyanobacteria to Pesticides 265
Binata Nayak, Shantanu Bhattacharyya, and Jayanta K. Sahu

Citations 289

Index 291

List of Abbreviations

ANN	Artificial Neural Network
BER	Base Excision Repair
CA	Chromosomal Aberrations
CEI	Cumulative Exposure Index
CNS	Central Nervous System
EPA	Environmental Protection Agency
GTS	Green Tobacco Sickness
HAT	Hours After Treatment
HBA	H-Bond Acceptors
HEA	Homogeneous Ecological Area
IARC	International Agency for Research on Cancer
IPM	Integrated Pest Management
MIP	Molecular Imprinted Polymer
MLA	Mouse Lymphoma Assay
MW	Molecular Weight
PPE	Personal Protective Equipment
QCM	Quartz Crystal Microbalance
QEF	Quantitative Estimates of Fungicide
QEH	Quantitative Estimates of Herbicide
QEI	Quantitative Estimates of Insecticide
RB	Rotatable Bonds
ROC	Receiver Operating Curve
ROS	Reactive Oxygen Species
SCE	Sister Chromatid Exchange
SVM	Support Vector Machine

List of Contributors

Davor Zeljezic
Institute for Medical Research and Occupational Health Croatia

Marin Mladinic
Institute for Medical Research and Occupational Health Croatia

Neiva Knaak
University of Vale do Rio dos Sinos, Laboratory of Microbiology and Toxicology, São Leopoldo, RS, Brazil

Diouneia Lisiane Berlitz
University of Vale do Rio dos Sinos, Laboratory of Microbiology and Toxicology, São Leopoldo, RS, Brazil

Lidia Mariana Fiuza
University of Vale do Rio dos Sinos, Laboratory of Microbiology and Toxicology, São Leopoldo, RS, Brazil

Danieli Benedetti
Laboratório de Genética Toxicológica, Universidade Luterana do Brasil - ULBRA, Canoas, Brazil

Fernanda Rabaioli Da Silva
Departamento de Genética, Instituto de Biociências, Laboratório de Imunogenética. Universidade Federal do Rio Grande do Sul - UFRGS, Porto Alegre, Brazil

Kátia Kvitko
Departamento de Genética, Instituto de Biociências, Laboratório de Imunogenética. Universidade Federal do Rio Grande do Sul - UFRGS, Porto Alegre, Brazil

Simone Pereira Fernandes
Departamento de Genética, Instituto de Biociências, Laboratório de Imunogenética. Universidade Federal do Rio Grande do Sul - UFRGS, Porto Alegre, Brazil

Juliana da Silva
Laboratório de Genética Toxicológica, Universidade Luterana do Brasil - ULBRA, Canoas, Brazil

Audrey Sassolas
Laboratoire IMAGES EA 4218, Université de Perpignan via Domitia, Perpignan, France

Beatriz Prieto-Simón
Nanobioengineering Group, Institute for Bioengineering of Catalonia, Barcelone, Spain

Jean-Louis Marty
Laboratoire IMAGES EA 4218, Université de Perpignan via Domitia, Perpignan, France

Sorin Avram
Department of Computational Chemistry, Institute of Chemistry of Romanian Academy Timisoara, 24 Mihai Viteazul Avenue, 300223 Timisoara, Romania.

Simona Funar-Timofei
Department of Computational Chemistry, Institute of Chemistry of Romanian Academy Timisoara, 24 Mihai Viteazul Avenue, 300223 Timisoara, Romania.

Ana Borota
Department of Computational Chemistry, Institute of Chemistry of Romanian Academy Timisoara, 24 Mihai Viteazul Avenue, 300223 Timisoara, Romania.

Sridhar Rao Chennamaneni
GVK Biosciences Pvt. Ltd., S1, Phase-1, Technocrats Industrial Estate, Balanagar, Hyderabad 500 037, India.

Anil Kumar Manchala
GVK Biosciences Pvt. Ltd., S1, Phase-1, Technocrats Industrial Estate, Balanagar, Hyderabad 500 037, India.

Sorel Muresan
Food Control Department, Banat's University of Agricultural Sciences and Veterinary Medicine, Calea Aradului 119, 300645 Timisoara, Romania.

Umesh Kumar Acharya
Central Queensland University, Institute for Resource Industries and Sustainability, Plant Science Group, Rockhampton, Australia

Phul Prasad Subedi
Central Queensland University, Institute for Resource Industries and Sustainability, Plant Science Group, Rockhampton, Australia

Kerry Brian Walsh
Central Queensland University, Institute for Resource Industries and Sustainability, Plant Science Group, Rockhampton, Australia

Dr. Rachid Rouabhi
Doctor of Toxicology and Ecotoxicology Larbi Tebessi University, Biology department, Tebessa Algeria

Sonia Soloneski
Faculty of Natural Sciences and Museum, National University of La Plata, Argentina

Marcelo L. Larramendy
Faculty of Natural Sciences and Museum, National University of La Plata, Argentina

M. Butinof
Faculty of Medical Sciences, National University of Córdoba, Córdoba, Argentina

R. Fernández
Faculty of Medicine, Catholic University of Córdoba, Córdoba, Argentina

M.J. Lantieri
Faculty of Medical Sciences, National University of Córdoba, Córdoba, Argentina

M.I. Stimolo
Faculty of Economics, National University of Córdoba, Córdoba, Argentina

M. Blanco
Faculty of Agricultural Sciences, National University of Córdoba, Córdoba, Argentina

A.L. Machado
Faculty of Psychology, National University of Córdoba, Córdoba, Argentina

G. Franchini
Faculty of Medical Sciences, National University of Córdoba, Córdoba, Argentina

M. Gieco
Faculty of Medical Sciences, National University of Córdoba, Córdoba, Argentina

M. Portilla
Faculty of Medical Sciences, National University of Córdoba, Córdoba, Argentina

M. Eandi
Faculty of Medical Sciences, National University of Córdoba, Córdoba, Argentina

A. Sastre
Faculty of Psychology, National University of Córdoba, Córdoba, Argentina

M.P. Diaz
Faculty of Medical Sciences, National University of Córdoba, Córdoba, Argentina

Binata Nayak
School of Life Sciences, Sambalpur University Jyoti Vihar, Burla, Odisha, India

Shantanu Bhattacharyya
School of Life Sciences, Sambalpur University Jyoti Vihar, Burla, Odisha, India

Jayanta K. Sahu
Trust Fund College, Bargarh, Odisha, India

Preface

Pesticides are used to repel, kill or control certain forms of pests, e.g. animals or plants. These chemical compounds can be divided into three main classes: insecticides, which are used to control insects; herbicides, which are used to destroy unwanted vegetation; and fungicides, which are used to control fungi and their spores, preventing them from damaging plants. Pesticides are employed extensively around the world and in recent years their use has even increased. On one hand, extensive use of pesticides in farming has lead to a higher production of pests that damage crops and, on the other hand, pesticide-resistant pests have emerged. Pesticide Chemistry and Toxicology should serve as a reference work for agriculturists, environmentalists and industry professionals. First chapter focuses on novel approaches in genetic toxicology of pesticides applying fluorescent in situ hybridization technique. In second chapter, we address the toxicological aspects of microbial and botanical biopesticides that act by ingestion, with emphasis on histopathological analysis of tissues and cells in the alimentary channel of the Lepidoptera as well as the specificity of the *B. thuringiensis* bacteria, aqueous extracts and oil essential of medicinal and forest. Third chapter focuses on genotoxicity induced by occupational exposure to pesticides. Fourth chapter presents a state-of-the-art update in pesticide biosensors. The aim of fifth chapter is to develop quantitative estimates of herbicide- (QEH), insecticide- (QEI), fungicide- (QEF), and, finally, pesticide-likeness (QEP). In sixth chapter, the dry-extract system for (near) infrared (DESIR) technique was implemented using reflectance near-infrared spectroscopy in context of detection of contact pesticide residues on fruit. The fungicides and their toxicity on biological and ecological systems are presented in seventh chapter. Eighth chapter gives an emphasis on genetic toxicological profile of carbofuran and pirimicarb carbamic insecticides. Ninth chapter offers a comparative analysis of two widely different agricultural settings (extensive and horticultural crops) and characterizes the pesticide applicator populations in each, including the health conditions associated with occupational pesticide use. Tenth chapter presents experimental results to illustrate the effects of Monocrotophos and Endosulfan in time and concentration dependent manner on growth, pigments and photosynthesis of these two alga.

Chapter 1

NOVEL APPROACHES IN GENETIC TOXICOLOGY OF PESTICIDES APPLYING FLUORESCENT IN SITU HYBRIDIZATION TECHNIQUE

Davor Zeljezic and Marin Mladinic

Institute for Medical Research and Occupational Health Croatia

INTRODUCTION

The use of chemical substances in the pest control has been known since ancient times. Records have been found indicating that 2.500 years BC ancient Sumerians applied sulphur in mite control (Price, 1973). In ancient China, some 1.200 years BC inorganic mercury and arsenic compounds have been used in lice and bug treatment (Smith & Kennedy, 2002). First records of primitive fungicide and rodenticide use reach back to the time of Roman Empire where copper compounds were used in plant protection against moulds, and hellebore (lat. Helleborus nigra) containing poisonous baits against mice and rats (Smith & Secoy, 1975). Pyrethrium, which has remained in use as insecticide till nowadays, and is derived from Chrysanthemum cinerariaefolium flowers was brought to Europe from Persia by Crusaders (Wandahwa et al., 1996). However, the real revolution in use of chemicals in pest management started in early 1940-ties with the discovery of insecticidal properties of DDT by Paul Müller and the beginning of its massive production. Within next decade pesticidal properties of various synthetic chemical compounds had been discovered (hexachlorocyclohexane, 2,4-D, dithiocarbamates, chloradane, organophosphorous compounds, etc.) and they have been introduced in agricultural practice and household pest control (Ware & Whitacre, 2004). In the year 2009, more than 1.500 different active ingredients in more than 2.500 formulations were present in the world market (BCPC, 2009) with annual use exceeding 1.000.000 tones on global scale.

It was in 1962, after the publication of three parts serial in The New Yorker magazine and book entitled „Silent Spring" by Rachel Carson, when the general public became aware of possible adverse effects of unsustainable use of pesticides. The book has been based on records of the adverse effects of DDT on birds reproductive system and deaths of adults resulting from pesticide exposure. Furthermore, Carson pointed out the potential of pesticides to circulate within the ecosystems, accumulate within the organisms and affect all links in the food chain including humans (Carson, 2002). It has been ten years later when Environmental Protection Agency (EPA) estimated that evidences of adverse effect of DDT to the environment are sufficient to ban its use in the USA. Soon its prohibition was extended to European countries.

CARCINOGENIC RISK OF LONG-TERM PESTICIDE EXPOSURE

Acute effects of pesticide poisoning are being recorded since their commercial use began (Green, 1949). However, more concern has been raised regarding the adverse effects that long-term exposure to pesticides might have on human organism, specifically genomic material. The damage may be accumulating over the years without any noticeable health effects, but silently mediating cancer development. First studies indicating connection between exposure to arsenic insecticides in regular application and increased incidence of melanoma and bronchial cancer among European grape growing farmers appeared in late 1960-ties (Jungmann, 1966). However, since carcinogenic potential of arsenic had been documented already at the beginning of 20th century, these epidemiological studies did not raise much concern. In 1968, Klayman published first article suggesting association between chemically synthesized active ingredients and risk of carcinoma. The author reported several cases of larynx carcinoma in "never-smoking, never- or rarely-alcohol drinking" men with the record of more than 10 years of exposure to insecticides malathion or lindane working in either greenhouse or landscape (Klayman, 1968). Later, a case report indicating connection between lindane exposure and leukemia has been published (Hoshizaki et al., 1969). First more complex descriptive epidemiological studies aiming to evaluate carcinogenic risk arising due to occupational exposure to pesticides started in middle 1970- ties. In a study comprising the railroad workers applying herbicides amitrol or phenoxy acids, elevated occurrence of tumor deaths from stomach and lung cancer was observed (Axelson et al., 1974). With advances in knowledge on carcinogenesis it became evident that many life-style factors may elevate the risk of developing neoplasia by inducing genome damage. The profound effect on human cancer risk of smoking habits, alcohol consumption, diet, diagnostic procedures involving ionizing radiation or ultrasound, some medications, exposure to dyes and solvents may additionally pronounce or even prevail over the effect of pesticide exposure. Thus, beside simple recording of the type of pesticide exposure, all possible confounding factors should be considered in statistical analysis of data obtained by the study. Furthermore, the study should include population of individuals without any record of pesticide exposure that matches examinees by age, residence region, and life-style factors (Becher, 2005). In descriptive epidemiological studies there are three major epidemiological measures of cancer risk that are calculated. 1) Risk ratio or relative risk (RR) is the ratio of percentages of individuals developing cancer among the examinees and controls. If RR is above 1 then exposure increases the risk. 2) Odds ratio (OR) representing the ratio of the odd of cancer for the examinees and control. 3) Confidence interval (CI) represents sampling error inherent thus, statistical uncertainty of extrapolating the risk observed at the level of study group to the true population. These epidemiological measures should be adjusted for potential confounders by using Mantel-Haenszel estimation, regression methods (Cox, Poisson, multiple), or fractional polynomials. To overcome the problem of reduced statistical power in studies with small sample sizes meta-analysis is performed. It combines the results of several studies that address a set of related research hypotheses by using a form of meta-regression models. Additionally, to deduce the exact contribution of pesticides to the observed incidence of malignant disease multivariate approach in comparison between exposed and control group should be applied taking into consideration all confounding factors (e.g. MANOVA with post hoc comparison). Multiple regression analysis will identify the effect of each of confounding factors on the observed effect and canonical correlations analysis could be helpful (Ahrens et al., 2005). However, most of the individuals comprised by epidemiological studies are subjected to multiple pesticide exposures making impossible to separate the contribution of each individual pesticide to observed adverse health effect. Nevertheless, based on results of epidemiological studies, higher risk from developing neoplasia has been suspected for long term exposure to EPTC, pendimethalin, aldicarb, alachlor, chlorpyrifos, cyanazine, carbofuran, glyphosate and others (Dich et al., 1997). Short overview of indicated associations between pesticide exposures and increased risk of developing neoplasia published within last 3 years is presented in Table 1.

Table 1. Most recent studies indicating association between pesticide exposure and cancer.

Exposure	Population	Cancer type/site	OR/RR (95% CI)	Reference
EPTC	Farmers	Colon	RR 2.09 (1.26–3.47)	van Bemmel et al., 2008
Trifluralin	Farmers	Colon	RR 1.76 (1.05–2.95)	Kang et al., 2008
Phenoxy herbicides	Plant workers	Myeloid leukemia	OR 6.99 (1.96–24.90)	van Maele-Fabry et al., 2008
EPTC	Farmers	Pancreas	OR 1.8 (1.0–3.3)	Andreotti et al., 2009
Pendimethalin			OR 1.7 (0.8–3.3)	
Metribuzin	Farmers	NHL	RR 2.42 (0.82-7.19)	DeLancey et al., 2009
Butylate	Farmers	Prostate	RR 2.09 (1.27–3.44)	Lynch et al., 2009
		NHL	RR 3.44 (1.29–9.21)	
Triazole fungicides	Farmers	HL	OR 8.4 (2.2–32.4)	Orsi et al., 2009
Urea herbicides			OR 10.8 (2.4–48.1)	
Organochlorines		Hairy cell leukemia	OR 4.9 (1.1–21.2)	
Phenoxy herbicides			OR 4.1 (1.1–15.5)	
Organophosphates	Residentials	Acute lymphoblastic leukemia	OR 2.5 (0.4–14.8)	Rull et al., 2009
Triazines			OR 4.1 (1.5–11.1)	
Permethrin	Farmers	Multiple myeloma	RR 5.72 (2.76–11.87)	Rusiecki et al., 2009
Terbufos	Farmers	Leukemia	RR 2.38 (1.35–4.21)	Bonner et al., 2010
		NHL	RR 1.94 (1.16–3.22)	
		Lung	RR 1.45 (0.95–2.22)	
Phenoxy herbicides	Plant workers	Urinary cancers	RR 4.20 (0.99–17.89)	Boers et al., 2010
		Genital cancers	RR 2.93 (0.61–14.15)	

However, there are certain shortcomings of epidemiological studies that may lead to a possible bias in estimation of exposure risk and which results in studies reporting inconsistent results due to exposure to a specific pesticide (Eastmond & Balakrishnan, 2001). Due to poorly defined exposure levels, combinatorial exposure to other potentially carcinogenic agents, small study groups, or maladjustment of data for possible confounders, regulatory agencies and organizations authorized for carcinogenic classification of chemicals consider most of the reported associations of pesticide exposure with risk of cancer inconclusive. Because of these, in addition to epidemiological studies they turn to the results of long- and short-term animal and in vitro assays. Consequently, International Agency for Research on Cancer (IARC) and Environmental Protection Agency (EPA) have recognized less than 10 active ingredients as proved human carcinogens. Yet, many substances have been assigned as likely to be carcinogenic to humans or with suggestive evidence for carcinogenic potential.

Relevance of Epidemiological Studies: Pro and Con

There is much controversy about the relevance of epidemiological studies in risk assessment of pesticide exposure. Since individuals occupationally or residentially exposed to pesticides are simultaneously affected by several substances it is not possible to determine contribution of the specific agrochemical to the observed health effect. As already mentioned this issue has been recognized by regulatory agencies. Their decisions mostly rely on surrogate short- and long-term testing performed under controlled laboratory conditions, and results of epidemiological studies may provide supportive information. There are potential endogenous confounding factors such as gender, age, genetic polymorphism, and exogenous such as smoking, alcohol intake, medications that may significantly influence observed adverse health effect and have to be considered in analysis and interpretation of the results (Anderson, 2000). Additional bias in estimation of risk may evolve due to poorly described exposure conditions, inconsistent level of exposure with time, lack of air quality measurements and data on personal protective equipment usage. Also, lot of skepticism has been raised about the value of short-term genotoxicity tests. As the results of acute testing had accumulated over the past years less and less correlation with results of chronic carcinogenicity tests on rodents has been observed (Casciano, 2000). Furthermore, though testing of active substances under strictly controlled conditions using different cell lines may provide valuable knowledge on adverse effect toward human genome, interpretation of obtained results is challenging. There is the likelihood of (a) false positive results mediated by rather high concentrations tested (which may also lead to increase in osmolality), pH value decline, or (b) false negative results mediated by neglecting the need for metabolic transformation or by testing highly pure active ingredients. The latest may also diminish the relevance of results obtained on animal models. Moreover, the results of carcinogenicity testing obtained on animals may not always univocally be extrapolated on humans since there are some distinctive differences in metabolic pathways between animals and humans. That may be crucial in activation of carcinogenic potential of substance of concern as it was reported for insecticide carbosulfan (Abass et al., 2009) and also raised controversy over saccharine risk assessment (Chappel, 1992). Major discrepancy between results obtained by using different short- and long-term experiments on both cell and animal models, and effects occurring in real-case exposures arises from the fact that conditions used in laboratory may differ significantly from those likely to be encountered in occupational and residential exposure to pesticides. By using active ingredients of high purity in experimental evaluations two important factors that occur in real exposure of humans to pesticide are completely

omitted:
   a. Possible effect of impurities contained within pesticide formulations that are byproducts of active substance synthesis,
   b. Possible effect of "inert" ingredients (solvents, potentiators, surfactants, emulsifiers, stabilizers) as standard components of pesticide formulations.

There are several examples of contaminants with adverse effect on human genome exceeding those of active ingredients in whose synthesis they are by-produced. Low levels of 2,3,7,8-tetrachlorodibenzo-para-dioxin (TCDD) known as a potent human carcinogen, are present in herbicide formulations containing chlorophenoxyacetic acids (Eastmond & Balakrishnan, 2001). Carbendazim may be found in sulphur pesticide formulations that are approved for application in organic agriculture (Balayiannis et al., 2009). It further contains 2,3 diaminophenazine (DAP) that is suspected to be responsible for carbendazim's carcinogenic effect in mice. Malaoxon impurities contained within malathion formulations are associated with its genotoxic effect (Blasiak et al., 1999) while pyrethroids contain contaminants that exhibit more sever toxic effects than active ingredient itself (Hadnagy et al., 1999). Inert ingredients are routinely added to pesticide formulation in order to facilitate the application and to secure rapid and efficient transport of the active ingredient to the target site within the pest organism. Organic solvents used as inert ingredients have been associated with elevated risk of non-Hodgkin's lymphoma (Blair & Zahm, 2008). It has been shown that toxic effects of herbicide formulations with atrazine, glyphosate, and fungicide vinclozoline may be increased by present inert ingredients (Cox & Surgan, 2006).

Consequently, as stated by Hill (2010), since detecting the effect of multiple chemical exposure epidemiological studies provide us with limited knowledge on the genotoxic mechanism of a single pesticide. However, their importance lies in the fact that they look directly at human risk in situ, and estimate impact of a specific exposure type on cytogenetic status of the population in question. Epidemiological studies of pesticide exposure and experimental evaluations mutually supplement each other, and both approaches represent inevitable segments of risk assessment mosaic of exposure to pesticide.

GENETIC TOXICOLOGY

Findings that cancer is triggered and promoted by occurrence and accumulation of genome damage induced by a series of physical and chemical agents from residential and occupational environment, genetic toxicology as a subfield of toxicology developed. Pioneer assays aiming to detect changes at the level of cell genome as the consequence of exposure to pesticides were conducted in 1970-ies. Though Kaszubiak (1968) was first to isolate Rhizobium mutants from the cultures treated with herbicides linuron, dinoseb, and dimethylurea, the first experiment intended to evaluate mutagenic effect was conducted in 1970 and it gave positive results for herbicide 3',4'-dichloropropionanilide (Prasad, 1970). The exponential growth of genotoxicity testing occurred in the middle of 1970-ties with the introduction of Ames test on mutant strains of bacterium Salmonella typhimurium and mouse lymphoma assay (MLA) on L5178Y cell line. Within a decade more than 100 genotoxicity tests for evaluation of genetic potential of chemicals emerged. Since most of them were flawed in providing results relevant for carcinogenic risk assessment they gradually disappeared (Casciano, 2000). The term "genetic toxicology" appeared already in 1975 in the title of the review paper of Legator & Zimmering (1975). This toxicological discipline aims to:

- Identify biomarkers of the exposure or effect at the level of the cell genome that would be affected by exposure to carcinogen in a dose-response manner and would highly correlate with risk of cancer development, and
- Interpret obtained results to deduce the mechanism of chemical-genome interaction, role of the observed effects in carcinogenesis induction, and their impact on human health.

Adverse effects that are subject of study of genetic toxicology arise at the exposure levels far beyond the concentrations that would induce observable toxic effects on cells, organs or organism, which is somehow in disagreement with other fields of toxicology. These effects could even hardly be classified as toxic since they do not have any short-term impact on human health, and in most cases they are efficiently repaired. Nevertheless, under conditions of long-term exposure their occurrence may prevail over the repair or they may be misrepaired. Accumulation of these genomic lesions may induce cell transformation, immortalization and neoplastic growth.

Nowadays less than 15 assays are commonly applied in biomonitoring and even less are officially accepted for regulatory genotoxicity testing. Basically, assays applied in genetic toxicology may be classified in three major groups:

- Assays detecting mutagenic potential (Ames test, MLA, Drosophilla wing-spot test, mouse dominant lethal assay)
- Assays detecting genotoxic potential (micronucleus assay, structural chromosomal aberration analysis, sister chromatid exchange assay)
- Assays detecting nonspecific primary DNA lesions (alkaline comet assay).

Over the last decade, chromosomal aberration analysis, micronucleus, and comet assay have emerged as most reliable in evaluation of genotoxic effect of human exposure to pesticides. Among them, chromosomal aberrations and micronuclei as biomarkers of the effect were also proved to be good predictors in cancer risk assessment (Fenech, 2007; Rossi et al., 2009).

Chromosomal Aberration Analysis

Alterations in chromosome structure as the consequence of exposure to external agents has been known for more than 50 years. Their occasional application in health surveillance programs of individuals occupationally exposed to potential carcinogens started in 1960-ies. Soon structural chromosomal alterations in peripheral blood lymphocytes have been accepted as surrogate effect that reflects events triggered in the precursor cells for carcinogenesis under the exposure conditions of the issue (Hagmar et al., 2000).

Formation of structural aberrations of chromosomes is a rather complex event, involving DNA replication process in S phase of the cell-cycle, and misrepair of induced DNA strand breaks in post-replication phases. Under physiological conditions the lymphocytes are mainly in "resting" G_0 phase. Most of the genomic lesions induced by chemical agents will be efficiently repaired, especially if they occurred in transcriptionally active regions of chromosomes. Nevertheless, unrepaired lesions will interfere with DNA replication process forming DNA strand breaks thus, chromatid breaks, but also chromosome breaks which may result from DNA breaks due to additional topoisomerase II impairment (Maynard et al., 2009). Post-replicate repair mechanisms, through specific error prone pathways may convert chromosome breaks that occurred within "rejoining distance" into more complex rearrangements in chromosome structure such as chromatid exchanges and dicentric chromosomes (Obe et al., 2002). Insect chemosterilants tepa and apholate were first pesticides proved to affect morphology of human chromosomes in 1968 (Chang & Klassen, 1968). They were followed by insecticides propane sultone, aldicarb, malathion, fungicides ziram, thiram, herbicides 2,4-D, symazine and others.

Due to their good correlation with the level of exposure to chemicals in dose-dependent manner, structural chromosomal aberrations have been accepted as valuable cytogenetic biomarker of effect in epidemiological studies and risk assessment. Additional support of their application in human biomonitoring lies in their predictive value in cancer epidemiology – elevated frequency of aberrations indicates a population in increased cancer risk (Rossner et al., 2005).

Micronucleus Assay

Within the last 5 years micronucleus assay has been recognized as most reliable and efficient cytogenetic test in detection of potential carcinogens. It took over the primacy of being the most relevant biomarker of effect from chromosomal aberrations (Cavallo et al., 2009). Micronuclei as manifestations of adverse effect of physical and chemical agents on cell genome have been known for more than 40 years (Matter & Schmid, 1971). They are small chromatin structures visible in cytoplasm of interphase cells, with maximum of 1/3 of nuclear diameter in size.

Micronuclei originate from:

- Chromosomal fragments formed as the result of induced DNA strand breaks (as discussed in 3.1) that lagged in anaphase for not possessing the centromere to be attached to mitotic spindle and pulled to one of the mitotic poles, or
- Whole chromosomes that lagged in the anaphase due to spindle or kinetochore protein damage and that remained unsegregated (Fenech, 2007).

Following mitosis, one of the newly formed cells will be deficient in the genetic information within the lagged chromosome/fragment, while the other micronuclei containing cell will be in surplus. Micronucleus assay owes its preference over chromosomal aberration analysis to the ability to detect two different mechanisms of genotoxicity:

- Clastogenic mechanism meaning direct interaction of genotoxic agent with DNA molecule. It mostly results in micronuclei harboring chromosomal fragments
- Aneugenic mechanism meaning interaction of genotoxins with mitotic spindle proteins not DNA molecule itself, leading to genomic instability by loss and malsegregation of chromosomes. It results in micronuclei harboring whole chromosomes (Muller et al., 2008).

Additional efficiency of micronucleus assay in detecting genotoxic chemical has been gained by implementation of scoring criteria proposed by HUman MicroNucleus (HUMN) project group (Fenech et al., 2003). Beside micronuclei (MN), other aberrant chromatin structures such as nuclear buds (NB) and nucleoplasmic bridges (NPB) are considered in evaluating genotoxic potential. NBs are chromatin structures observed as nuclei extensions. They are formed in the process of elimination of (a) genomic regions harboring the genes related to metabolism of or resistance to exogenous chemical substance that have been amplified due to chronic exposure, or (b) DNA-repair complexes. NPBs, as chromatin structures connecting newly formed nuclei in telophase, are mostly manifestations of dicentric chromosomes, telomeric fusion of chromosomes, or union of sister chromatids (Fenech, 2007). In 1974, metepa was one of the first pesticides reported to induce micronuclei formation (Richardson, 1974). Soon, potential for micronuclei induction was reported for herbicidal phenylalkylureas, fungicide thiram, insecticide malathion etc.

Hence, micronuclei formation has been accepted as valuable cytogenetic biomarker of exposure to chemicals and there are ever more scientific evidences of its possible applicativity in epidemiological studies estimating the cancer risk of exposure to chemicals (Fenech, 2007; El-Zein et al., 2008).

Fluorescent in situ Hybridization

In 1969 for the first time hybridization of small radioactively labeled RNA fragments has been successfully applied in microscopic localization of specific genes (Buongiorno-Nardelli & Amaldi, 1969). Some 20 years later, RNA molecules have been replaced with DNA probes, and radioactive labeling with antigen labeling and immunocytochemical detection of sequences. Finally, when in early 1990-ties a fluorochrome labeled DNA probe has been produced, fluorescent in situ hybridization (FISH) as a powerful cytogenetic technique with high potential of discovering new biomarkers of effect in epidemiological studies of exposure to carcinogens was introduced. Basically, FISH technique enables visualization of specific genes, chromosome regions (centromeres, telomeres) or whole chromosomes within the cell genome in both, interphase nuclei and metaphase chromosomes. Thus, it provides us with the information regarding

- The copy number of a specific gene or chromosome in detection of aneuploidy that may arise as the result of chromosome malsegregation or loss of broken fragments in exposure to genotoxic chemical,
- The chromosome regions, or a specific chromosome harbored by micronuclei,
- The occurrence of translocations as stable chromosomal rearrangements (Raap, 1998).

First attempts to categorize content of micronuclei occurred in 1990 (Becker et al., 1990). Several years later began its occasional application in human biomonitoring (TitenkoHolland et al., 1994), but it has not been before 2003 that micronuclei content has been indicated as a valuable parameter in characterization of the effect of exposure to chemicals on human genome (Norppa & Falck, 2003). The presence of whole chromosomes in micronuclei may be detected by anti-kinetochore antibodies that will target kinetochore proteins in centromeric region. However, it has been deduced that many micronucleated chromosomes may possess disrupted kinetochore or it may be detached thus, exhibiting no signal following immunocytochemical detection. Therefore detection of pan-centromeric regions within the micronuclei by hybridization of fluorescent DNA probe will render more relevant results in determining the micronuclei content. A year after the first application of pan-centromeric FISH to analyze the content of micronucleus in evaluation of the effect of exposure to pesticides was reported, and three years later according to the HUMN recommendations other aberrant chromatin structures in relation to pesticide exposure have been categorized regarding the centromere content.

Translocations are persistent alteration in chromosome morphology that significantly affect genome integrity. They are of similar ethiology as dicentric chromosomes, originating from double strand DNA breaks (DSB) that are misrepaired mostly by non-homologous end joining (NHEJ) and ectopic homologous repair. Translocations are considered as most valuable biomarker in cancer risk assessment (Obe et al., 2002). High translocation frequencies have been observed in all tumor cells, some type of them being highly associated with a specific type of cancer thus, being etiologic for the neoplasia in question (Mitelman et al., 1997). Due to their high correlation with risk of developing cancer, translocations are considered as a valuable cytogenetic biomarker of effect in evaluating human exposure to genotoxic agents. Their application in biomonitoring began in early 1990-ties, but it has been restricted to studies of exposure to ionizing radiation (Tucker et al., 1993). Several years later, Steenland et al. (1997) introduced application of translocations as biomarkers of effect in occupational exposure to pesticides by evaluating the effect of ethylenebis(dithiocarbamate) (EBDC) fungicides. However, until 2009 no further use of chromosome painting by FISH in evaluation of pesticide genotoxicity has been reported. The application of FISH technique in translocation analysis and in revealing the content of aberrant chromatin structures provides us with more precise knowledge regarding the extent of genome affected by pesticide-induced damage and its relevance to carcinogenicity.

FISH IN GENETIC TOXICOLOGY OF PESTICIDES

Background Epidemiological Studies

Concept of using structural chromosomal aberrations in peripheral blood lymphocytes in cytogenetic biomonitoring of subjects occupationally or residentially exposed to potential carcinogens has been based on the finding that level of genetic damage in lymphocytes reflects the effects occurring in precancer cells of target tissue. Until the late 1990-ties it has been generally accepted that occupational exposure to chemicals may influence chromosome structure mostly by inducing chromatid-type of aberrations such as gaps and chromatid breaks. It has been assumed that chromosome breaks may also occur but at much lower frequency than chromatid breaks, while more complex rearrangements as dicentric and ring chromosomes have been considered as cytogenetic biomarkers of exposure to ionizing radiation or a small group of radiomimetic chemicals (e.g. antineoplastic drug irinotecan) which interact with DNA in pathways that resembles the one of ionizing radiation (IAEA, 2001). Accordingly, epidemiological studies evaluating cytogenetic effects of occupational exposure to pesticides have reported increased level of alterations in chromosome morphology, but chromosome breaks being the most serious reported lesion. As discussed in section 2.1, all such studies published within last 2 decades comprised groups of examinees in multiple pesticide exposure. Significant effect on induction of chromatid and chromosome breaks in farmers has been published by Hoyos et al. (1996) due to exposure to mixture of dithiocarbamates, carbamates and organophosphates, by Garry et al. (1996) in applicators of broad spectrum of insecticides and herbicides, by Antonucci & de Syllos Cólus (2000) in applicators of organophosphates, carbamates, and some herbicides (Mann-Whitney U-test, P<0.05). Although Hoyos et al. (1996) presented detailed information regarding exposure conditions and adjusted their statistical analysis for confounders, none of that has been done by Garry et al. (1996) and Antonucci & de Syllos Cólus (2000). In the later study on chlorophenoxy herbicide applicators and exposed foresters Garry et al. (2001) showed significant increase in chromatid breaks but only in lymphocytes of applicators using more than 3,785 liters of herbicides per season (Wilcoxon rank-sum test, P=0.017). Again, no adjustment for confounders has been done. Conversely, Lander et al., (2000) analyzed results using a multiple log-linear Poisson regression model for smoking, age, and coffein intake as possible confounders. The authors monitored chromosomal aberrations in greenhouse workers exposed to residues of 10 different insecticides, 6 herbicides, and 3 growth regulators prior to and after the spraying season. Number of cells harboring structural chromosomal aberrations significantly increased after the spraying season (P=0.05). However, among all recorded structural alterations the effect has been significant only for chromatid gaps (P=0.001), and most prominent increase has been observed for non-glove using smokers (P=0.04). In spite of excellent study design, aberration nomenclature used by the authors does not follow the IAEA (2001) recommendations nor has been elaborated, which bias comparison of results with those reported in previously discussed studies.

Studies reporting complex rearrangements are summarized in Table 2. Contrary to previously cited papers, Kourakis et al. (1996) were first to report the presence of dicentric and even ring chromosomes in the absence of significant increase in number of chromatidtype aberrations, in group of both, outdoor and greenhouse spraying workers exposed to complex mixture of pesticides (organophosphoric and organochlorinic compounds, carbamates, dithiocarbamates). However, occurrence of complex alterations was not statistically significant. Since the workers did not use any personal protection equipment the complex cytogenetic effects may be attributed to high level of exposure due to direct pesticide intake by inhalation, through the skin and eye contact. Non-significant presence of complex structural alterations accompanied by significantly increased chromatid breaks was observed in vineyard growers mostly exposed to insecticide diazinon and fungicide dithiocarbamate (Joksic et al., 1997). Unfortunately, though results were presented at the level of subgroups regarding smoking habits and gender, in both studies statistics has been done without considering life-style factors as possible confounders. Although Amr (1999) did report significant increase of dicentrics (Student's t-test, $P<0.001$) among pesticide handling workers exposed mostly to mix of DDT, chlorinated hydrocarbons, organophosphates, pyrethroids and carbamates, due to major lack of data regarding the group characteristics, and inclusion criteria, reported results are considered inconclusive. For instance, of 300 included examinees that were subjected to biochemical analyses and medical diagnostic procedures, only 30 of them were chosen for chromosomal aberration analysis without any explanation of criteria for selecting them. Furthermore, most of the examinees had medical history of suffering from various disorders. Again, no multivariate analysis of obtained data has been applied. Other study reporting elevated frequency of unstable chromosome rearrangements comprised pesticide plant workers primarily exposed to atrazine and cyanazine, with minor exposure to alachlor, 2,4-D, and malathion (Zeljezic & Garaj Vrhovac, 2001). Production of pesticides was organized seasonally and workers have been exposed to pesticides during 8 months of their production. Until next production season they were transferred to the working places out of the exposure zone. For the first time it has been reported that following exposure period, beside chromatid and chromosome breaks, dicentric chromosomes may also be significantly increased (MANOVA, $P_{ScheffePostHocc}<0.01$). Further, significant occurrence of chromatid exchanges in the form of quadriradials has been observed (MANOVA, $P_{ScheffePostHoc}<0.01$).

Table 2. Studies reporting complex alterations in chromosome structures in pesticide exposed subjects. Results presented as average per 100 cells ± S.E., N/R data not reported by authors, *P<0.05, **P<0.01.

Exposure type	Chromosome rearrangements exposed vs. control			Reference
	Dicentric	Ring	Exchange	
Organophosphates, carbamates, dithiocarbamates, organochlorines	0.07±0.03 N/R	0.03±0.02 N/R	0.00±0.00 N/R	Kourakis et al., 1996
Ethofumesate, diazinon, vinclozolin, 2,4-D, dithiocarbamate, metalaxyl etc.	0.02±0.00 0.00±0.00	0.19±0.00* 0.02±0.00	N/R	Joksic et al., 1997
Mancozeb, methamidophos, captan, chlorpyrifos	0.08±0.30 0.05±0.09	0.02±0.05 0.03±0.07	N/R	Steenland et al., 1997
Atrazine, cyanazine, alachlor, 2,4-D, malathion	0.42±0.95** 0.00±0.00	0.00±0.00 0.00±0.00	0.10±0.40* 0.00±0.00	Zeljezic & Garaj Vrhovac, 2001
Captan, mancozeb, endosulfan, methiocarb, glyphosate, linuron etc.	Dicentric and ring chromosomes reported to be observed without providing quantitative data			Costa et al., 2006

At the end of the non-exposure period frequency of dicentric chromosomes significantly decreased, and no quadriradials were observed confirming the unstable nature of those chromosome-type aberrations (IAEA, 2001). Residual dicentrics were only detected in individuals with more than 18 years of employment in pesticide production. Conversely to previously cited studies, the authors applied multivariate analysis of variances considering smoking, gender, and age. None of them did significantly influence intergroup variations in aberration frequency. Furthermore, to avoid possible effect of x-ray diagnostics on induction of chromosome-type aberrations only workers without the record of being subjected to such procedures were allowed to participate in the study. Still, alcohol intake, nutrition, medications and other life-style factors have not been considered. Publishing of those results on pesticide genotoxicity coincided with the review article of Obe et al. (2002). The authors proposed the mechanism by which chemical genotoxins including pesticides may give rise to chromosome-type aberrations that had been previously considered as cytogenetic biomarkers of exposure to ionizing radiation. Accordingly, chemicals may induce DNA lesions such as single-strand breaks as the consequence of interaction with DNA that would result in hindered DNA replication, alkylations, bulky adducts formation, or oxidative DNA damage. During DNA replication or by error-prone DNA repair pathways, these lesions may be transduced into DSB. Further, improperly repaired DSB may give rise to complex alterations in chromosome structure such as dicentric chromosomes and exchanges. As stated by Obe et al. (2002) there are three major pathways of DSB repair: (a) ectopic homologous recombination repair (EHRR), (b) non-homologous end joining (NHEJ), and (c) single-strand annealing. Error-prone activity of two of them, NHEJ and EHRR are responsible for dicentric formation. Conversely to NHEJ that requires two initial DSB within "rejoining distance", for EHRR to form a dicentric a single DSB is needed. Furthermore, EHRR may involve homologous sequences of different chromosomes resulting in dicentric or translocation. However, it has to be stated that the repair of chemically induced DNA damage may be slower than that of ionizing radiation; thus, the probability of misrepair induced aberrations would be low (Preston, 2000) which could also be observed from the results summarized in Table 2.

More recent studies also reported appearance of dicentrics and quadriradials in lymphocytes of workers exposed to pesticides. Costa et al. (2006) detected complex structural alterations in lymphocytes of sprayers applying 33 different active ingredients without specifying the type. Ergene et al., (2007) detected dicentrics detected in residentially exposed subjects living in region contaminated mostly with organochlorines, organophosphates, carbamates, pyrethroids and benzoyl ureas. Both groups of authors did adjust statistical analysis for confounding factors. The incidence of chromosomal aberrations has been significantly affected (MANOVA, $P<0.01$) by pesticide exposure only in the study of Ergene et al., (2007) indicating that residential exposure to pesticides may also adversely affect genome integrity.

Nevertheless, findings of studies documenting the presence of dicentric chromosomes and other complex chromosome-type aberrations in lymphocytes showed necessity of using novel cytogenetic approaches that may provide more detailed knowledge regarding the potential of pesticides to affect genome integrity and induce complex genome rearrangements considered as a driving force for carcinogenicity.

FISH in Analysis of Pesticide Induced Aberrant Chromatin Structures

Generally, micronuclei represent the most known aberrant chromatin structure, and within the last 3 decades they have been used in evaluation of cytogenetic effect of exposure to pesticides (Bull et al., 2006). Etiologically, they may originate either from chromosomal fragments as result of unrepaired chromosome breaks or whole chromosomes that lagged in anaphase due to damage of mitotic spindle or kinetochore. Identifying each of these two types of micronuclei provides us more detailed knowledge regarding the mechanism of genotoxicity and enables distinguishing between clastogenic pesticides that directly interact with genome and aneugenic that give rise to genomic instability by damaging protein structures. The latest leads to malsegregation of chromosomes and may result in their loss and aneuploidy. Application of fluorochrome labeled DNA probes that would hybridize in pan-centromeric region of each of 46 chromosomes makes it possible do recognize micronuclei harboring whole chromosome. When analyzing lymphocyte preparations under epifluorescence microscope, such micronuclei will contain one or more fluorescent signals. Conversely, micronuclei harboring chromosomal fragments remain without any signal.

Bolognesi et al. (2004) were among first to hybridize all-chromosome pan-centromeric probes to micronucleus slides in elucidating origin of micronuclei induced by exposure to mixture of pesticides. Study comprised floriculturists mostly exposed to organophosphates, carbamates, organochlorines, benzimidazoles, thiophtalimides, pyrethroides, and bupirimate, exposure duration ranging form 2 to 70 years. Although no statistically significant difference has been observed between pesticide users and control subjects, MN frequency increased with amount of pesticides applied, number of formulations individuals were exposed to, involvement in pesticide preparation, years of exposure, and non-use of personal protective equipment. While in the control group ratio of centromere containing micronuclei (61.9%) did not significantly differ from the reference values (±60%), in floriculturists the ratio of whole chromosome harboring micronuclei (C+MN) prevailed over the fragment harboring ones (C-MN). Though not associated with pesticide exposure duration, ratio of C+MN positively correlated with non-use of gloves in pesticide preparation. As expected, proportion of C+MN increased in both groups with age, which could be attributed to micronucleation of X and Y chromosomes associated with ageing. The effect of lagging sex chromosomes is more pronounced in females (Norppa & Falck, 2003). The most interesting conclusion made by Bolognesi et al. (2004) is higher ratio of C+MN in benzimidazolic fungicide appliers (66.52±16.11) compared to floriculturists applying other pesticide classes (63.78±14.02), which does not surprise knowing that benzimidazoles are proved as spindle microtubule poisoning agents. Single limitation of the study arises from the lack of multivariate statistical analysis of obtained data that would adjust them for age, wide range of exposure duration, and other possible confounders.

Though it renders deeper insight in mechanism of genotoxicity, revealing micronuclei content does not provide any information regarding the ability of pesticides to affect genome integrity by inducing complex rather unstable chromosome rearrangements as it has been indicated by application of chromosomal aberration analysis in section 4.1. To obtain that kind of knowledge, HUMN recommendations for considering other aberrant chromatin formations, especially nucleoplasmic bridges, had to be implemented in biomonitoring study coupled with all-chromosome pan-centromeric FISH analysis (Fenech et al., 2003).

Such an approach has been applied in evaluation of cytogenetic effect of occupational pesticide exposure in carbofuran production workers (Zeljezic et al., 2007). Though the use of carbofuran had been banned in the EU countries, in 2008 the application for its inclusion in Annex I of Council Directive 91/414/EEC concerning the placing of plant protection products on the market thus, reallowance of its use, has been resubmitted. In 2003 in some EU countries carbofuran was still among top 5 active substances applied to vegetable crops. Epidemiological studies reported that individuals exposed to carbofuran may have increased risk for lung cancer (Usmani et al., 2004) and non-Hodgkins lymphoma (Zheng et al., 2001). In study of Zeljezic et al. (2007) lymphocytes of carbofuran production plant workers micronuclei, nuclear buds and nucleoplasmic bridges were analyzed for centromere content. Acethylcholinesteraze (AChE) activity in both, whole blood and plasma also has been recorded, as the biomarker of acute exposure to carbamate and organophosphorus insecticides. As shown in the Fig. 1. considering smoking, alcohol intake, gender, difference in age, medical procedures and exposure duration as possible covariates the incidences of MN, NBs, and NPBs have been significantly increased ($P_{DuncanPostHoc}$=0.0040; 0.0089; 0.046, respectively) in carbofuran exposed examinees. Although the ratio of centromere harboring MNs (69.6±12.5%) did not exceed referent values suggested by Norppa & Falck (2003), it has been increased compared to the unexposed group (54.3±10.3%). Statistical significance in the difference ($P_{DuncanPostHoc}$=0.0084) suggests that carbofuran, beside interacting with DNA to induce breaks in chromosome structure may also exhibit aneugenic effect by damaging mitotic spindle, and consequently resulting in chromosome loss and aneuploidy. Evident increase in the presence of heteromorphic sites of chromosomes 1, 9, 15, 16, and Y in C+MN of carbofuran handling workers ($P_{DuncanPostHoc}$=0.0036), suggests that micronucleation of whole chromosomes is not restricted to sex chromosomes as it may be expected in elderly subjects. Rather, it may indicate higher susceptibility of genomic regions rich in A-T base pairs toward aneugenic effect of carbofuran and/or its metabolites. Additional support for both clastogenic and aneugenic activity of carbofuran could be found in literature data. Stehrer-Schmid & Wolf (1995) reported adverse effect of the insecticide on formation of microtubuli and impairment of their function during chromosome segregation. Clastogenic effect of carbofuran observed as induction of C-MN may find its support in the paper of Zhang et al. (2005) who deduced carbofuran's ability to intercalate between base pairs and produce DNA-carbofuran adducts. Contrary to general perception that ACh-sensitive receptors are specific for cells of nervous tissue, they are also present on the membrane surface of most non-neuronal cells, including the periferal blood lymphocytes. Their specific physiological role has mostly remained unclear though they are confirmed to be involved in immune functions, control of gene expression, cytoskeletal organization, secretion, absorption etc. Consequently, as discussed by Rull et al. (2009) changes in AChE activity due to chronic exposure to anti-ChE compounds may induce amplification of AChE gene, and may affect cell proliferation and differentiation (Vidal, 2005). Both processes are closely associated with carcinogenesis, which may also provide a support for the results of previously cited epidemiological studies suggesting a correlation between carbamate exposure and increased risk of lymphoma. Possible amplification of AChE and butyrylcholinesterase (BuChE) genes under the chronic burden of exposure to carbamates may be the reason of elevated occurrence of NBs in lymphocytes of pesticide plant workers ($P_{DunvanPostHoc}=0.0089$). Amplification of AChE gene under chronic exposure conditions may be the reason of increased NB formation. Observed NBs might be also formed as the result of the elimination of DNA-repair complexes that have been overrepresented due to increased level of DNA damage induced (a) directly by genotoxic activity of pesticide or (b) indirectly by endogenous reactive oxygen species formed in the cells as the result of carbofuran-produced oxidative stress (Calviello et al., 2006). Although we did observe sporadic centromere signals in NBs of exposed subjects (5.8±2.21%) C+NB/C-NB ratio did not differ from the control subjects ($P_{DuncanPostHoc}=0.72$).

Unlike for C+MN, centromere signal in NBs would indicate budding of epicentric interstitial fragments or joining of chromosome harboring MNs with nucleus. Less likely scenario would be that whole chromosome may be expelled from the nucleus forming the NBs by triggering aneusomy rescue mechanism, which would be possible for highly aneugenic pesticides. Lagged chromosomes that have been incorporated within daughter nucleus instead of forming MN may cause distortion in chromosome territories and be eliminated in the budding process (Lindberg et al., 2007). However, this theory has not been proved yet nor did carbofuran exposure exhibit such prominent aneugenic potential considering observed C+MN/C-MN ratio. Finally, ratio of centromere signals has been significantly elevated ($P_{DuncanPostHoc}=0.046$) in NPBs of pesticide plant workers (20.3±12.3%) compared to the control subjects where none of detected bridges harbored the pancentromeric region. Elevated occurance of NPBs as biomarkers of dicentric and ring chromosomes, additionally supports results discussed in section 4.1 indicating that long-term exposure to pesticides is capable of inducing complex alterations in chromosome morphology. Since detected NPBs were mostly not accompanied by the presence of micronuclei (98.2±1.5%) it may be suggested that they are formed by misrepair of DSB by NHEJ or by telomere-end fusion (TEF). The latter occurs as a consequence of telomere shortening which has been commonly associated with aging, but also may be prematurely mediated by chronic influence of chemicals affecting cell proliferation and is observed in precancer cells (Fenech, 2006). Both NHEJ and TEF are characterized by the lack of acentric formation that would result in micronucleus formation. Further, increased frequency of NPBs has been recognized as a biomarker of elevated risk of lung cancer (Multivariate logistic regression analysis OR=29.05, CI=7.48-112.80, P<0.001; El-Zein, et al. 2006), for which an association has been indicated with occupational carbofuran exposure (Usmani et al., 2004).

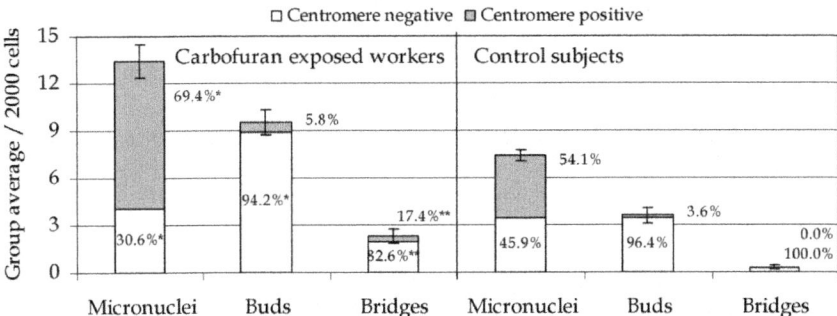

Figure 1. Aberrant chromatin structures regarding to presence of FISH centromere signals in carbofuran production line workers and controls. *P<0.05, **P<0.01 compared to the control.

AChE activity, as a biomarker of exposure to carbamates, will not reflect possible cumulative effect of pesticides in evaluation of their adverse impact on human health. Consequently in presented study (Zeljezic et al., 2007) we did not find any correlation between plasma and whole blood AChE activity and years spent handling carbofuran formulations as assessed by multiple regression analysis (R=0.023). Average whole blood AChE activity was 94.5±1.33% (81-100%), and plasma activity 99.2±0.55% (86- 100%).

Thus, to relevantly assess potential carcinogenic risk of long-term exposure to low-doses of carbamate and organophosphorous compounds, appropriate cytogenetic biomarker of exposure that will reflect cumulative effect should be determined. Interestingly, centromere containing NPBs (Fig. 2.) correlated with employment/exposure years (R=0.77, β=0.68), confirming previously reported finding that long-time occupational exposure to pesticides is capable of inducing complex chromosomal rearrangements. Additionally, multiple regression analysis considering age, gender, medical procedures and life-style factors as possible confounders revealed significant influence of carbofuran exposure duration on incidence of C+MN (R=0.83, β=0.76),

NBs (R=0.86, β=0.79) suggesting that aberrant chromatin structures may be applied as nonspecific biomarkers in evaluation of cytogenetic effect of long-term carbamate exposure.

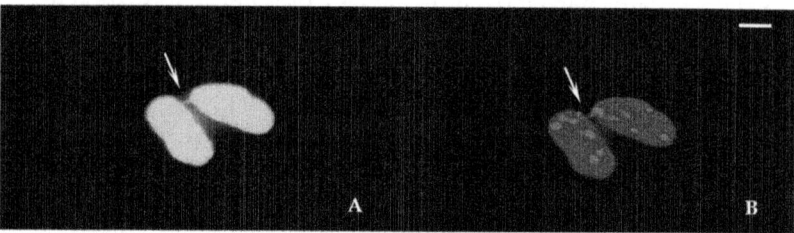

Figure 2. Binucleated lymphocyte of a carbofuran plant worker with nucleoplasmic bridge indicated by arrow: A) DAPI staining, B) centromeric DNA dyed red (Texas Red-conjugated DNA probe), and nuclei blue (DAPI). Objective 100x, bar 1 μm.

Dynamics by which discussed cytogenetic biomarkers are being affected by pesticide exposure has been reported based on a case of acute carbofuran intoxication (Zeljezic et al., 2009). A single male worker on Furadan production line who also participated in previously discussed cytogenetic monitoring, has been transferred to medical facility with symptoms of acute anti-ChE poisoning after accidental inhalation of pesticide containing dust. The patient experienced cephalalgia, disorientation, suffocation, perspiration, weakness, fatigue, abdominal pain, and vomited. Measured whole blood AChE activity was 57% which has been significantly decreased (χ^2=16.10, P=0.0001) compared to the earlier records obtained for the same worker (83%). As shown in the Fig. 3., cytogenetic analysis revealed that 180 minutes upon intoxication frequencies of MN, C+MN, and NPBs, remained unaffected compared to the records obtained in the study prior to intoxication (P_{χ^2}=0.847; 0.683; 0.180, respectively). Only the number of C-NBs has been elevated significantly (χ^2=11.93, P=0.0006). It seems unlikely that observed NBs represent amplified AChE or BuChE genes that are being extruded from the cells. More reasonable explanation would be that these early NBs harbor expelled DNA repair complexes or which is more likely, broken chromatid/chromosome fragments as the result of genotoxicity. Latest theory might be supported by results of the alkaline comet assay. The level of primary DNA damage detected 3 hours following intoxication has been significantly increased compared to referent value for the same worker detected prior to the accident ($_{PMann-Whitney}$=0.0019). Nevertheless, C-NBs may be indicated as possible early cytogenetic biomarkers of the effect of exposure to carbamate insecticides. In the next 72 hours, frequency of C-NB continued to increase insignificantly (χ^2=2.50, P=0.114), followed by significant occurrence of C+NBs χ^2=15.25, P=0.00009). Simultaneously, total MN, and proportion of MN originating from whole chromosomes were significantly increased ($\chi^2=10.40$, P=0.0013; $\chi^2=10.30$, P=0.0013) which categorizes them as late biomarkers of the effect.

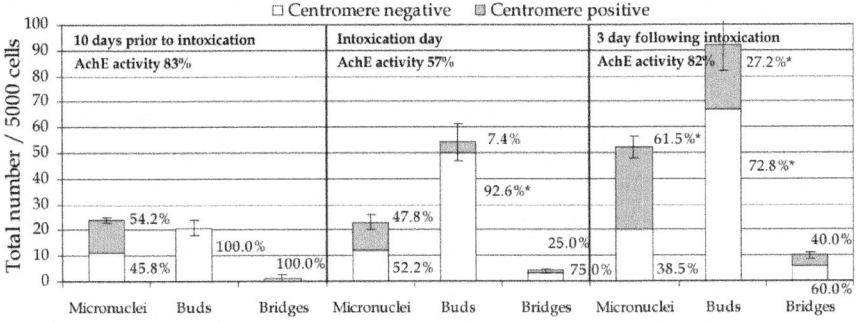

Figure 3. Dynamics of formation of aberrant chromatin structures regarding to presence of FISH centromere signals in carbofuran intoxicated production plant worker. *P<0.05 compared to values prior to intoxication.

Conversely, even 3 days upon intoxication the level of NPBs was not significantly elevated ($\chi^2=2.58$, P=0.1086). As already discussed, NPBs did significantly correlate with exposure to carbamate. However, NPBs are formed by dicentric or ring chromosomes whose formation is mediated through a cascade of misrepairs or significant telomere shortening. These cytogenetic events need longer period of time to occur and to form a critical pool of aberrant chromosomes that will be manifested as NPBs. Time consuming aspect of formation of chromosomes aberrant in their structure is even more pronounced in chemical than ionizing radiation genotoxicity. Some bridges will be lost due to their breakage when the daughter cells separate which also negatively reflects on their manifestation as NPBs. Thus, NPBs might be only considered as biomarkers of the effect in long-term exposure to pesticides. However, their presence in lymphocytes of individuals under pesticide exposure together with the reports indicating their presence in lymphoma patients stressed the need to look after the possible induction of translocations due to pesticide exposure. Their increased presence would indicate populations with elevated risk of tumor development.

Pesticide Exposure and Translocation Yield Assessed by FISH Chromosome Painting

Chromosomal translocations indicate severe impairment of genome integrity and are associated with the carcinogenic transformation of the cell, even more used as specific biomarkers in cancer diagnostics, classification, prognosis, deciding on the treatment and evaluating its efficacy. Thus, evaluation of translocation yield in population of concern is applied as a surrogate approach in epidemiological studies for estimation of cancer risk.

First attempts aiming to assess the level of translocations in peripheral blood lymphocytes of individuals exposed to pesticides started in early 1990-ties by applying G-banding technique. Method is based on treatment of metaphase spreads with trypsine and Giemsa staining that reveals a pattern of bands being specific for each of chromosome pairs. Any change in the band pattern indicates chromosomal rearrangement. Based on the determination of patterns proceeding and following the rearrangement site chromosomes involved in the translocation may be identified. By using G-banding technique, Garry et al. (1996) detected increased (Wilcoxon rank sum test, $P=0.003$) frequency of rearrangements in phosphine mixing or applying subjects (1.7 ± 0.5 per 100 metaphases) compared to the control group (0.5 ± 0.1). However, 12 months upon some subjects ceased using the phosphine lymphocyte translocation frequency decreased to the control values. Conversely, it remained increased in individuals who continued with pesticide application. Translocations at 1p13, and 14q32 were observed in majority of applicators. Since these rearrangements are accepted to be related to non-Hodgkin's lymphoma (NHL) the authors assumed that longterm exposure to phosphine poses a risk of developing NHL. Similar results have been reported for mixed pesticide applicators (Garry et al., 1996). In applicators of multiple insecticides significantly elevated rearrangement frequency (Wilcoxon rank sum test, $P=0.005$), has been reported (1.4 ± 0.3) compared to the control individuals (04 ± 0.1). Though elevated (1.0 ± 0.3) translocation yield in mixed herbicides applicators was not significantly affected ($P=0.096$). Again, 14q32 position has been involved in translocations common to most examinees indicating elevated risk of NHL. Unfortunately, both studies are characterized by rather poor pesticide exposure characterization and lack of multivariate statistical approach considering lifestyle factors as possible confounders; though exposed and control groups were matched by age and smoking habits. In their third study, Garry at el. (2001) applied G-banding in detection of translocations in forest and areal pesticide applicators with most prominent exposure to chlorophenoxy herbicides. Translocation, deletion and insertion yield (exposed 2.22 ± 0.38 vs. control 0.65 ± 0.30) has been significantly affected only in subjects applying more than 3.785 liters of herbicides per season (Wilcoxon rank sum test, $P=0.003$). Since Poisson regression analysis adjusted for smoking status revealed insignificant negative correlation between rearrangement frequency and measured 2,4-D urinary concentrations observed impact on induction of translocations could should be attributed to other than chlorophenoxy herbicides.

To obtain relevant results in assessing translocation frequency by analyzing G-band patterns on metaphase chromosomes requires well trained and experienced scorer, able to recognize any change in the banding sequence or width of bands that may indicate chromosomal rearrangement. Even then, due to significant variability in banding quality between metaphases of the same donor, and different donors, as well as due to terminal interchromosomal rearrangements that may not affect banding pattern, many translocations remain undetected which bias sensitivity and reliability of the technique. All these limitations classify G-banding technique laborious and inadequate for detection of cytogenetic effects resulting from low level pesticide exposures. Referred drawbacks may be circumvented by the use of FISH where single, several or all chromosomes are hybridized with DNA probes specific for each of them, and labeled with various combinations of flourochromes. As the result, each of the 22 of pairs of autosomes and 2 sex chromosomes are dyed in different color which makes easier to spot interchromosomal rearrangements and enhances the sensitivity of the method in detection of translocations. Table 3. provides an overview of studies applying FISH in translocation analysis in pesticide exposed subjects. More thorough evaluation of translocation frequency by applying chromosome painting FISH in regards to duration of pesticide exposure considering various confounders has been conducted on a group of pesticide plant workers exposed to mix of carbofuran, chlorpyrifos, metalaxyl and dodine (Zeljezic et al., 2009). In this study approach used in ionizing radiation biodosimetry has been applied, painting chromosomes 1, 2, and 4 by FISH in red, green, and yellow and analyzing their mutual rearrangements, and those with other chromosomes (IAEA, 2001).

Table 3. FISH in detection of translocations associated with long-term pesticide exposure.

Exposure type	FISH probe	Translocations per 100 cells vs. control	Significance	Correlation to exposure years	Reference
Mancozeb, methamidophos, captan, chlorpyrifos	Chr #1,2,4	1.21±0.97 0.92±0.82	$P=0.05$	N/A	Steenland et al., 1997
Carbofuran, chlorpyrifos, metalaxyl, dodine	Chr #1,2,4	1.63±0.78 0.51±0.23	$P=4\times10^{-5}$	$R^2=0.356$, $P=0.0003$	Zeljezic et al., 2009
Dieldrin, toxaphene, lindane, atrazine	LSI IGH/BCL2 genes loci on Chr #14 & 18	N/A; OR(CI$_{95\%}$) 2.1 (0.8-5.4) 1.0 (referent)	$P=0.04$	Yes; only descriptive	Chiu & Blair, 2009

Pairs of chromosomes 1, 2, and 4 represent 22.34% of the DNA content of the female genome and 22.70% of the male translocations detected painting them in 3 colors will represent 39.4% of total translocations involving all other chromosomes. Due to different gene densities along these chromosomes efficiency of DNA repair varies between them classifying translocations involving Chr 1 among most persistent, and those involving Chr 4 among least stable. Formula derived by Lucas & Sachs (1993) is used to extrapolate observed translocations frequencies for painted chromosomes to total genomic translocation frequency. In lymphocytes of pesticide plant workers genomic translocation frequency was significantly higher than in matching controls ($P_{ScheffePostHoc}=0.000004$), being higher in females (0.0062 ± 0.0027 per cell) than in males (0.0043 ± 0.0016) regardless of adjustment for the age difference. Most of detected translocations were reciprocal (2 bicolor chromosomes in metaphase; Fig. 4.), though their frequency in examinees (67.8 ± 1.34) has been lower than among the controls (80.1 ± 1.83) indicating involvement of rather small chromosomal fragments into rearrangements that are beyond resolution of the technique. Furthermore, while among the controls no complex rearrangements have been detected, among pesticide handling workers $8.5\pm0.51\%$ translocations involved 3 or more chromosomes. Multiple regression analysis adjusted for confounders showed significant correlation between translocation yield and age ($R^2=0.274$, $P=0.021$) and pesticide exposure duration ($R^2=0.356$, $P=0.0003$) the effect of latest being more prominent (exposure $\beta=0.623$, age $\beta=0.524$). Although distribution of translocations between chromosomes was random, involvement of chromosome 4 in rearrangements positively correlated with years of employment in pesticide production ($R^2=0.48$, $p=0.0008$). Since no significance in dependence of translocations upon age has been detected among controls ($R^2=0.10$, $P=0.088$), and since age and years of exposure significantly correlated among examinees ($R^2=0.50$, $P=0.0000$), and finally due to higher impact of exposure it may be concluded than translocations detected in pesticide plant workers are a consequence of impaired genomic stability due to long-term exposure. No cytogenetic effect of dodine or metalaxyl was revealed thus, observed genotoxicity may be attributed mainly to mixed carbofuran and chlorpyrifos exposure. As discussed in section 4.1, chemically induced single-strand breaks, abasic sites, oxidative damage lesions and alkylated bases may be transferred into DSBs by error-prone base excision repair (BER; Maynard et al., 2009). Both, strand breaks induced in the course of BER, and oxidized DNA bases due to their topoisomerase II poisoning activity may be converted into DSBs (Khan et al., 2009). By error-prone activity either of original or non-classic pathways of NHEJ in its standard or modified repair, translocations may be formed (Weinstock et al., 2006). Indications for such cascade of error-prone damage transformation may be indicated by results of multiple regression analysis revealing high correlation of translocations with chromatid-type aberrations ($R^2=0.26$, $P=0.003$) in pesticide manufacturing workers being highly correlated with years of employment ($R^2=0.479$, $P=0.00001$).

Steenland et al. (1997) also reported significant increase in translocations by FISH painting of chromosomes 1, 2 and 4 in EBDC fungicide applicators (age adjusted Poisson regression $P=0.05$). Analyses restricted to reciprocal translocations nonsignificant exposure effect ($p = 0.24$) which may be mediated by significantly lower resolution of earlier FISH applications. However, total, and reciprocal translocations significantly correlated ($R=0.28$, $P=0.02$; $R=0.27$, $P=0.02$, respectively) with incidence of sister-chromatid exchanges that are manifestations of DSB repair.

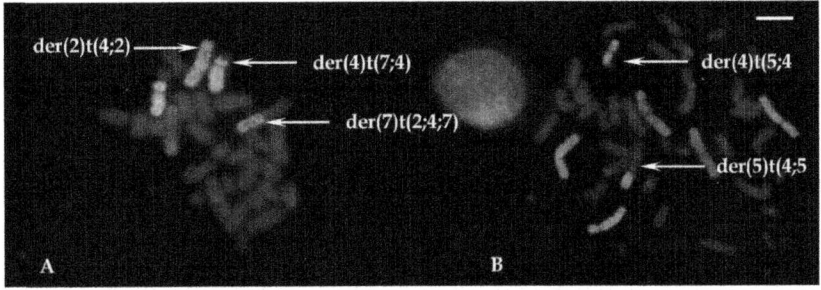

Figure 4. Translocations in lymphocytes of the pesticide plant workers: A) complex translocations, B) reciprocal translocation t(4;5). Chr 1 – red; Chr 2 – green; Chr 4 – yellow; Objective 100x, bar 1 μm.

Similar but gene specific approach of FISH use in estimation of cancer risk due to pesticide exposure has been reported by Chiu & Blair (2009). Gene locus specific LSI IGH/BCL2 dualfusion probe has been applied to evaluate association of t(14;18) rearrangement in NHL patients with their previous long-term pesticide exposure. Translocation t(14;18) represents the hallmark of follicular lymphoma as one of the most common adult NHLs. Exposure to crop and animal insecticides and herbicides have been associated with t(14;18) positive NHL (multivariate regression $P=0.01$). The risk of NHL for dieldrin applicators has been estimated to OR=2.4 ($CI_{95}= 0.8–7.9$), toxaphene OR=3.2 ($CI_{95}=0.8–12.5$), lindane OR=3.5 (CI=1.4–8.4), and atrazine OR=1.7 ($CI_{95}=1.0–2.8$) (Chiu & Brian, 2009).

CONCLUSION

Presented data indicate that within the last decade by applying the FISH technique in monitoring studies of subjects exposed to pesticides, a new set of cytogenetic biomarkers has been introduced providing us with more detailed insight in the effect of long-term occupational pesticide exposure on genome integrity. Some biomarkers such as chromosomal rearrangements and translocation had been over a long period of time considered irrelevant in chemical carcinogenesis and cancer risk assessment due to pesticide exposure. Valuable knowledge regarding the error-prone effect of distinct DNA repair mechanisms helped us to understand how primary genome lesions induced by agrochemicals may be transformed into chromosomal rearrangements. The use of FISH in evaluation of the impact of occupational exposure to pesticides on human genome may provide us with more precise insight in the extent and type of genome damage, indicate possible specificity of pesticides of concern toward specific chromosomes or regions and help us in evaluation of pesticide exposure relevant for cancer risk.

REFERENCES

1. Abass, K.; Reponen, P.; Mattila, S. & Pelkonen, O. (2009). Metabolism of carbosulfan. I. Species differences in the in vitro biotransformation by mammalian hepatic microsomes including human. Chemico-biological Interactions, Vol. 181, No. 2, pp 210-219, 00092797
2. Ahrens, W.; Krickeberg, K. & Pigeot, I. (2005). An Introduction to Epidemiology, In: Handbook of epidemiology, Ahrens, W. & Pigeot, I. (Ed.), pp 1-42, Springer Verlag, 354000568, Berlin
3. Amr, M.M. (1999). Pesticide monitoring and its health problems in Egypt, a Third World country. Toxicology Letters, Vol. 107, pp 1-13, 03784274
4. Anderson, D. (2000). Examination of various biomarkers after occupational exposure to different chemicals, In: Human Monitoring after Environmental and Occupational Exposure to Chemical and Physical Agents, Anderson, D.; Karakaya, A.E. & Šrám, R.J. (Ed.), pp 94-100, IOS Press, 9051995958, Amsterdam
5. Andreotti, G.; Beane Freeman, L.E.; Hou, L.; Coble, J.; Rusiecki, J.; Hoppin, J.A.; Silverman, D.T. & Alavanja, M.C.R. (2009). Agricultural pesticide use and pancreatic cancer risk in the Agricultural Health Study. International Journal of Cancer, Vol. 124, 2495–2500, 0020-7136
6. Antonucci, G.A. & de Syllos Cólus, I.M. (2000). Chromosomal aberrations analysis in a Brazilian population exposed to pesticides. Teratogenesis, Carcinogenesis, and Mutagenesis, Vol. 20, pp 265–272, 02703211
7. Axelson, O.; Rehn, M. & Sundell, L. (1974). Exposure to herbicides--mortality and tumor incidence: An epidemiologie investigation of

Swedish railroad workers. Lakartidningen, Vol. 71, No. 24, pp 2466-2470, 00237205

8. Balayiannis, G.P.; Anagnostopoulos, H. & Kellidou, I (2009). Facile and rapid determination of contamination in sulphur pesticide formulations by liquid chromatographytandem mass spectrometry. Bulletin of Environmental Contamination and Toxicology, Vol. 82, No. 2, pp 133-136, 00074861

9. BCPC - The British Crop Protection Council (2009). The pesticide Manual (Incorporating the Agrochemicals Handbook) a world compendium, Tomlin, C., (Ed.), pp 21, Datix International, 1901396188, Bungay

10. Becher, H. (2005). General Principles of Data Analysis, In: Handbook of epidemiology, Ahrens, W. & Pigeot, I. (Ed.), pp 595-624, Springer Verlag, 354000568, Berlin

11. Becker, P.; Scherthan, H. & Zankl, H. (1990). Use of a centromere-specific DNA probe (p82H) in nonisotopic in situ hybridization for classification of micronuclei. Genes Chromosomes Cancer, Vol. 2, No. 1, pp 59-62, 10452257

12. Blair, A. & Zahm, S.H. (2008). Agricultural exposures and cancer. Environmental Health Perspectives, Vol. 103, pp 205-208, 00916765

13. Blasiak, J.; Jaloszynski, P.; Trzeciak, A. & Szyfter, K. (1999). In vitro studies of the organophosphorous insecticide malathion and its two analogues. Mutation Research, Vol. 445, No. 2, 275-283, 00275107

14. Bolognesi, C.; Landini, E.; Perrone, E. & Roggieri, P. (2004). Cytogenetic biomonitoring of a floriculturist population in Italy: micronucleus analysis by fluorescence in situ hybridization (FISH) with all-chromosome centromeric probe. Mutation Research, Vol. 557, pp 109-117, 00275107

15. Boers, D.; Portengen, L.; Bueno-de-Mesquita, H.B.; Heederik, D. & Vermeulen, R. (2010). Cause-specific mortality of Dutch chlorophenoxy herbicide manufacturing workers. Occupational and Environmental Medicine, Vol. 67, No. 1, pp 24-31, 13510711

16. Bonner, M.R.; Williams, B.A.; Rusiecki, J.A.; Blair, A.; Beane Freeman, L.E.; Hoppin, J.A.; Dosemeci, M.; Lubin, J.; Sandler, D.P. & Alavanja, M.C.R. (2010). Occupational exposure to terbufos and the incidence of cancer in the Agricultural Health Study. Cancer Causes and Control, Vol. 21, No. 6, pp 871-877, 09575243

17. Bull, S.; Fletcher, K.; Boobis, A.R. & Battershil, J.M. (2006). Evidence for genotoxicity of pesticides in pesticide applicators: a review. Mutagenesis, Vol. 21, No. 2, pp 93–103, 02678357

18. Buongiorno-Nardelli, M. & Amaldi, F. (1969). Autoradiographic detection of molecular hybrids between tRNA and DNA in tissue sections. Nature, Vol. 225, pp 946-947, 00280836
19. Calviello, G.; Piccioni, E.; Boninsegna, A.; Tedesco, B.; Maggiano, N.; Serini, S.; Wolf, F.I. & Palozza, P. (2006). DNA damage and apoptosis induction by Mancozeb in rat cells: Involvement of the oxidative mechanism. Toxicology and Applied Pharmacology, Vol. 211, pp 87-96, 0041008X
20. Carson, R. (2002). Silent Spring, Mariner Books, 0618249060, New York, NY
21. Casciano, D.A. (2000). Introduction: Historical Perspectives of Genetic Toxicology, In: Genetic Toxicology, Li, A.P. & Heflich, R.H. (Ed.), pp 1-12, CRC Press, 0849388153, Boca Raton, FL
22. Cavallo, D.; Ursini, C.L.; Rondinone, B. & Iavicoli, S. (2009). Evaluation of a suitable DNA damage biomarker for human biomonitoring of exposed workrs. Environmental and Molecular Mutagenesis, Vol. 50, No. 9, pp 781-790, 08936692
23. Chang, T.H. & Klassen, W. (1968). Comparative effects of tretamine, tepa, apholate and their structural analogues on human chromosomes in vitro. Chromosoma, Vol. 24, No. 3, pp 314-323, 00095915
24. Chappel, CI.. (1992). A review and biological risk assessment of sodium saccharin. Regulatory Toxicology and Pharmacology, Vol. 15, No. 3, pp 253-270, 02732300
25. Chiu, B.C.-H. & Blair A. (2009). Pesticides, chromosomal aberrations, and Non-Hodgkin's lymphoma. Agromedicine, Vol. 14, No. 2, pp 250–255, 1059-924X
26. Costa, C.; Teixeira, J.P.; Silva, S.; Roma-Torres, J.; Coelho, P.; Gaspar, J.; Alves, M.; Laffon, B.; Rueff, J. & Mayan, O. (2006). Cytogenetic and molecular biomonitoring of a Portuguese population exposed to pesticides. Mutagenesis, Vol. 21, No. 5, pp 343– 350, 02678357
27. Cox, C. & Surgan, M. (2006). Unidentified Inert Ingredients in Pesticides: Implications for Human and Environmental Health. Environmental Health Perspectives, Vol. 114, No. 12, pp 1803-1806, 00916765
28. DeLancey, J.L.; Alavanja, M.C.R.; Coble, J.; Blair, A.; Hoppin, J.A.; Austin, H.D. & Beane Freeman, L.E. (2009). Occupational exposure to metribuzin and the incidence of cancer in the Agricultural Health Study. Annals of Epidemiology, Vol. 19, No. 6, pp 388-395, 10472797

29. Dich, J.; Hoar Zahm, S.; Hanberg, A. & Adami, H-O. (1997). Pesticides and cancer. Cancer Causes and Control, Vol. 8, pp 420-443, 09575243
30. Eastmond, D.A. & Balakrishnan, S. (2001). Genetic Toxicity of pesticides, In: Handbook of Pesticide Toxicology, Krieger, R. (Ed.), pp 747-767, Academic Press, 0124262627, San Diego, CA
31. El-Zein, R.A.; Fenech, M.; Lopez, M.S.; Spitz, M.R. & Etzel, C.J. (2008). Cytokinesis-blocked micronucleus assay biomarkers identify lung cancer amongst smokers. Cancer Epidemiology, Biomarkers & Prevention, Vol. 17, No. 5, pp 1111-1119, 10559965
32. El-Zein, R.A.; Schabath, M.B.; Etzel, C.J.; Lopez, M.S.; Franklin, J.D. & Spitz, M.R. (2006). Cytokinesis-blocked micronucleus assay as a novel biomarker for lung cancer risk. Cancer Research, Vol. 66, No. 12, pp 6449-6456, 00085472
33. Ergene, S.; Çelik, A.; Çavaş, T. & Kaya, F. (2007). Genotoxic biomonitoring study of population residing in pesticide contaminated regions in Göksu Delta: Micronucleus, chromosomal aberrations and sister chromatid exchanges. Environment International, Vol. 33, pp 877–885, 01604120
34. Fenech, M. (2006). Cytokinesis-block micronucleus assay evolves into a "cytome" assay of chromosomal instability, mitotic dysfunction and cell death. Mutation Research, Vol. 600, pp 58–66, 00275107
35. Fenech, M. (2007). Cytokinesis-block micronucleus cytome assay. Nature Protocols, Vol. 2, pp 10084-1104, 17542189
36. Fenech, M.; Chang, W.P.; Kirsch-Volders, M.; Holland, N.; Bonassi, S. & Zeiger, E. (2003). HUMN project: detailed description of the scoring criteria for the cytokinesis-block micronucleus assay using isolated human lymphocyte cultures. Mutation Research, Vol. 534, pp 65–75, 00275107
37. Garry, V.F.; Tarone, R.E.; Kirsch, I.R.; Abdallah, J.M.; Lombardi, D.P.; Long, L.K.; Burroughs, B.L.; Barr, D.B.; & Kesner J.S. (2001). Biomarker correlations of urinary 2,4-D levels in foresters: genomic instability and endocrine disruption. Environmental Health Perspectives, Vol. 109, No. 5, pp 495-500, 00916765
38. Garry, V.F.; Tarone, R.E.; Long, L.; Griffith, J.; Kelly, J.T. & Burroughs, B. (1996). Pesticide appliers with mixed pesticide exposure: G-banded analysis and possible relationship to non-Hodgkin's lymphoma. Cancer Epidemiology, Biomarkers and Prevention, Vol. 5, pp 11-16, 10559965
39. Green, R.E. (1949). Poisoning by the organic phosphate insecticides. Wisconsin Medical Journal, Vol. 48, No. 11 p 1007, 00436542

40. Hadnagy, W.; Seemayer, N.H.; Kühn, K.H.; Leng, G. & Idel, H. (1999). Induction of mitotic cell division disturbances and mitotic arrest by pyrethroids in V79 cell cultures. Toxicology Letters, Vol. 107, No. 1-3, pp 81-87, 03784274

41. Hagmar, L.; Tinnerberg, H.; Mikoczy, Z.; Strömberg, U.; Bonassi, S.; Huici Montagud, A.; Hansten, I.-L.; Knudsen, L.E. & Norppa, H. (2000). Do cytogenetic biomarkers, used for occupational health surveillance predict cancer?, In: Human Monitoring after Environmental and Occupational Exposure to Chemical and Physical Agents, Anderson, D.; Karakaya, A.E. & Šrám, R.J. (Ed.), pp 1-6, IOS Press, 9051995958, Amsterdam

42. Hill, M.K. (2010). Understanding Environmental Pollution, pp 89-116, Cambridge University Press, Cambridge

43. Hoshizaki, H.; Niki, Y.; Tajima, H.; Terada, Y. & Kasahara, A. (1969). A case of leukemia following exposure to insecticide. Nippon Ketsueki Gakkai Zasshi, Vol. 32, No. 4, pp 672-677, 00015806

44. Hoyos, L.S.; Carvajal, S.; Solano, L.; Rodriguez, J.; Orozco, L.; Lopez, Y. & Au W.W. (1996). Cytogenetic monitoring of farmers exposed to pesticides in Colombia. Environmental Health Perspectives, Vol. 104, Supp. 3, pp 535-538, 00916765

45. IAEA (2001). Cytogenetic analysis for radiation dose assessment: a manual, IAEA, 9201021011, Vienna

46. Joksic, G.; Vidakovic, A. & Spasojevic-Tisma, V. (1997). Cytogenetic monitoring of pesticide sprayers. Environmental Research, Vol. 75, pp 113–118, 00139351

47. Jungmann, G. (1966). Arsenic cancer in vintagers. Landarzt, Vol. 42, pp 1244-1247, 00237728

48. Kang, D.; Park, S.K.; Beane-Freeman, L.E.; Lynch, C.F.; Knott, C.E.; Sandler, D.P.; Hoppin, J.A.; Dosemeci, M.; Coble, J.; Lubin, J.; Blair, A. & Alavanja, M. (2008). Cancer incidence among pesticide applicators exposed to trifluralin in the Agricultural Health Study. Environmental Research, Vol. 107, pp 271–276, 00139351

49. Kaszubiak, H. (1968). The effect of herbicides on Rhizobium. 3. Influence of herbicides on mutation. Acta Microbiologica Polonica, Vol. 17, No. 1, pp 51-57, 01371320

50. Khan, F.; Sherwani, A.F. & Afzal, M. (2009). Chromosomal aberration and micronucleus studies of two topoisomerase (II) targeting anthracyclines. Journal of Environmental Biology, Vol. 30, No. 3, pp 409-412, 02548704

51. Klayman M.B., Exposure to Insecticides, Arch Otolaryngol. 1968;88(1):116-117
52. Kourakis, A.; Mouratidou, M.; Barbouti A. & Dimikiotou M. (1996). Cytogenetic effects of occupational exposure in the peripheral blood lymphocytes of pesticide sprayers. Carcinogenesis, Vol. 17, No. 1, pp 99-101, 01433334
53. Legator, M. & Zimmering, S. (1975). Genetic Toxicology. Annual Review of Pharmacology, Vol. 15, pp 387-408, 00664251
54. Lindberg, H.K.; Wang, X.; Jarventaus, H.; Falck, GC.-M.; Norppa, H. & Fenech, M. (2007) Origin of nuclear buds and micronuclei in normal and folate-deprived human lymphocytes. Mutation Research, Vol. 617, pp 33-45, 00275107
55. Lucas, J.N. & Sachs, R.K. (1993). Using three-color chromosome painting to test chromosome aberration models. Proceedings of National Academy of Science USA, Vol. 90, pp 1484–1487, 00278424
56. Lynch, S.M.; Mahajan, R.; Beane Freeman, L.E.; Hoppin, J.A. & Alavanja, M.C.R. (2009). Cancer incidence among pesticide applicators exposed to butylate in the Agricultural Health Study. Environmental Research, Vol. 109, pp 860-868, 00139351
57. Matter, B. & Schmid, W. (1971). Trenimon-induced chromosomal damage in bone-marrow cells of six mammalian species, evaluated by the micronucleus test. Mutation Research, Vol. 12, No. 4, pp 417-425, 00275107
58. Maynard, S.; Schurman, S.H.; Harboe, C.; de Souza-Pinto, N.C. & Bohr, V.A. (2009). Base excision repair of oxidative DNA damage and association with cancer and aging. Carcinogenesis, Vol. 30, pp 2–10, 01433334
59. Mitelman, F.; Mertens, F. & Johansson, B. (1997). A breakpoint map of recurrent chromosomal rearrangements in human neoplasia. Nature Genetics, Vol. 15, pp 417–474, 10614036
60. Muller, J.; Decordier, I.; Hoet, P.H.; Lombaert, N.; Thomassen, L.; Huaux, F.; Linson, D. & Kirsh-Volders, M. (2008). Clastogenic and aneugenic effects of multi-wall carbon nanotubes in epitheliail cells. Carcinogenesis, Vol. 29, No. 2, pp 427-433, 01433334
61. Norppa, H. & Falck, G.C-M. (2003). What do human micronuclei contain? Mutagenesis, Vol. 18, No. 3, pp. 221–233, 02678357
62. Obe, G.; Pfeiffer, P.; Savage, J.R.K.; Johannes, C.; Gödecke, W.; Jeppesen, P.; Natarajan, A.T.; Martínez-López, W.; Folle, G.A. & Drets

M.E. (2002). Chromosomal aberrations: formation, identification and distribution. Mutation Research, Vol. 504, pp 17–36, 00275107

63. Orsi, L.; Delabre, L.; Monnereau, A.; Delval, P.; Berthou, C.; Fenaux, P.; Marit, G.; Soubeyran, P.; Huguet, F.; Milpied, N.; Leporrier, M.; Hemon, D.; Troussard, X. & Clavel, J. (2009). Occupational exposure to pesticides and lymphoid neoplasms among men: results of a French case-control study. Occupational and Environmental Medicine, Vol. 66, No. 5, pp 291-298, 13510711

64. Prasad, I. (1970). Mutagenic effects of 3›,4›-dichloropropionanilide and its degradation products. Canadian Journal of Microbiology, Vol. 16, No. 5, pp 369-372, 00084166

65. Preston, R.J. (2000). Mechanisms of Introduction of Chromosomal alterations and sister chromatid exchange, In: Genetic Toxicology. Li, A.P. & Heflich, R.H. (Ed.), pp 41-66, CRC Press, 0849388153, Boca Raton, FL

66. Price, J.D. (1973). Agricultural Entomology, In: History of Entomology, Smith, R.F.; Mittler, T.E. & Smith, C.N., (Ed.), pp 307-372, Annual Reviews, 0686092988, Palo Alto, CA

67. Raap, A.K. (1998). Advances in fluorescence in situ hybridization. Mutation Research, Vol. 400, pp 287-298, 00275107

68. Richardson, J.C. (1974). A preliminary assessment of cytogenetic effects of metepa on mose bone marrow using the micronucleus test. Mutation Research, Vol. 26, No. 5, pp 391- 394, 00275107

69. Rossi, A.M.; Hansteen, I-L.; Skjelbred C.F.; Ballardin, M.; Maggini, V.; Murgia, E.; Tomei, A.; Viarengo, P.; Knudsen, L.E.; Barale, R.; Norppa, H. & Bonassi, S. (2009). Association between frequency of chromosomal aberrations and cancer risk is not influenced by genetic polymorphisms in GSTM1 and GSTT1. Environmental Health Perspectives, Vol. 117, No. 2, pp 203-208, 00916765

70. Rossner, P.; Boffeta, P.; Ceppi, M.; Bonassi, S.; Smerhovsky, Z.; Landa, K.;, Juzova, D. & Šrám, R.J. (2005). Chromosomal aberrations in lymphocytes of healthy subjects and risk of cancer. Environmental Health Perspectives, Vol. 113, No. 5, pp 517-520, 00916765

71. Rull, R.P.; Gunier, R.; von Behren, J.; Hertz, A.; Crouse, V.; Buffler, P.A. & Reynolds, P. (2009). Residential proximity to agricultural pesticide applications and childhood acute lymphoblastic leukemia. Environmental Research, Vol. 109, pp 891-899, 00139351

72. Rusiecki, J.A.; Hou, L.; Lee, W.J.; Blair, A.; Dosemeci, M.; Lubin, J.H.; Bonner, M.; Samanic, C.; Hoppin, J.A.; Sandler, D.P. & Alavanja,

M.C.R. (2009). Cancer incidence among pesticide applicators exposed to metolachlor in the Agricultural Health Study. International Journal of Cancer, Vol., 118, pp 3118–3123, 0020-7136

73. Smith, A.E. & Secoy, D.M. (1975). Forerunners of Pesticides in Classical Greece and Rome. Agricultural and Food Chemistry, Vol. 23, No. 6, pp 1050, 00218561

74. Smith, E.H. & Kennedy, G.G. (2002). History of Pesticides, In: Encyclopedia of Pest Management, Pimentel, D. (Ed.), 0824706326, pp 376-372, Taylor & Francis, London

75. Steenland, K.; Cedillo, L.; Tucker, J.; Hines, C.; Sorensen, K.; Deddens, J. & Cruz, V. (1997). Thyroid hormones and cytogenetic outcomes in backpack sprayers using ethylenebis(dithiocarbamate) (ebdc) fungicides in Mexico. Environmental Health Perspectives, Vol. 105, No. 10, pp 1126-1130, 00916765

76. Stehrer-Schmid, P. & Wolf, H.U. (1995). Genotoxic evaluation of three heterocyclic Nmethylcarbamate pesticides using the mouse bone marrow micronucleus assay and the Saccharomyces cerevisiae strains D7 and D61. Mutation Research, Vol. 345, pp 111–125, 00275107

77. Titenko-Holland, N.; Moore, L.E. & Smith, M.T. (1994). Measurement and characterization of micronuclei in exfoliated human cells by fluorescence in situ hybridization with a centromeric probe. Mutation Research. Vol. 312, No. 1, pp 39-50, 00275107

78. Tucker, J.D.; Ramsey, M.J.; Lee, D.A. & Minkler, J.L. (1993). Validation of chromosome painting as a biodosimeter in human peripheral lymphocytes following acute exposure to ionizing radiation in vitro. International Journal of Radiation Biology, Vol. 64, pp 27–37, 09553002

79. Usmani, K.A.; Hodgson, E. & Rose, R.L. (2004). In vitro metabolism of carbofuran by human, mouse, and rat cytochrome P450 and interactions with chlorpyrifos, testosterone, and estradiol. Chemico-biological Interactions, Vol. 150, pp 221–232, 00092797

80. van Bemmel, D.M.; Visvanathan, K.; Beane Freeman, L.E.; Coble, J.; Hoppin, J.A. & Alavanja, M.C.R. (2008). S-ethyl-N,N-dipropylthiocarbamate exposure and cancer incidence among male pesticide applicators in the agricultural health study. Environmental Health Perspectives, Vol. 116, No. 11, pp 1541-1556, 00916765

81. van Maele-Fabry, G.; Duhayon, S.; Mertens, C. & Lison, D. (2008). Risk of leukaemia among pesticide manufacturing workers: A review and meta-analysis of cohort studies. Environmental Research, Vol. 106, pp 121–137, 00139351

82. Vidal, C.J. (2005). Expression of cholinesterases in brain and non-brain tumours. Chemicobiological Interactions, Vol. 157–158, pp 227–232, 00092797
83. Wandahwa, P.; van Ranst, E. & van Damme, P. (1996). Pyrethrum (Chrysanthemum cinerariaefolium Vis.) cultivation in West Kenya: origin, ecological conditions and management. Industrial Crops and Products, Vol. 5, No. 4, pp. 307-322, 09266690
84. Ware, G. & Whitacre, D. (2004). The pesticide book, pp 128-132, Meister Media Worldwide, 1892829118, Willoughby, OH
85. Weinstock, D.M.; Richardson, C.A.; Elliott, B. & Jasin, M. (2006). Modeling oncogenic translocations: distinct roles for double strand break repair pathways in translocation formation in mammalian cells. DNA Repair, Vol. 5, pp 1065-1074, 15687864
86. Zeljezic, D. & Garaj Vrhovac, V. (2001). Chromosomal aberration and single-cell gel electrophoresis (comet) assay in longitudinal risk assessment of occupational exposure to pesticides. Mutagenesis, Vol. 16, No. 4, pp 359-363, 02678357
87. Zeljezic, D.; Lucic Vrdoljak, A.; Lucas, J.N.; Lasan, R.; Fucic, A.; Kopjar, N.; Katic, J.; Mladinic, M. & Radic, B. (2009). Effect of occupational exposure to multiple pesticides on translocation yield and chromosomal aberrations in lymphocytes of plant workers. Environmental Science & Technology, Vol. 43, pp 6370–6377, 0013936X
88. Zeljezic, D.; Lucic Vrdoljak, A.; Radic, B.; Fuchs, N.; Berend, S.; Orescanin, V. & Kopjar, N. (2007). Comparative evaluation of acetylcholinesterase status and genome damage in blood cells of industrial workers exposed to carbofuran. Food and Chemical Toxicology, Vol. 45, pp 2488–2498, 02786915
89. Zhang, L.J.; Min, S.G.; Li, G.X.; Xiong, Y.M. & Sun, Y. (2005). The mechanism of carbofuran interacts with calf thymus DNA. Guang Pu Xue Yu Guang Pu Fen Xi, Vol. 25, No. 5, pp 739-742, 10000593
90. Zheng, T.; Zahm, S.H.; Cantor, K.P.; Weisenburger, D.D.; Zhang, Y. & Blair, A. (2001). Agricultural exposure to carbamate pesticides and risk of non-Hodgkin lymphoma. Journal of Occupational and Environmental Medicine, Vol. 43, pp 641–649, 10762752

Chapter 2

TOXICOLOGY OF THE BIOINSECTICIDES USED IN AGRICULTURAL FOOD PRODUCTION

Neiva Knaak, Diouneia Lisiane Berlitz, and Lidia Mariana Fiuza

University of Vale do Rio dos Sinos, Laboratory of Microbiology and Toxicology, São Leopoldo, RS, Brazil

INTRODUCTION

As populations grow in numbers, the demands for food production increase and are generally met by the intensification of livestock breeding and the increase of agricultural activities. This in turn increases the quantity of chemical pesticides required to control the losses in production caused by insect pests preying on the food plants and disturbing the animals. Once applied these pesticides may cause resistance to the synthetic molecules, contaminate biotic and abiotic components like plants, soil, water and/or the local water network and can also effect non-targeted organisms, such as fish, small mammals, birds and so on.

Contamination of rivers with chemical pesticides is almost always due to excess material being carried away by rain or irrigation waters or by erosion of contaminated soil particles [1, 2]. Therefore, as this generalized contamination of the environment increases, researchers are seeking alternative methods for controlling the pests but which cause less damage than the chemical products [3].

In the biological control system, the bacteria of the *Bacillus* genus have considerable potential for use as control agents, because, as well as being lethal to the insect pests, they remain viable for long periods in storage [4]. Amongst these bacteria, *Bacillus thuringiensis* has achieved commercial-scale success in controlling various insect pests, plant pathogens, nematodes and mites, mainly because this micro-organism has a high specificity to the pest-targets [5].

In addition to the use of microorganisms, various botanical pesticides are being studied and can be associated with Integrated Pest Management (IPM). The products originating in plants refer to plant species that, over a long time developed defense mechanisms against herbivores, pathogens and other stress agents. Among the toxins produced by these plants are found nitrogenous substances such as non-protein amino acids, cyanogenic glycosides, certain peptides and proteins, and several alkaloids. The toxicity of a substance is related to the dose taken by the insect, its age, the absorption mechanism and the manner of excretion [6].

In this chapter we address the toxicological aspects of microbial and botanical biopesticides that act by ingestion, with emphasis on histopathological analysis of tissues and cells in the alimentary channel of the Lepidoptera as well as the specificity of the *B. thuringiensis* bacteria, aqueous extracts and oil essential of medicinal and forest.

HISTOLOGY OF LEPIDOPTERA

The alimentary channel of insects is composed of three main regions with different embryological origins: stomodeum or foregut, middle or midgut and hindgut or proctodeu. This canal represents a contact area between the insect and the environment and is the focus of much of the applied research concerning pest control [7, 8], especially the midgut region where the epithelial cells are involved in the processes of absorption and secretion of enzymes (columnar cells), ion homeostasis (goblet cells), endocrine function (endocrine cells) and the renovation of the epithelia (regenerative cells) [9, 10, 11]. A defense mechanism in this region is the peritrophic membrane, whichplays a fundamental role in the biology of the midgut, being positioned between the food contents and epithelial layer, performing the function of protecting the epithelium from mechanical damage, and in addition acts as a barrier against toxins and chemicals harmful to the insects [12, 13].

In Lepidoptera, the midgut epithelium is composed of four cell types (Figure 1) which are involved in the processes of absorption and secretion of enzymes (columnar cells), ion homeostasis (goblet cells), endocrine function (endocrine cells) and the renewal of the epithelium (regenerative cells) [14, 15]. The regenerative cells are undifferentiated andare responsible for the renovation ofthe midgut epithelium, substituting the cells that wear out and are lost during the digestive process – they also make it possible for the alimentary channel to grow larger at each ecdysis. They are found at the base of the midgut, alone or in groups, and there are no differences in their abundance along the midgut [16, 17, 18].

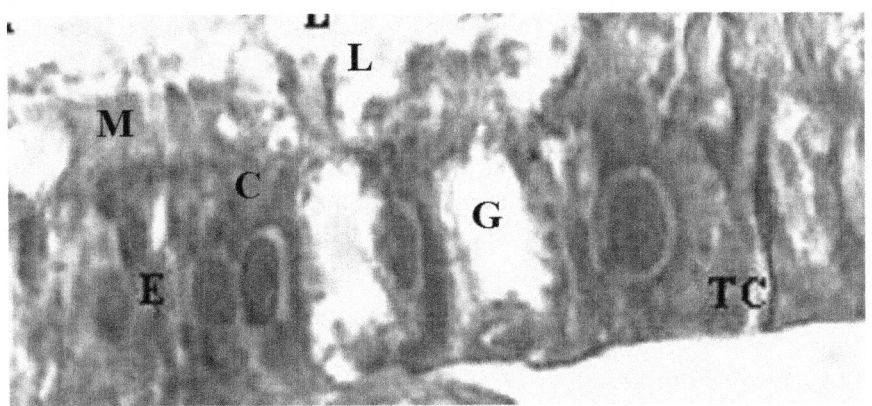

Figure 1. Types of cells found in Lepidoptera. L=lumen; TC=connective tissue; E=epithelium, C=columnar cells; G=globet cells;

Changes in the alimentary canal, especially in the region of the midgut, can affect the growth and development of insects, as well asall the physiological events, because these processes depend on adequate food, on its absorption and transformation in the alimentary canal [9, 11].

Among the insects, the Lepidoptera order causes the greatest economic losses in crops – in maizeand ricewhich are attacked principally by *Spodoptera frugiperda,* in sugar-caneattacked by the stem-borer *Diatraeasaccharallis* and in soybeans mainly by the Lepidoptera pest *Anticarsiagemmatalis.*

All regions of the epithelial layer of the alimentary canal of *S. frugiperda* caterpillars are coated with a single layer of cells, with a flat morphology in the stomodeum region and cubic morphology in the proctodeus [19]. The muscles are arranged along the channel in a uniform manner [7]. According to Cavalcante & Cruz-Landim [20] and Pinheiro et al. [16], in the midgut epithelium four cell types predominate: columnar, goblet, regenerative, and endocrine cells. These cells predominate along the *S. frugiperda* midgut epithelium and are considered responsible for the absorption of digested food, and demonstrate morphology similar to that of other Lepidoptera [8, 21]. Studies by Harper & Hopkins [22] and Harper & Granados [23], indicate that these epithelial cells are also responsible for secretion of a micro fiber net soaked in a matrix of proteins and glycoproteins -denominated a peritrophic membrane – which performs various functions such as: the protection of the epithelium against chemical and mechanical damage caused by the alimentation, the creation of a physical barrier against micro-organisms and digestion division [18].

Pinto & Fiuza [24], analyzing the histology of the midgut of *A. gemmatalis* caterpillars with an optical microscope, observed that the alimentary canal is divided basically into three portions, which are: the anterior intestine or stomodeum, the middle intestine or mesentero, and the posterior intestine or proctodeus. The cells do not have a cuticle coating, the intestinal wall being constituted by an epithelium of approximately 40m in height, which is separated from the haemolymph only by a thin layer of loose connective tissue. The cylindrical and goblet cells are evenly distributed along the intestine section. In the intestinal lumen is a thin membrane, called the peritrophic membrane, which surrounds the bolus of food, separating it from the epithelial cells of the midgut. The apical surface of the cylindrical cells has dense microvilli measuring 3m in height. The goblet cells show invaginations of the intestinal lumen as far as the nucleus.

MICROBIAL BIOPESTICIDES

Among the microbial insecticides, the *Bacillus thuringiensis* bacterium is highlighted because it shows great promise for development in the line of organic products and in the area of genetically modified plants. Besides being a sporulent bacteria found naturally in the soil, it is distinguished by its production of protein inclusions with insecticidal activity against several orders of insects and phytonematodes [25,26]. According to Shelton et al. [27], the firstbiopesticide containing subspecies of *B. thuringiensis* were sold in France in 1930. In 1995 more than 180 products based on this bacterial species were recorded by the Environmental Protection Agency (EPA) in the United States of America. In a study by Fiuza & Berlitz [28] they stated that *B. thuringiensis* formulas were then being marketed in Brazil. These are described in Table 1.

Table 1. Products made from *Bacillus thuringiensis* for controlling agricultural pest*

Products	Companies	*B. thuringiensis (Bt)*	Target insects
Dipel	Abbott	*Bt kurstaki*	Lepidoptera
Thuricide	Sandoz	*Bt kurstaki*	Lepidoptera
Agree	Mitsui	*Bt aizawai*	Lepidoptera
Bactur	Milenia Agrociências	*Bt kurstaki*	Lepidoptera
Ecotech Pro	Bayer	*Bt kurstaki*	Lepidoptera
Bactospeine	Solvay	*Bt kurstaki*	Lepidoptera
Javelin	Sandoz	*Bt kurstaki*	Lepidoptera
Foray	Novo-Nordisk	*Bt kurstaki*	Lepidoptera
Biobit	Novo-Nodisk	*Bt kurstaki*	Lepidoptera

Foil/Condor	Ecogen	*Bt kurstaki*	Lepidoptera
Delfin	Sandoz	*Bt kurstaki*	Lepidoptera
Cutlass	Ecogen	*Bt kurstaki*	Lepidoptera
LarvoBt	Fermone	*Bt kurstaki*	Lepidoptera
Nubilacid	Radonja	*Bt kurstaki*	Lepidoptera
MVP	Mycogen	*Bt kurstaki*	Lepidoptera
Bac-control	Agricontrol	*Bt kurstaki*	Lepidoptera
XenTari	Abbott	*Bt aizawai*	Lepidoptera
M-One	Mycogen	Bt san diego Bt tenebrionis	Coleoptera
Di-Terra	Abbott	Bt san diego Bt tenebrionis	Coleoptera
Trident	Sandoz	Bt san diego Bt tenebrionis	Coleoptera
Novodor	Novo Nordisk	Bt san diego Bt tenebrionis	Coleoptera
M-One Plus	Mycogen	Bt san diego Bt tenebrionis	Coleoptera
Foil	Ecogen	*Bt* (recombinant)	Coleoptera

Case Studies

A study by Berlitz & Fiuza [29] evaluated the toxicity of *B. thuringiensis aizawai* on *S. frugiperda* and demonstrated that, 6 hours after the application of the treatments, an increase in the cell volume occurred and the intestinal microvilli ruptured. Significant differences in the cell volume of the treatment as compared with that of the control caterpillars were observed 12 hours after the application.

In the histopathological analysis of the midgut of *S. frugiperda* caterpillars treated with the *B.thuringiensis thuringiensis* 407 (pH 408) strain, structural changes were observed six hours after application of the treatment (HAT), where there were cells in the intestinal lumen and elongation of the microvilli, as compared to the control. After nine applications of the treatment (HAT) the action was intensified with vacuolization of the cytoplasm, and the beginning of the degradation of the peritrophic membrane – this was entirely absent after 12 HAT.

Treatment with *B. thuringiensis kurstaki* HD-73 strain was similar, except that rupture of the microvilli (BBMV's) and vacuolization of the cytoplasm began at 12 HAT. The results of the histopathological analysis of the midget of *S. frugiperda* caterpillars demonstrate that treatment with the *B. thuringiensis thuringiensis* strain was more efficient, because the degradations of the microvilosities started 9 hours after treatment application (HAT), while in the *B. thuringiensis kurstaki* the same effect was noticed only after 12 HAT[30].

Knaak and Fiuza [31] tested the nuclear polyhedrosis virus of *Anticarsia gemmatalis* (VPNAg) and *B.thuringiensis kurstaki* HD-1 (Dipel$_®$) in 2_{nd} instar caterpillars of *A. gemmatalis* (Lepidoptera, Noctuidae), and observed that when both entomopathogens are utilized simultaneously they are more efficient, because they caused alterations in the intestinal cells after 6 HAT while when used separately they produced the alteration only after 12 HAT.

If the dose of toxin administered does not cause the death of the insect, its cells are substituted permitting normal alimentation and the recuperation of the insect [32]. Several studies report on the cell changes produced in the middle intestine of caterpillars intoxicated with Cry proteins from *B. thuringiensis* - for example: the increase in the volume of the epithelium cells, the rupture of the microvilli, cytoplasmic vacuolization, the changes in cytoplasmic organelles and cell hypertrophy[30,33-35].

Studies of the mode action of the *B. thuringiensis* Cry proteins seek to clarify the mechanisms by which these proteins produce their entomopathogenic effects and to elucidate the specificity of the various toxins.

Mode Action of the *Bacillus Thuringiensis* Cry Proteins

When the crystals of *B. thuringiensis* are ingested by susceptible insects they are solublized under alkaline conditions in the middle intestine and then broken into smaller fragments by proteases [5, 36,37]. Binding occurs because of the association of the activated toxin with specific proteins located in the microvilli of the epithelial cells of the middle intestine [38] and is followed by the formation of the pore [39] – the ion flux from this pore leads to cell lyses and consequently, to the death of the susceptible insect [5].

The solubility of the protein and crystals in liquids with alkaline pH values such as those in the middle intestine of the insects liberates the protoxins of 130-140kDafor Cry1 and 70kDa for Cry2. This phase determines the specificity of the *B. thuringiensis* isolate to the target species, both because of the alkalinity of the digestive system and because of the composition of the *B. thuringiensis* crystals.

- The protoxins are activated by the digestive enzymes, forming toxic fragments of 60-65kDa. At this stage both the photolytic composition and the structure of the crystal protein are important. The toxins recognize specific receptors in the microvilli of the epithelial cells of the middle intestine of susceptible larvae to which they bind. Studies made with BBMV (*Brush Border Membrane Vesicles*) isolated from the larvae of Lepidoptera show that the strong binding affinity between the toxin and the receiver is considered an important factor in determining the insecticidal spectrum of the Cry proteins [40]. Research data demonstrates that there is a positive correlation between binding, in vitro, of the toxin in the intestinal receptor and the toxicity, in vivo. On the other hand, other studies describe that while the recognition of the receptor is necessary it is not sufficient in itself to provoke the toxicity, which suggests the existence of other factors related to the mode of action of the Cry proteins. In 1994, Knight et al. [41] isolated from the BBMV of *Manduca sexta*(Lep., Sphingidae) larvae an aminopeptidase N implicated in the interaction of the toxin Cry1Ac. The receptor models now described demonstrate that an insect may present, in variable quantities, various kinds of receptors which could be recognized by different toxins. Research data show that these models may explain the specificity of the *B. thuringiensis* toxins.

- After recognition of the receptor, the toxins induce the formation of pores in the cellular membrane of the intestinal epithelia.
- The formation of pores in the cellular membrane provokes an ionic imbalance between the cytoplasm and the cell's external environment. Histopathological analysis performed after the intoxication of insects reveal destruction of the microvilli, epithelial cell hypertrophy,cytoplasmic vacuolization and cellular lyses which cause paralysis and death of the insect.

When selecting insecticidal proteins, synthesized by this bacteria,a preliminary in vitro analyzes of membrane receptors can facilitate a rapid determination of the range of action of Cry proteins against the target species. The evaluation of the toxicity in vivo can then be limited to those isolates which were pre-selected as active in vitro.

Detection of Receptors in Tissues of Insects

Preparation of the Tissues of Insects

The insect's digestive tubes are dissected and fixed for 24h in *Bouin Hollande Sublimé*10%, washedindistilled water for 12 hours and dehydrated in an ethanol mixture increasing in strength gradually from 70 to 100% [42]. Thetissues are thenimpregnated in mixed baths (ethanol/toluene/Paraplast) and embedded in 100% Paraplast at 58°C. The longitudinal or transverse sections, 7-10 mm thick, prepared using LKB microtome are mountedon glass slides, tanadaswithpoly-l-lysine at 10% and stored at 4°C for subsequent analysis of the histology of the receptors.

Labeling of Cry Proteins with Biotin

Cry proteins can be biotinylated, according to the method described in Bayer & Wilcheck [43], where the incorporation of biotin at the N-terminal of the protein is done using BNHS (biotinyl-N-hydroxysuccinimide) ina buffer of sodium bicarbonate, pH 9.

The reaction product should be purified with Sephadex G-25 (Sigma), and the fractions biotinylated and identified by dot-blot, which usesa nitrocellulose membrane, the conjugate of streptavidin alkaline phosphatase diluted in TST buffer (Tris-Triton-Saline, pH 7.6) and the revelation substrate(BCIP and NBT in a Tris buffer, pH9.5).

The concentration of biotinylated Cry proteins can be determined by the Bradford method [44] using BSA as the protein standard. The purity and integrity of the labeled proteins can be evaluated by western blot using a nitrocellulose membrane (Sigma) and Towbin buffer, pH 8.3 with 10% ethanol. The membranes can be developed usingthe same technique described in the dot-blot.

In Vitro Detectionofmembrane Receptors

Pre-treatment of Tissues

The in vitro analyses of the receptors of Cry proteins of *B. thuringiensis* are performed with histological sections of the digestive systems of larvae being studied. The actual detection results from incubation with the tissue proteins, wich have been previously dew axed and rehydrated as shown in the drawing.

Reactions with Biotinylated Proteins [45]

In the analyses with biotinylated Cry proteins, histological sections are incubated at room temperature for 1 hour with the biotinylated proteins. Proteins not bound to the receptor sites are removed with TST, pH 7.6.

At thenext step, the tissuesare treated with streptavidin conjugated to an enzyme (peroxidase or alkaline phosphatase) or fluorochrome (fluorescein or phycoerythrin), diluted in a TST buffer. The resulting "protein-receptor" complex reaction, using the enzyme conjugate, can be developed with the

DAB substrate for peroxidase (Figure 2A) and with BCIP/NBT for alkaline phosphatase (Figure 2B). The sections are fixed with Pertex mounting medium between the slide and the glass cover slip.

To develop tissues treated with fluorescein the histological sections are mounted with *Mowiol* and stored at 4°C for analysis by optical microscopy (OM - Figure 3A) or by laser scanning microscopy (LSM – Figure 3B).

Immune Detection Reactions of Protein Receptors

In immunohistochemical analyzes with unlabeled proteins (native preoteins), the receptors are developed with the primary antibody (AC_1 specifically against the Cry protein) and secondary antibody (AC_2, directed against the AC_1) conjugated to an enzyme or fluorochrome, which are developed and assembled according to the method described above.

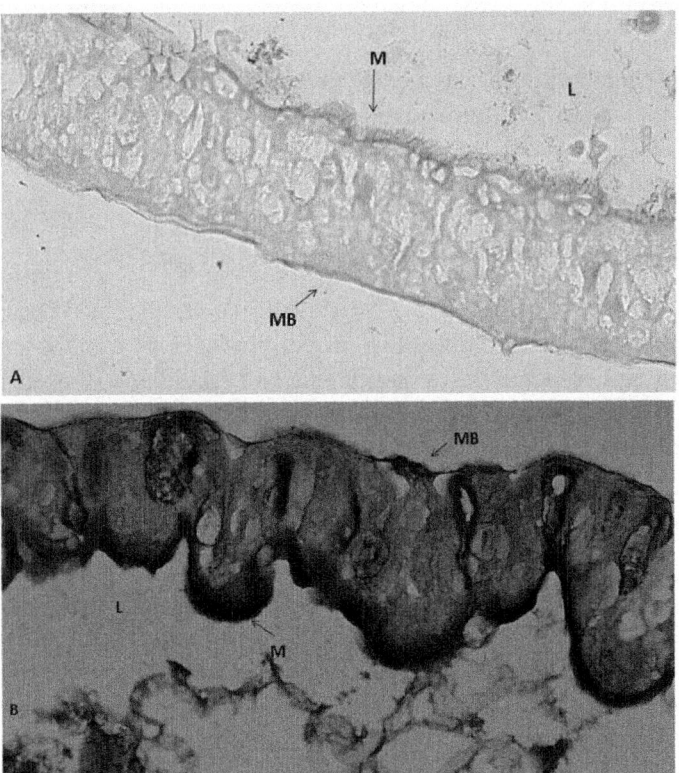

Figure 2. Cry proteins biotinylated receptor, in Lepidoptera insect tissue revealed with peroxidase (A) and alkaline phosphatase (B). (MB) Basal Membrane; (L) Lumen; (M) microvilli.

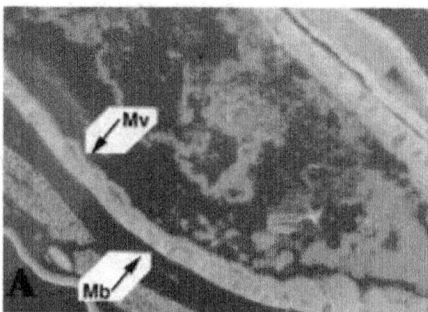 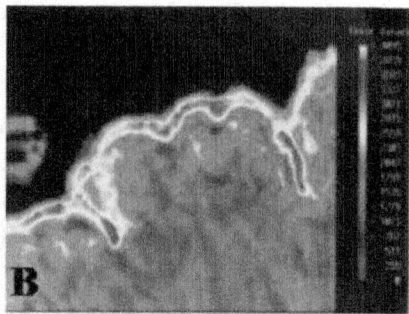

Figure 3. Receivers biotinylated Cry proteins, detected in the tissues of insects with fluorescein and observed in MO (A) and MVL (B) [45].

In immuno localization, the caterpillars are previously treated in vivo with the Cry proteins and after wards the tissues and the immunohistochemical reactions are prepared.

In both methods, the controls are prepared by alternative omission of each step of the reaction, to eliminate the possibility of false-positive reactions. The samples developed with enzymes like peroxidase and alkaline phosphatase can be evaluated by optical microscopy.

PLANT BIOINSECTICIDES

Until quite recently the use of chemical insecticide has been the most widely used method of controlling insect pests. However, the products are expensive and in some cases inefficient and hazardous if used intensively and/or incorrectly. However, some success has been achieved in programs of Integrated Pest Management (IPM) with an alternative to chemicals extracted from various secondary metabolites present in roots, leaves and seeds of plants as an alternative to chemicals called "plant pesticides" [46-48].

The natural products obtained from vegetable raw materials present a wide variety of molecules, with great diversity in structure and biological activity [49]. This wide range of new sites of action on target organisms can be considered another reason for the growing interest in phytotoxins, because, even if not commercially available able, they may suggest lines for the synthesis of entirely new products [50]. This is important if we consider the speed with which the insects and micro-organisms have developed resistance to chemicals commonly used as biological control agents of target species.

According to Mello & Silva-Filho [51], the components of insecticides can be divided into the following groups: (i) derivatives of chemical compounds (tannins, terpenoids, flavonoids, alkaloids, quinones, linomoides, phenols), (ii) molecules produced from the processing proteins (chitinase, lectins, inhibitors alpha-amylase and proteinase inhibitors) and (iii) volatile compounds of plants such as essential oil. In this case, the chemical components can be divided into two classes. The first, based on biosynthesis, in which are found the derivates of the terpenoids, is formed via the mevalonic acetate, and the second, by derivatives which are located in the phenylpropanoid, aromatic compounds formed by way of shikimic acid [52].

Normally, the oily essential in the leaves and resins contain some constituents in high concentration, and about 30-40 minor compounds at concentrations less than 1% [53]. These substances are found in plants in the form of complexes, whose components are integrated and reinforce its action on the organism. Even when the plant has only one active principle, this has a beneficial effect superior to that produced by the same substance produced by chemical synthesis. However, the use of botanical insecticides depends on identification of the active compounds, their mode of action, production, formulation, stability, dose, action on natural enemies, field persistence, toxicity tests for record, among others [54].

The Laboratory of Microbiology and Toxicology at the UNISINOS University uses two models for histopathologic analysis:

- Following treatment with plant extracts or oily essences 10 larvae of the insect target are collected in periods of 1, 3, 6, 9, 12, 24 and 48 hours after treatment (HAT).The specimens are fixed in aqueous paraformaldehyde and then dehydrated in ethanol solutions of increasing strength [50, 70, 90 and 100%). After dehydration, the larvae are embedded in resin (Leica Historesin)for 12 h. After that the specimens are put into molds of solid polypropylene using the same resin with a polymerize. The resin blocks are mounted on supports after polymerization, and cut into 3mm thick sections in a microtome with a glass blade. The histological sections are stretched out onto glass slides and stained with Schiff-Naphtol Blue Black periodic acid, dehydrated, and mounted between the glass slides and the cover slips with *Entellan*.

- After application of the treatments with plant extracts or essential oil, 10 target insect larvae in each treatment, are collected in periods of 3, 6, 9, 12, 24 and 27HAT. After fixation in *Bouin Hollande Sublime-BHS* for 24 hours [38], the larvae are subjected to dehydration in increasingly strong ethanol solutions, followed by rapid baths of xylene and impregnation in paraffin. Longitudinal histological sections of the midgut are realized at a thickness of 5m. Coloration is applied with *Azul de Heidenhain* [41], which differentiates the microvilli of the middle intestine by the presence of glycoproteins which can be observed by the system of comparative histology in an optical microscopy.

Knaak et al. [55] found that the toxicity of the extracts of *Petivesia alliacea, Zingiber officinale, Ruta graveolens, Malva silvestris, Baccharis genistelloides* and *Cymbopogon citratus* caused damage such as: vacuolization of the cytoplasm, disruption of microvilli, peritrophic membrane destruction and cell changes in the midgut of *S. frugiperda* (Figure 4). This study the changes mentioned in the previous work were not observed until 48HAT, which was the maximum time rated.

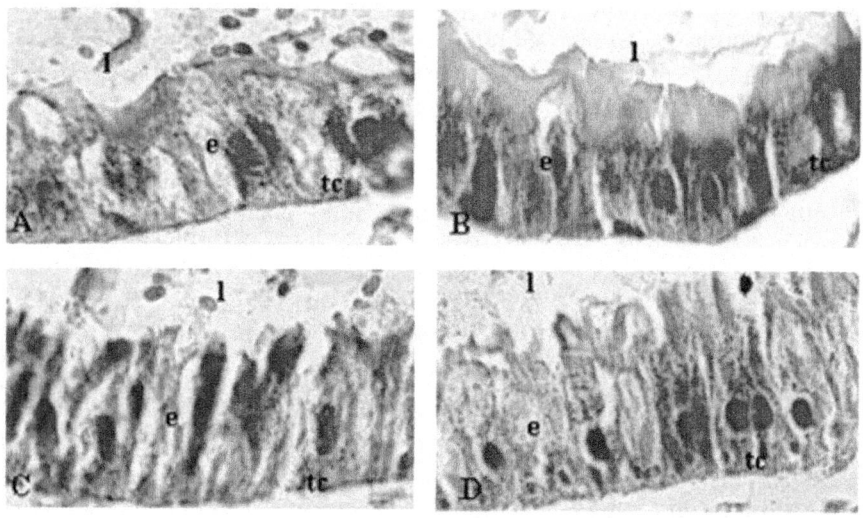

Figure 4. Longitudinal sections of the midgut of *Spodoptera frugiperda* caterpillars treated with plants extrats, (A) *Petivesia aliacea* (24 HAT), (B) *Petivesia aliacea* (24 HAT); (C) *Zingiber officinale* (6 HAT); (D) *Zingiber officinale* (27 HAT); (E) *Z. officinale* (6 HAT); (F) *Z. officinale* (24 HAT); (G) Controle. Increase 400X; bar=2.44m; =changes; TC=connective tissue; M=microvilli; E=epithelium; L= lumen.

The chemical compounds (tannins, terpenoids, flavonoids, alkaloids, quinones, linomoides,phenols), molecules produced from the processing of proteins (chitinase, lectins, alfa-amylaseinhibitor and proteinase inhibitors) and volatile compounds from the plant [56, 57], present in plantextracts, undergo different changes according to the physico-chemical conditions along the digestivetract of insects.

In treatments with the essential oil of *R. graveolens* and *Malva* sp., cellular projections were observed in the intestinal lumen at 9 HAT, and the oil of *R. graveolens* caused cell elongation at 12 HAT. Thus, it appears that changes in the structure of the midgut of *S. frugiperda* may increase during the treatment with essential oils. The intensity of the pathological effects is dependent on time period and the concentration used [58].

In the treatments with essential oil of *Z. officinale* and *C. citratus*, was observed projections of epithelial cells in the lumen, at 3 and 6HAT respectively (Figure 5A, D and E). After 24 hours treatment with *C. citratus* (Figure 5B), we observed elongation of the microvilli, while in the treatment with *Z. officinale*cell elongation occurred (Figure 5G). At 24HAT with the *Z. officinale* oil several morphological changes occurred, such as: cellular disorganization, destruction of the epithelium, of the microvilli and the peritrophic membrane. It can still be observed that the changes are dependent on the time period and the concentrations used [56].

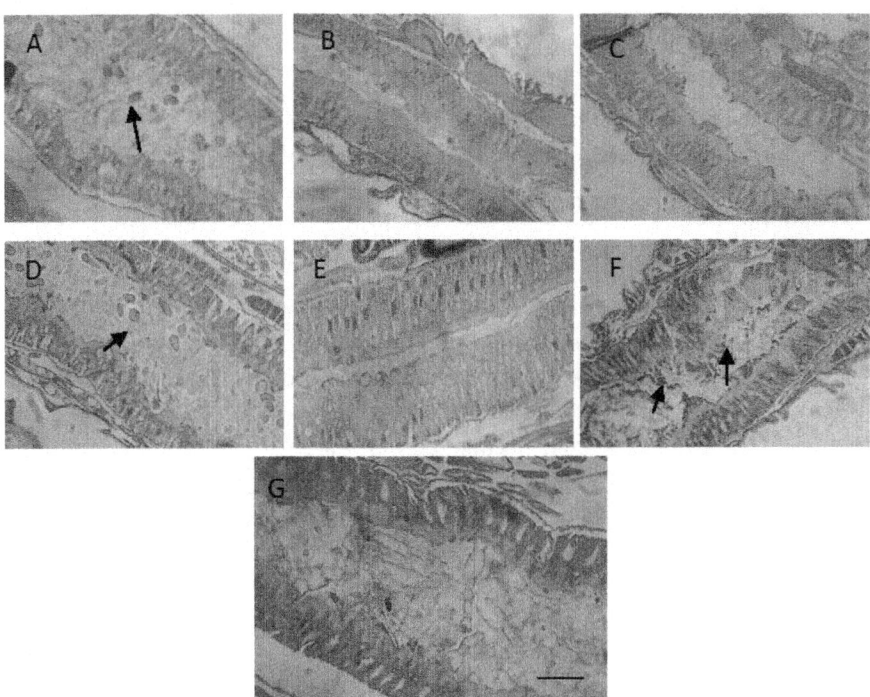

Figure 5. Longitudinal sections of the midgut of *Spodoptera frugiperda* caterpillars treated with essential oils, (A) *Cymbopogon citratus* (6 HAT) (B) *C. citratus* (24 HAT); (C) *C. citratus* (48 HAT); (D) *Zingiber officinale* (3 HAT); (E) *Z. officinale* (6 HAT); (F) *Z. officinale* (24 HAT); (G) Controle. Increase 400X; bar=2.44m; =changes; TC=connective tissue; M=microvilli; E=epithelium; L= lumen.

Knaak et al. [55] evaluated the effectof the interaction of various plant extracts with Xentari,*B. thuringiensis aizawai*, in the midgut of *S. frugiperda*, demonstrating that the histopathological effects of*Z. officinale, M. silvestris, R. graveolens* and *B. genistelloides,*in the midgut of*S. frugiperda*were more intense when compared to extracts of *P. alliacea and C.citratus*, which showed a positive interaction with Xentari, accelerating the process of destruction of intestinal cells, which represents a reduction in the lethal time of the target species,*S. frugiperda*.

Some plants are outstanding, for instance, *Melia azedarach*, a plant similar to*Azadirachtina indica*which produces azadirachtine efficiently for more than 400 species of insects [58], its active ingredient is already a commercial product called Neemix 4.5. This causes different reactions in insects,acting as an alimentary inhibitor retarding growth, reduces fertility, and causes morphogenetic and behavioral changes [59].

The toxicity of cinnamon is related to the presence of different compounds, such as salanalina, meliaterina and meliacarpinina E, from which additionally other derivatives can be obtained which show insecticidal action against the Coleopteran orders (Curculionidae, Tenebrionidae and Chrysomelidae) and Lepidoptera [60-62], amongst others.

Correia et al. [63] evaluated the histology of the alimentary canal of larvae of *S. frugiperda* treated with neem (Neemseto®) at concentrations 0.5 and 1.0%,on non-treated leaves. They found that regions of the stomodeum and proctodeus showed no morphological changes. However, the middle intestinal region, showed morphological changes that varied in intensity according to the exposure time and concentration (0.5 and 1.0%). After 48h and 96h of treatment, tissue changes were observed in the larvae treated with neem – the cells of the epithelial lamina became slender and elongated. It was not possible to distinguish cell types beyond reducing significantly reducing the secretion activity of the two concentrations studied, in comparison with the control.

CONCLUSION

In toxicological analysis of microbial biopesticides,with *Bacillus thuringiensis* as the object of analysis, thelabeling detected in the region of the microvilli of the cells of the intestinal epithelial cells revealed the presence of membrane receptors of Cry proteins in the midgut of the larvae of the insects when compared to the controls representing the alternative omission of different components of the reactions, as can be seen by the absence of staining on the microvilli of intestinal epithelial cells of the insects used here as a model of this approach to the study. The imunedetections and the detection of biotinylated Cry proteins were used by various authors, for identifying the intestinal membrane receptors, in larvae of different species of Lepidoptera [33, 45, 64-65], Diptera [66] and Coleoptera [33,67]. These authors proved that the binding of the Cry proteins to the microvilli of the insect's midgut indicate the existence of a receptor specific to the cited protein in the target insect.

Considering therefore, the studies of receptors in vitro and the toxicological analyses in vivo, it can be confirmed that the methods for detection of membrane receptors can be applied in selecting the toxins of *B. thuringiensis* that are active against insects, and that there is a positive correlation between the in vitro and the in vivo analysis. However, to determine the median lethal concentration (LC_{50}) of a Cry protein a bioassay is necessary since the binding of proteins to the receptors may vary in concentration and affinity, as previously described by several authors for different insect species [64, 68, 69, 70]. Other studies show that the Cry proteins that are toxic to insects correspond to those which bind irreversibly to the epithelial cell receptors of the target insects [71].

In the case of plant insecticides, chemical compounds, (tannins, terpenoids, flavonoids, alkaloids, quinones, phenols and linomoides), the molecules produced from the processing of proteins (chitinase, lectins, inhibitors alpha-amylase and proteinase inhibitors) andvolatile compounds from plants, such as the oily essences present in plants [57]undergo different changes according to the physical and chemical conditions along the digestive tract of the insects.

In recent years,in the search for alternative methods to control agricultural pests without using artificial chemical substances, research was expanded to obtain greater knowledge on the mode action of microbial toxins and plant insecticides. Studies realized in vitro on the cells and tissues of the alimentary canal of insects, especially of Lepidoptera, have identified and located the receptors in the target species. Thus,for the application of plant and microbial toxins as biopesticides, it is fundamental to evaluate the range of action and the specificity of the active ingredient. In this context, Fiuza [72] mentions that the analysis in vitro of membrane receptors can be considered an indispensable tool due to the large number of botanical compounds plus the isolates, the strains and proteins of B. thuringiensis that have been identified, and which have considerable potential for pest management.

REFERENCES

1. O. Cáceres, et al.1981Resíduos de pesticidas clorados em água das cidades de São Carlos e Araraquara. Ciência e Cultura, São Paulo, 33
2. L. H. Aguiar, G. Moraes, 1999Hepatic alanine and aspartic amino transferases of the freshwater teleost. Bryncephalus (Matrinchã) exposed to the organophosphorous methyl parathion (folidol 600

registred), Fish response totoxic-environments, Kennedy, Canadá. 145152

3. C. R. Carlini, M. F. Grossi-de-Sá, 2002Toxicology of the Bioinsecticides Used in Agricultural Food ProductionToxicon, 401515 EOF39 EOF
4. S. B. Alves, 1998Toxicology of the Bioinsecticides Used in Agricultural Food Productioned. Piracicaba: FEALQ, 1163p.
5. E. Schnepf, N. Crickmore, J. Van Rie, D. Lereclus, J. Baum, J. Feitelson, D. R. Zeigler, D. H. Dean, 1998Bacillus thuringiensis and its pesticide crystal proteins. Microbiology Molecular Biology Reviews, 62
6. M. L. Saito, F. Lucchini, 1998Substâncias obtidas de plantas e a procura por praguicidas eficientes e seguros ao meio ambiente. Jaguariúna: EMBRAPA-CNPMA. 46p.
7. R. F. Chapman, 1998Toxicology of the Bioinsecticides Used in Agricultural Food Productionth ed., Cambridge, Cambridge University Press, 788p.
8. S. M. Levy, A. M. F. Falleiros, E. A. Gregório, N. R. Arrebola, L. A. Toledo, 2004Toxicology of the Bioinsecticides Used in Agricultural Food ProductionBrazilian Journal of Biology64
9. . Mordue, Luntz, A. J. , A. Blackwell, 1993Toxicology of the Bioinsecticides Used in Agricultural Food ProductionJournal of Insect Physiology39903 EOF924 EOF
10. A. S. Chiang, D. F. Yen, W. K. Peng, 1986Toxicology of the Bioinsecticides Used in Agricultural Food ProductionJournal of Invertebrate Pathology47333 EOF339 EOF
11. . Mordue, Luntz, A. J. , A. J. Nisbet, 2000Azadirachtin from the neem tree Azadirachta indica: its action against insects. An. Soc. Entomol. Brasil, 29
12. M. J. Lehane, 1997Toxicology of the Bioinsecticides Used in Agricultural Food ProductionAnn Ver.Entomol, 42525 EOF50 EOF
13. W. R. Terra, 2001Toxicology of the Bioinsecticides Used in Agricultural Food ProductionArch. Insect Biochem. Physiol., 4747 EOF61 EOF
14. W. R. Terra, C. Ferreira, 2005Biochemistry of digestion. In Comprehensive Molecular Toxicology of the Bioinsecticides Used in Agricultural Food Production171224Elsevier, Oxford.
15. D. O. Pinheiro, I. Quagio-Grassiotto, E. A. Gregório, 2008Toxicology of the Bioinsecticides Used in Agricultural Food ProductionNeotropical Entomology37413 EOF419 EOF

16. D. O. Pinheiro, R. J. Silva, I. Quagio-Grassiotto, E. A. Gregório, 2003Toxicology of the Bioinsecticides Used in Agricultural Food ProductionNeotropical Entomology32
17. V. Wanderley-Teixeira, A. A. C. Teixeira, F. M. Cunha, M. K. C. M. Costa, A. F. S. L. Veiga, J. V. Oliveira, 2006Histological description of the midgut and thepy loric valve of Tropida criscollaris (Stoll, 1813) (Orthopetera: Romaleidae). Brazilian Journal of Biology,66
18. G. F. Martins, L. A. O. Neves, J. E. Serrão, 2006Toxicology of the Bioinsecticides Used in Agricultural Food ProductionMicron, 37161 EOF8 EOF
19. A. A. Correia, V. Wanderley-Teixeira, A. A. C. Teixeira, J. V. Oliveira, J. B. Torres, 2009Toxicology of the Bioinsecticides Used in Agricultural Food ProductionNeotropical Entomology38
20. V. M. Cavalcante, C. Cruz-Landim, 19 EOF40 EOF1999Toxicology of the Bioinsecticides Used in Agricultural Food ProductionNaturalia, 24
21. S. M. Levy, A. M. F. Falleiros, E. A. Gregório, N. R. Arrebola, L. A. Toledo, 2004Toxicology of the Bioinsecticides Used in Agricultural Food ProductionBraz J Biol, 64
22. M. S. Harper, T. L. Hopkins, 1997Toxicology of the Bioinsecticides Used in Agricultural Food ProductionTissue Cell, 29463 EOF75 EOF
23. M. S. Harper, R. R. Granados, 1999Toxicology of the Bioinsecticides Used in Agricultural Food ProductionTissue Cell, 31202 EOF
24. L. M. N. P. Pinto, Histologia do intestino médio de lagartas de Anticarsia gemmatalis Hübner 1818Lepidoptera, Noctuidae). Acta Biologica Leopoldensia, 22
25. N. Crickmore, D. R. Zeigler, E. Schnepf, J. Van Rie, D. Lereclus, J. Baum, A. Bravo, D. H. Dean, 2005Bacillus thuringiensis toxin nomenclature. (Online.) http://www.lifesci.su
26. L. D. Marroquin, D. Elyasnia, J. S. Griffitts, J. S. Feitelson, R. V. Aroian, 2000Toxicology of the Bioinsecticides Used in Agricultural Food Productionetics, 155
27. A. M. Shelton, J. Z. Ahao, R. T. Roush, 2002Economic, ecologycal, food safety and social consequences of Toxicology of the Bioinsecticides Used in Agricultural Food Production
28. L. M. Fiuza, D. L. Berlitz, 2009Produtos de Bacillus thuringiensis: registro e comercialização. Biotecnologia Ciência & Desenvolvimento, 38

29. D. L. Berlitz, L. M. Fiuza, 2004Avaliação toxicological de Bacillus thuringiensis aizawai para Spodoptera frugiperda (Lepidoptera: Noctuidae), em laboratório. Biociências, 12
30. N. Knaak, A. R. Franz, G. F. Santos, L. M. Fiuza, 2010Histopathology and the lethal effect of Cry proteins and strains of Bacillus thuringiensis Berliner in Spodoptera frugiperdaJ.E. Smith Caterpillars (Lepidoptera, Noctuidae). Brazilian Journal Biology, 70
31. N. Knaak, L. M. Fiuza, 2005Toxicology of the Bioinsecticides Used in AgriculturalFoodProductionBrazilianJournalofMicrobiology362195199
32. A. F. Spies, K. D. Spence, 1995Toxicology of the Bioinsecticides Used in Agricultural Food Productionand Cell, 17
33. A. Bravo, K. Hendrickx, S. Jansens, M. Peferoen, 1992Toxicology of the Bioinsecticides Used in Agricultural Food ProductionJournal of Invertebrate Pathology60247 EOF
34. V. M. Griego, L. J. Fancher, K. D. Spence, 1980Scanning electron microscopy of the disruption of tobacco horn worm, Manduca sexta, midgut by Bacillus thuringiensis endotoxin. Journal of Invertebrate Pathology, 350
35. S. Mathavan, P. M. Sudha, S. M. Pechimuthu, 1989Effect of Bacillus thuringiensis on the midgut cells of Bombyxmori larvae: A histopathological and histochemical study. Journal of Invertebrate Pathology, 53
36. Z. Shao, Y. Cui, H. Yi, J. Ji, Z. Yu, 1998Processing of D-endotoxin of Bacillus thuringiensis subsp. Kurstaki HD-1 inHeliothis armigera midgut juice and the effect of proteases inhibitors. Journal of Invertebrate Pathology, 72
37. A. Tojo, K. Aizawa, 1983Dissolution and degradation of Bacillus thuringiensis δ endotoxin by gut juice protease of the silkworm Bombyx mori. Appl. Environ. Microbiology, 45
38. J. L. Schwartz, M. Juteau, P. Grochulski, M. Cygler, G. Prefontaine, R. Brousseau, L. Masson, 1997Toxicology of the Bioinsecticides Used in Agricultural Food ProductionFEBS Lett., 410
39. L. Masson, B. E. Tabashnik, Y. B. Liu, J. L. Schwartz, 1999Helix 4 of the Bacillus thuringiensis Cry1Aa toxin lines the lumen of the ion channel. Journal of Biology Chemical, 274
40. L. M. Fiuza, C. Nielsen-Leroux, E. Goze, R. Frutos, J. F. Charles, 1996Binding of Bacillus thuringiensis Cry1 toxins to the midgut brush border membrane vesicles of Chilosuppressalis (Lepidoptera, Pyralidae):

evidence of shared binding sites. Applied Environmental Microbiollogy, 62

41. P. Knight, N. Crickmore, D. Ellar, 1994The receptor for Bacillus thuringiensis CryIAc delta-endotoxin in the brush border membrane of the lepidopteran Manduca sexta is aminopeptidase N. Molecular Microbiology, 11

42. P. Brandtzaeg, 1982Tissue preparation methods for imunocytochemistry. In: Bullock, G. & Petruz, P., Techniques in imunocytochemistry. Academic Press, London. 4951

43. E. Bayer, M. Wilcheck, 1990Protein biotinylation. Methods in Enzymology, 184

44. M. Bradford, 1976Toxicology of the Bioinsecticides Used in Agricultural Food ProductionAnalytical Biochemistry, 72

45. L. M. Fiuza, 1995Etude des sites récepteurs et de la toxicité des delta-endotoxines de Bacillus thuringiensis Berliner chez les larves de la Pyrale du riz, Chilo suppressalis Walker. Thèse de doctorat en Sciences Agronomiques, ENSA-M, Montpellier, France. 180p.

46. H. Schmutterer, 1990Toxicology of the Bioinsecticides Used in Agricultural Food ProductionAnnu. Rev. Entomology, 35

47. A. R. Roel, J. D. Vendramim, R. T. S. Frighetto, N. Frighetto, 2000Atividade tóxica de extratos orgânicos de Trichilia pallida Swartz (Meliaceae) sobre Spodoptera frugiperda (J.E. Smith). An. Soc. Entomol. Brasil, 29

48. A. R. Roel, 2001Utilização de plantas com propriedades inseticidas: uma contribuição para o desenvolvimento rural sustentável. Ver Int Desenv Local 1

49. M. Reigosa, N. Pedrol, 2002Toxicology of the Bioinsecticides Used in Agricultural Food ProductionPlymouth: Science Publishers, 316p.

50. S. O. Duke, F. E. Dayan, A. M. Rimando, 2000Natural products and herbicide discovery. IN: Cobb, A.H.; Kirkwood, R.C. (Ed.). Toxicology of the Bioinsecticides Used in Agricultural Food ProductionSheffield: Academic Press, 105133

51. M. O. Mello, M. C. Silva-Filho, 2002Plant-insect intections: na evolutionary arms race between two distinct defense mechanisms. Brazilian Journal of Plant Physiology, 14

52. J. O. Strapazzon, 2004Composição química e análise antimicrobiana do óleo volátil de Annona squamosa L. (Ariticum.) Chapecó, SC. Monografia de graduação. Universidade Comunitária Regional de Chapecó, 53 p.

53. J. Takabayashi, M. Dicke, M. A. Posthumus, 1994Volatile herbivore-induced terpenoids in plant-mite interactions: variation caused bybiotic and abiotic factors. Journal of Chemical Ecology, 20http://dx.doi.org/10.1007/BF02059811
54. S. S. Martinez, H. F. Emdem, 2001Redução do crescimento, deformidades e mortalidade de Spodoptera littoralis (Boisduval) (Lepidoptera: Noctuidae) causadas por Azadiractina. Neotropical Entomology, 30
55. N. Knaak, M. S. Tagliari, L. M. Fiuza, 2010Histopatologia da interação de Bacillus thuringiensis e extratos vegetais no intestino médio de Spodoptera frugiperda (Lepidoptera: Noctuidae). Arquivos do Instituto Biológico, São Paulo, 77
56. N. Knaak, 2011Potencial dos óleos essenciais e proteínas vegetais, obtidos de plantas medicinais, no controle de pragas da cultura do arroz irrigado.Tese de Doutorado, Programa de Pós-Graduação em Biologia da Universidade do Vale do Rio dos Sinos. 107p.
57. C. M. O. Simões, E. P. Schenkel, G. Gosmam, J. C. P. Mello, L. A. Mentz, P. R. Petrovick, 2003Toxicology of the Bioinsecticides Used in Agricultural Food Productioned., Porto Alegre, Universidade, 1102 p.
58. S. S. Martinez, 2002O nim- Azadirachta indica: natureza, usos múltiplos, produção. Londrina, Instituto Agronômico do Paraná, 142p.
59. M. Breuer, B. Hoste, A. De Loof, S. N. H. Naqvi, 2003Toxicology of the Bioinsecticides Used in Agricultural Food ProductionPestic. Bioch. Physiology, 7699 EOF103 EOF
60. Y. Huang, J. M. W. L. Tan, R. M. Kini, S. H. Ho, 1997Toxicology of the Bioinsecticides Used in Agricultural Food ProductionJournal of Stored Products Research4289 EOF298 EOF
61. F. I. Bohnenstengel, V. Wray, L. Witte, R. P. Srivastava, P. Proksch, 1999Inseticidal meliacarpins (C-seco limnoids) from Melia azedarach. Phytochemistry, 50
62. M. C. Carpinella, M. T. Defago, G. Valladares, S. M. Palacios, 2003Antifeedant and inseticide proprieties of a limnoid from Melia azedarach (Meliaceae) with potential use for pest management. J. Agric. Food Chemistry, 51
63. A. A. Correia, V. Wanderley-Teixeira, A. A. C. Teixeira, J. V. Oliveira, J. B. Torres, 2009Toxicology of the Bioinsecticides Used in Agricultural Food Productionm. Neotropical Entomology3883 EOF91 EOF
64. P. Denolf, S. Jansens, M. Peferoen, D. Degheele, J. Van Rie, 1993Toxicology of the Bioinsecticides Used in Agricultural Food Production

65. U. Estada, J. Ferre, 1994Toxicology of the Bioinsecticides Used in Agricultural Food Production60
66. O. Ravoahangimalala, J. Charles-F, J. Schoeller-Raccaud, 1993Immunological localization of Bacillus thuringiensis serovar israelensis toxins in midgut cells of intoxicated Anopheles gambie larvae (Diptera: Culicidae). Research of Microbiology, 144
67. A. Boets, S. Jansens, P. Denolf, M. Peferoen, D. Degheele, J. Van Rie, 1994Sequential observations of toxin distribution and histopathological effects of CryIIIA in the gut of intoxicated Leptinotarsa decemlineata larvae. XXVIIth Annual Meeting of the Society for Invertebrate Pathology, 377
68. J. Van Rie, S. Jansen, H. Höfte, D. Degheeled, H. Van Mellaert, 1990Receptors on the brush border membrane of the insect midgut as determinants of the specificity of Bacillus thuringiensis δ-endotoxins. Applied Environ. Microbiology, 56
69. M. K. Lee, T. H. You, B. A. Young, J. A. Cotrill, A. P. Valatis, D. H. Dean, 1996Aminopeptidase N purified from Gypsy moth brush border membrane vesicles is a specific receptor for Bacillus thuringiensis Cry1Ac toxin. Applied Environmental Microbiollogy,62
70. G. Hua, L. Masson, J. L. Jurat-Fuentes, G. Schwab, M. J. Adang, 2001Binding Analyses of Bacillus thuringiensis Cry delta-endotoxins using brush border membrane vesicles of Ostrinia nubilalisApplied Environmental Microbiollogy, 67
71. Y. Liang, S. S. Patel, D. H. Dean, 1995Irreversible binding kinetics of Bacillus thuringiensis Cry1A delta-endotoxins to Gypsy moth brush border membrane vesicles is directly correlated to toxicity. Journal of Biology Chemical, 270
72. L. M. Fiuza, 2009Mecanismo de ção de Bacillus thuringiensis. Biotecnologia Ciência & Desenvolvimento, 38

Chapter 3

GENOTOXICITY INDUCED BY OCUPATIONAL EXPOSURE TO PESTICIDES

Danieli Benedetti[1], Fernanda Rabaioli Da Silva[2], Kátia Kvitko[2], Simone Pereira Fernandes[2], and Juliana da Silva[1]

[1]Laboratório de Genética Toxicológica, Universidade Luterana do Brasil - ULBRA, Canoas, Brazil

[2]Departamento de Genética, Instituto de Biociências, Laboratório de Imunogenética. Universidade Federal do Rio Grande do Sul - UFRGS, Porto Alegre, Brazil

INTRODUCTION

Pesticides are used to repel, kill or control certain forms of pests, e.g. animals or plants. These chemical compounds can be divided into three main classes: insecticides, which are used to control insects; herbicides, which are used to destroy unwanted vegetation; and fungicides, which are used to control fungi and their spores, preventing them from damaging plants (Maroni et al., 2000). Pesticides are employed extensively around the world and in recent years their use has even increased. On one hand, extensive use of pesticides in farming has lead to a higher production of pests that damage crops and, on the other hand, pesticide-resistant pests have emerged. Increased crop production demands increased use of pesticides (Mostafalou and Abdollahi, 2013).

The widespread use of agricultural chemicals in the food production and public health sectors has released large amounts of potentially toxic substances into the environment, most of which are unspecific and therefore potentially also target the human organism (Bolognesi, 2003; Dyk and Pletschke, 2011). Humans are exposed to the ubiquitous pesticides, e.g. in form of food contaminations through the production line, but also in the household, workplace, hospitals and schools (Bolognesi, 2003; Aprea et al., 2012).

Exposure to pesticides can induce two kinds of toxic effects: acute and chronic. The acute effects are immediate and include headache, nausea and/or other more serious effects, even death. Chronic health effects occur, when individuals are exposed continuously or repeatedly to foreign substances. In the scientific literature, the effects of acute exposure are more clearly described. In contrast to that, effects of chronic exposure still need to be further investigated, especially how they are triggered (Ray and Richards, 2001; Sanborn et al., 2007; Kortenkamp et al., 2007).

The degree of danger associated with chemical exposure can be evaluated by health risk assessments. Chemical exposure can be evaluated with respect to a single compound or to complex mixtures. Mixtures of toxins may influence and even amplify the toxicity of individual components through synergies, potentiation, antagonism, inhibition or additive effects (Muntaz, 1995; Reffstrup et al., 2010). The assessment of chronic exposure to mixtures of pesticides should improve the understanding of underlying intoxication mechanisms (Bond and Medinsky, 1995; Sanborn et al., 2007; Reffstrup et al., 2010). Indeed, the number of studies involving chronic exposure to pesticides and their consequences to human health (Muntaz, 1995; Mostafalou and Abdollahi, 2013) in the scientific literature is increasing. Individuals, who are in direct contact with and exposed repeatedly to low levels of pesticides (e.g. agrochemicals) as part of their work (e.g. agricultural or cargo workers, etc.) may therefore provide a good opportunity to study the deleterious effects of chronic pesticide exposure to human health (Bolognesi, 2003).

OCCUPATIONAL EXPOSURE TO PESTICIDES: TOXICOLOGY AND ABSORPTION PATHWAYS

Pesticides can be classified according to their chemical structures: carbamates (CBM), dithiocarbamates (DTC), synthetic pyrethroids, organochlorines (OC), organophosphorous (OP) compounds, thiocarbamates, phenoxyacetates (PHE), quaternary ammonium compounds and coumarins (Maroni et al., 2000). The individual toxicity of these compound classes is expressed by the dose inducing lethality in 50% of the specimens in tests with laboratory animals (LD50). During these LD50 tests, usually mice are exposed to a single given dose (Maroni et al., 2000; Suiter and Scharf, 2012). In practice however, the toxicity of pesticides should not be evaluated on the basis of a single dose, but by the absorption of small doses over a given time period (Bolognesi, 2003; Kortenkamp et al., 2007; Reffstrup et al., 2010). In addition, agricultural workers are usually exposed to a mixture of pesticides (Bolognesi, 2003;Kortenkamp et al., 2007; Aprea, 2012) and fundamental aspects such as type and duration of the exposure can severely affect the toxicodynamics of the pesticides (Gammon et al., 2012).

In laboratory tests, toxicokinetic models are important in order to determine the kinetic parameters of the active components and to understand chemical interactions between pesticides (Bond and Medinsky, 1995). Toxicokinetics refer to the route a xenobiotic takes to get into, through, and out of the body.

It can be divided into several processes including absorption, distribution, metabolism, and excretion.

The effects of chronic exposure, which pesticides induce in humans, are highly sensitive to several parameters, e.g. dose, duration, and especially the absorption pathway (Aprea, 2012). In agricultural surroundings occupational exposure mostly involves absorption via dermal and/or respiratory routes (Leoni et al., 2012; Aprea, 2012). This type of exposure occurs predominantly during the period of the application of the toxins, e.g. through spraying (Ranjbar et al., 2002). The penetration of the skin itself depends on several factors: type of pesticide, temperature, relative humidity, type of exposed unprotected area of the body (e.g. the back of the hand, wrist, neck, foot, armpit or groin), contact time, and the presence of wounds or skin lesions, which greatly facilitate absorption. Cases of absorption through the gastrointestinal tract are also known, albeit less frequent, because larger pesticide particles tend to be deposited in the upper airways of the respiratory tract (Aprea, 2012).

The knowledge about absorption pathways allows a more apt description of real doses of absorption (together with corresponding toxic effects), rather than a description of the dose, which is considered potentially toxic. For example, Ortiz and Bouchard (2012) demonstrated toxicokinetic effects for the fungicide captan after absorption. Unfortunately, it was impossible to isolate the toxic effects resulting from exposure, because the absorption pathways and toxic doses of this compound in humans are not yet known exactly. Several studies have reported a rapid absorption of organophosphates (OPs) via dermal routes, e.g. through connective and mucous membranes, but gastrointestinal and respiratory absorption routes are also known (Stallones and Beseler, 2002). Gammon et al., (2011) reported minor toxicity for CBMs when absorbed throught he skin, but more serious toxic effects after gastrointestinal incorporation. Pyrethroids are generally unstable under environmental conditions and tend to be rapidly absorbed after degradation through hydrolysis, but don't accumulate in the body (Suiter and Scharf, 2012). In contrast, OCs are relatively stable under comparable conditions and can accumulate when absorbed; Absorption doses can moreover be cumulative, depending on the absorption route (George and Shukla, 2011).

Many of the toxicological effects of pesticides have been demonstrated to be mediated by induced redox signaling. Exposure to a wide variety of pesticides induces oxidative stress, as reflected in the accumulation of reactive oxygen species (ROS), lipid peroxidation and DNA damage. For some pesticides, the mechanisms leading to alterations in the cellular redox homeostasis are partially understood. Pesticides can alter cellular redox equilibria via different mechanisms, including their enzymatic conversion to secondary reactive products (e.g. ROS), depletion of cellular antioxidant defenses and/or impairment of antioxidant enzyme functions (Franco et al., 2009; Limon-Pacheco and Gonsebatt, 2009).

Nutrigenomics and nutrigenetics are recent research areas, which seek to understand the effects of diet and nutrients as genetic response modulators to pesticides. The effects of nutrient deficiencies or imbalances, as well as the toxic concentrations of some dietary compounds have been the subject of nutritional research. About 40 micronutrients are required in an optimal human diet, and their levels may vary depending on age, genetic predisposition, etc. (Ames, 1999; Ames 2001). Most interestingly, the genomic damage caused by moderate micronutrient deficiency is of the same order of magnitude as the levels of damage caused by exposure to high doses of environmental toxins (Kym, 2007; Dangour et al., 2010; Wald et al., 2010). Folates and other B-complex vitamins perform key functions in biological processes pivotal to a healthy constitution. Even moderate folate deficiencies may cause genomic damage in the general population. Folates maintain genome stability by regulating DNA synthesis and repair as well as methylation processes. Deficiencies in folic acid can therefore increase chromosomal instability (Beetstra et al., 2005). A major co-factor in the folate metabolism is vitamin B_{12} and clinical evidence suggests that the inappropriate intake of vitamin B_{12} may result in damage of the DNA. Moreover, chromosome repair mechanisms may be compromised, when vitamin B_{12} concentrations are too low (Swanson et al., 2001). Age and gender are other factors, which may possibly influence the level of DNA damage. Fenech and Bonassi (2011) showed that the damage to DNA increases with age, probably due to a combination of several factors such as inadequate nutrition, occupational or environmental exposure to genotoxins, and a wider variety of other unhealthy lifestyle factors.

PESTICIDE METABOLISM

Metabolism is one of the most important factors in the toxic profile of a pesticide. During the first steps of the metabolism, chemical compounds are bio-transformed by phase I enzymes, usually the cytochrome P450 (CYP) system. Phase II conjugating enzyme systems, which are present in the glutathione complex, subsequently transform these reaction products into more soluble and excretable forms (Guengerich and Shimada 1991; Eleršek and Filipič, 2011). These enzymatic reactions are generally beneficial, since they help to eliminate compounds from the body. Sometimes however, these enzymes transform otherwise harmless substances into highly reactive forms – a phenomenon known as "metabolic activation" (Guengerich 2001; Abass et al., 2010).

The metabolism reacts towards these xenobiotics in phase I by generating functional and/or polar groups, with the goal to create a substrate for enzymatic reactions in phase II (Hodgson and Goldstein, 2001; Parkinson, 2001). The CYP system comprises a large family of multigenes, which are important in the metabolic phase I of xeno- and endobiotics. Their reactions occur predominantly in the liver, which is the place of subsequent eliminations (Abass et al., 2010). Beyond the hepatic tissue, CYP multigenes can be found in the lung (Lawton et al., 1990), brain (Bergh and Strobel, 1992), and kidney tissue (Hjelle et al., 1986), as well as in the gastrointestinal tract (Peters and Kremers, 1989), and dermal (Khan et al., 1989) and mucous membranes (Eriksson and Brittebo, 1991). The result of these catalytic reactions depends on the type of pesticide and can range from induction to enzyme inhibition (Patil et al., 2003). The toxicity of OPs and CBMs for example, can be monitored after exposure by measuring the esterase activity (Chambers et al., 2001). The high toxicity of OPs and CBMs is attributed to their ability to mimic esters (natural compounds present in biological organisms), such as acetylcholinesterase (AChE) and butyrylcholinesterase (BChE) (Ray et al., 1998; Chambers et al., 2001). Metabolic phase I reactions take place in the liver, where chemical bonds between phosphorus and carbon atoms are cleaved by alkylations (methyl- or ethylation), resulting in the formation of active enzyme centers (Wild, 1975). Trough these phosphorylation processes in the esterase enzymes (AChE and BChE), complexes are formed between the enzymes and the pesticide (Ray et al., 1998; Kamanyire and Karalliedde, 2004; Gupta, 2006). Moreover, the phosphorylation of hydroxyl groups inactivates the enzymatic activity towards substrates and the esterase enzymes lose both stability and function (Pullman and Valdemoro, 1960; Wild, 1975; Ray et al., 1998; Kamanyire and Karalliedde, 2004; Costa, 2006). These interactions can result in the formation of reversible and irreversible complexes, depending on the pesticide and the recovery time of the esterase. OPs tend to form more stable, sometimes irreversible, complexes, whereas CBMs usually form less stable and reversible complexes. The resulting complexes can be depleted by enzymes known as "oximes" (Ray et al., 1998; Kamanyire and Karalliedde, 2004). These metabolic depletion transformations can generate metabolites, which a far more toxic than the original foreign species (Abass et al., 2009; Eleršek and Filipič, 2011). During phase I of the metabolism, OPs and CBMs are involved in oxidation and hydrolysis processes. The oxidation reaction is important for the neurotoxicity of CBMs and OPs, since a desulphurization generates metabolites know as "oxons" through CYP enzymes. Oxons are also known as oxygen analogues of pesticides (Eleršek and Filipič, 2011). Usually, CYPs are relatively specific in the detoxification of chemical compounds, e.g.: diazinon is metabolized by CYP2C19; parathion by CYP3A4 / 5 and CYP2C8;

chlorpyrifos by CYP2B6 (Eleršek and Filipič, 2011), or by CYP3A4 and CYP2C9 (Leoni, et al., 2012); atrazine, terbuthylazine, ametryn and terbutryn by CYP1A2 (Lang et al., 1997); endosulfan and carbosulfan by CYP2B6 (Abass et al., 2010).

Other metabolic enzymes, such as paraxonases facilitate hydrolysis reactions. Their function is to eliminate OP/CBM-generated oxons, which is achieved by cleavage of a dialkyl phosphate group. However, through this elimination reaction, highly reactive metabolites (e.g. ROS) can be generated. Eleršek and Filipič (2011) considered these to be genotoxic, since they can interact with DNA molecules. Paraxonases play a protective role against the toxic metabolite oxon, but the potential protection is specific to the type of pesticide and depends on the individual's genotype for the PON gene, which expresses these enzymes. Recent studies on animals, demonstrated an increased expression of PON1 as a result of the promotion, signal transduction and transcription factors on the expression of paraoxomase during the metabolism of OPs. However, there are no known relationships between genotypes, which efficiently detoxify through paraxomases, and/or the activities of AChE and BChE (Costa et al., 2012).

Chemical interactions between xenobiotics may cause saturation of enzymes involved in the metabolism (Bond and Medinsky, 1995). Moreover, evaluations involving low doses during exposure to pesticides may not alter metabolism enzymes, which operate without saturation, and therefore mask a possible effect of intoxication (Bolognesi, 2003; Dyk and Pletschke, 2011). Also, the efficiency of conjugation, a process involved in the glutathione complex during phase II of the metabolism, should be proportional to potential excretion (Eleršek and Filipič, 2011). Accordingly, individual genetic variability, involved in the metabolic transformations of pesticides, can influence the observed pathophysiological effects.

PATHOPHYSIOLOGY OF PESTICIDES

Pyrethroids, OPs, CBMs and OCs represent different classes of insecticides (George and Shukla, 2011). The main effect on human health they share can be attributed to neurotoxicity (Dyk and Pletsche, 2011). Pyrethroids, which lack (type I) or contain a cyano group (type II) in the phenoxybenzyl moiety of their chemical structure (Maroni et al., 2000; Nasuti et al., 2003), interfere with the opening and closing of sodium channels, extending the time of entry for Na^+ cations into the cell (Narahashi, 1996; Spencer et al., 2001). Type II pyrethroids interfere moreover with the chloride channels, blocking the neurotransmitter glutamate receptor (GABA) in the postsynaptic nerve. As a result, the binding of GABA at the receptor site is inhibited and the influx of Cl^- anions into the nerve cell is suppressed (Manna et al., 2006; Suiter and Scharf, 2012). GABA is the major inhibitory neurotransmitter in the central nervous system (CNS) of vertebrates and the absence of synaptic inhibition leads to a CNS hyperexcitability. The same effect can be observed through the incoporation of OCs, especially as the active ingredient in fipronil (Suiter and Scharf, 2012).

OPs and CBMs are neurotoxic due to their inhibition of cholinesterases (AChE, BChE), which interfere with the function of the neurotransmitter acetylcholine (ACh) and long-term effects can be observed (Maroni et al., 2000; Mansour, 2004). AChE and BChE are responsible for hydrolyzing ACh, which is widely distributed in the nervous system of vertebrates (Ray et al., 1998; Chambers et al., 2001). In order to regenerate cholinergic synapses, ACh must be rapidly hydrolyzed by AChE, producing choline and acetic acid after a neurochemical transmission (Namba and Hiraki, 1971).

The inhibition of AChE, caused by OP and CBM insecticides results in an accumulation of ACh at the cholinergic synapses and neuromuscular junctions, eventually causing various signs and symptoms (Maroni et al., 2000; Suiter and Scharf, 2012), Especially muscarinic and nicotinic sites as well as other areas of the CNS are severely affected. Usually, affected receptors are present on the surface of nerve cells (Ray et al., 1998; Kamanyire and Karalliedde, 2004). Due to the effects of AChE on muscarinic and nicotinic receptors, cardiac responses, such as tachycardia, sinus bradycardia, hypertension, hypotension, changes in heart rate and force of heart muscle contraction can be observed. Saadeh et al. (1997) also observed cyanosis and increased serum levels of creatinine and lactate dehydrogenase after OP poisoning. Cardiovascular symptoms occur most frequently after poisoning with pyrethroids, OPs and OCs. Cardiac sodium channel proteins are responsible for both rapid upstroke of the action potential and rapid propagation of nerve impulses through the heart tissue. Thus, their function is central to the origin of cardiac arrhythmias (Balser, 1999). Studies of ventricular myocytes in cats showed that deltamethrin increased the duration of the action potential. The kinetic changes produced in the cardiac sodium channels were similar to those induced by pyrethroids in the sodium channels of the nerve membranes (De La Cerda et al., 2002).

Neuropathy caused by exposure to pesticides is usually related to chronic poisoning cases, since the neurological damage in patients with acute intoxications can be reversed and controlled with adequate treatment (Ray et al., 1998; Ray and Richards, 2001; Costa, 2006; Jayasinghe and Pathirana, 2012). OPs are retained on the endoplasmic reticulum of the axons, promoting apoptosis and injury of the muscle spindle located in the center of the nervous system (involving the spine, spinal cord and cerebellum). This damage is manifested in symptoms such as lethargy, tingling, numbness and weakness of the hip (Ray et al., 1998). Furthermore, elevated risks of developing Parkinson's disease, psychiatric disorders and depressive memory disorders have been discussed (Calvert et al., 2007).

In a review, Rahimi and Abdollahi (2007) suggested hyperglycemia as another effect caused by chronic exposure to OP pesticides. OPs are able to alter the mechanisms involved in the glucose metabolism and thus potentially induce diabetes in exposed individuals. The risk of the general population to develop type 2 diabetes from exposure to environmental OP insecticides, especially in the form of residual contaminants of food supplies, has also been investigated by Rezg et al. (2010).

OP poisoning results in repeated stimulation of cholinergic nerves, which stimulate nerve fibers in the postganglionic parasympathetic muscarinic receptors. This can cause symptoms such as nausea, vomiting, abdominal pain, diarrhea, and tenesmus (Simpson and Schuman, 2002). The phenoxyacetic acid moieties of some herbicides have been associated with the development of gastric cancer. A study showed that chronic exposures to herbicides can result in a 70% chance to develop adenocarcinomas (Ekstrom et al., 1999). Xenobiotics are mainly metabolized in the liver and various types of enzymes, such as alanine aminotransferase, aspartate aminotrasferase (Gomes, 1999; Sarhan and Al-Sahhaf, 2011), and gammaglutamil transferase, as well as other amino acids and proteins (Gomes, 1999) may be affected by their presence. Forensic analysis in humans has also shown histopathological changes in the liver, e.g. necrosis, fat accumulation, and modified centrilobular sinusoidal dilatation (Seema and Tirpude, 2008). Studies conducted on rabbits showed that after the absorption of OPs, leukocyte infiltration occurred in the liver parenchyma, alongside cytoplasmic vacuolization, fatty degeneration and the emergence of pyknotic nuclei in the hepatocytes (Sarhan and Al-Sahhaf, 2011).

OPs are also able to inhibit enzymes, which are important for the metabolism of mitochondrial antioxidant defenses. These are in turn pivotal to the process of respiration and the generation of ATP (Kamanyire and Karalliedde, 2004; Shadnia et al., 2007). This way, pesticides can be directly linked to oxidative stress conditions via lipid peroxidation, which is a molecular mechanism involved in apoptosis (Rastogi et al., 2009). The mitochondrial ATP depletion leads to a stimulation of proteolytic enzymes and a subsequent DNA fragmentation, resulting in cellular death (Shadnia et al., 2007). Mutagenic effects could be observed through the frequency of micronucleus tests (MN), which - on average - were found to be increased after the exposure to OPs (Bolognesi, 2011). These results can be related to certain types of cancer such as Non-Hodgkin's Lymphoma and Leukemia (Bonner et al., 2010).

Mancozeb is a fungicide, commonly used for a wide spectrum of crops (especially soy) and contains a substance with important effects on human health: ethylene(bis)dithiocarbamate (EBCD). EBCD is easily metabolized into ethylenethiourea (ETU), which decreases the activity of tumor suppression proteins, thus facilitating tumor growth (George and Shukla, 2011; Paro et al., 2012). ETU has also been linked to congenital malformation and thyroid disorders (George and Shukla, 2011). Lower concentrations of ETU can affect the morphology and function of cells in the ovarian follicles of mammals (Paro et al., 2012). The effects of paraquat (1,1-dimethyl-4,4-chloride bipyridylium), which is a prototypical agricultural herbicide, have been described by Ranjbar et al. (2002). It promotes toxic effects mainly in the liver and kidney. The latter is predominantly affected by an unchanged excretion of paraquat in the urine (O'Leary et al., 2008). This results in increased tissue injury through lipid peroxidation, which is a secondary effect of the excessive generation of ROS and/or a depletion of the antioxidant defenses (Ranjbar et al., 2002; Samai et al., 2010). Nephrotoxicity processes can also be observed, usually as a result of oxidative stress and/or DNA damage (Samai et al., 2010). A recent study has investigated the effects of OPs on the neoplastic skin cells of rats. The main hypothesis was that the exposure to the herbicide glyphosate, a member of OP family, could lead to increased levels of oxidative stress and result in increased levels of DNA damage (George et al., 2010).

During or after the use of paraquat, triazines, OPs and thiocarbamates, the development of respiratory symptoms was observed among farmers (Hoppin et al., 2002). Paraquat can initiate pulmonary fibrosis through the generation of ROS, whereas glyphosate was associated with the development of chemical pneumonitis (Kirkhorn and Garry, 2000). After exposure to major dosage of OPs and CBMs, signaling in muscarinic receptors was affected and OPs were found in the nerves of post-ganglionic parasympathetic fibers, resulting in respiratory hypersecretion, rhinorrhea, bronchospasm, dyspnea, and cyanosis. These symptoms can evolve progressively and end - due to a complete lack of nerve signals - in apnea and respiratory paralysis (Gaspari and Paydarfar, 2007).

Pesticides interfere with the endocrine system and neurobehavioral development. Moreover, reproduction mechanisms are affected via the endocrine function of steroid hormones, which act as agonists/antagonists in the reproductive system (LeBlanc et al., 1996). The normal reproductive development depends on the interaction of steroid hormones with tissue specific receptors. Xenobiotics may affect the balance between androgen, estrogen and progesterone and hindered interactions between steroids and their receptors may have adverse endocrine effects. These interactions have been examined in studies regarding the exposure to CBMs, e.g. in fungicides. The biosynthesis of imidazole resulted in a deficient production of testosterone hormones (DiMattina et al., 1988), whereas chlordecone and endosulfan increased the testosterone metabolism (Le Blanc et al., 1966). These results contribute to the understanding of the causes of infertility in humans exposed to pesticides (Le Blanc et al., 1996). They also help to explain stillbirths, deformities during embryonic development, as well as congenital malformations caused by OPs (Garry et al., 2002; Maurizio et al., 2008). OCs may affect the function of the thyroid gland, especially regarding the level of thyroxine (T4) production (Le Blanc and Wilson, 1996).

Pesticides can also cause immune alterations, e.g. immunodeficiencies. However, these depend on various environmental factors, which are related to changes of cell functions, and the presence of sub-cellular and/or molecular components in the immune system. The immune response furthermore depends on the interaction between antigens and different cells in the immune system, e.g. lymphocytes or macrophages. Adverse effects can be triggered by direct or indirect immune responses, mainly because some pesticides are more selective than others and may not involve all cell types of the immune system. More specifically, an inhibition of esterases could induce a degranulation of mast cells, thus triggering the release of histamine, which could result in allergic reactions of exposed human individuals. The enzyme phospholipase A2 is involved in the signaling of inflammatory processes, interfering with the humoral and cellular immune system and with T-type lymphocytes (Li et al., 2000; Kamanyire and Karalliedde, 2004; Li, 2007; Li et al., 2009). It can also produce antibodies, autoantibodies and inhibit natural cell killers such as CD5 and CD26, which promote cytotoxicity (Li, 2007; Li et al., 2009). Most of the pesticide metabolites generate free radicals, which are involved in the generation of oxidative stress conditions. The main mechanisms of the immunotoxicity of pesticides therefore usually involve homeostasis of the pro-oxidant agents and antioxidant defenses.

GENOTOXIC DAMAGES OF PESTICIDES

Exposure to pesticides has been associated with an increase in the occurrence of non-Hodgkin's lymphoma (Hardell and Eriksson, 1999), multiple myeloma (Khuder and Mutgi, 1997), soft tissue sarcoma (Kogevinas et al., 1995), and lung sarcoma (Blair et al., 1983). Pancreatic, stomach, liver, bladder, and gall bladder cancer have also been reported (Ji et al., 2001; Shukla and Arora, 2001). Moreover, relations to Parkison's disease (Gauthier et al., 2001) and reproductive influences (Arbuckle et al., 2001) have been examined. Several reports are concerned with chromosomal aberrations (CA) (Au et al., 1999; Zeljezic and Garaj-Vrhovac, 2001; Jonnalagadda et al., 2012), sister chromatid exchange (SCE) (Shaham et al., 2001; Zeljezic and Garaj-Vrhovac, 2002), micronuclei (MN) (Falck et al., 1999; Pastor et al., 2003; De Bortolli et al., 2009; Da Silva et al., 2012a; Benedetti et al., 2013) and Comet cells (Grover et al., 2003; Zeljezic and Garaj-Vrhovac, 2001; Da Silva et al., 2012b; Benedetti et al., 2013) as a result of pesticide exposure. In general, significantly increased levels of these biomarkers were found, suggesting severe genotoxic effects of these pesticides.

Various studies have reported significant incidences of cytogenetic damage in agricultural workers, floriculturists, vineyard cultivators, cotton field workers and others (Bolognesi, 2011). Studies involving biomarkers of exposure are usually used in order to assess occupational exposure, i.e. to correlate exposure to chemical reagents with health effects (Aprea, 2012). For this purpose, different biomarkers regarding exposure, effect or susceptibility towards xenobiotics are used to express a specific measure of interaction between a given biological system and a genotoxin (Bolognesi, 2003; Aprea, 2012). The influence of genotypes on the cytogenetic damage is the specific ability of individuals to influence genotoxic biomarkers, i.e. to activate or detoxify substances with respect to their potential to induce mutations, cancer and other diseases (Hagmar et al., 1994; Hagmar et al., 1998). A variety of enzymatic isoforms have been suggested to influence the individual's risk of contracting cancer after exposure to genotoxins (Sulbatos, 1994; Clapper, 2000). Genomic stability has moreover been linked to several dietary micronutrients, nutrient imbalances, dietary deficiencies, as well as excessive exposure to environmental mutagens or carcinogens, all of which can potentially increase genetic damage. As previously discussed, deficiencies of folic acid or other vitamin B cofactors (e.g. B_{12} and B_6) may cause impaired DNA repair.

In view of these diverse and complex findings, the investigation of humans exposed to pesticides constitutes to be a highly important research topic. MN tests and comet assays are accurate and practical analysis tools, complying with most of the criteria used in human bio-monitoring (Fairbairn et al., 1995; Grover et al., 2003; Moller et al., 2000). In order to assess, if a prolonged exposure to complex mixtures of pesticides could lead to an increase in cytogenetic damage, our group has examined individuals occupationally exposed to agricultural pesticides in Rio Grande do Sul (Brazil), and the public health workers occupationally exposed to agricultural pesticides in Piauí (Brazil). We were also interested in the potentially important effects of gene polymorphisms, which encode proteins involved in the xenobiotic metabolization/detoxification of phase I or II. These should influence the DNA repair pathways, which should allow an evaluation of the genetic predisposition of individuals towards their xenobiotic metabolizing capacity, i.e. the individual susceptibility towards genotoxic effects of pesticides. Therefore, we also investigated the polymorphism of the PON, GSTM1, GSTT1, CYP2E1, OGG1 and XRCC1 genes. Apart from observing the occupational exposure of individuals towards pesticides, we were also interested in the influence of micronutrient intake (vitamin B_{12}, B_6 and folates), as well as the influence of MTHFR C677T polymorphism on the observed DNA damage.

In order to study all of the aforementioned aspects, our group conducted an investigation on vineyard workers, which involved a total number of 173 individuals (Rohr et al., 2011). Of these, 108 were agricultural workers exposed to pesticides and 65 were control individuals. As evident from MN tests, the individuals exposed to pesticides showed a high rate of DNA damage ($P<0.001$; Mann-Whitney U test), relative to the control group. In addition, some of the MN results of the exposure group suggested genetic polymorphisms of PON, GSTM1, GSTT1, and CYP2E1. OGG1 and XRCC1 are examples of important proteins in the base excision repair (BER) pathway (Au et al., 2004; Goode et al., 2002; Hao et al., 2004; Muniz et al., 2008). In another study, we evaluated two BER polymorphisms: OGG1 Ser326Cys: rs1052133 and XRCC1 Arg194Trp: rs1799782 as well as the combined genotypes of these polymorphisms with PON1 Gln192Arg. The modifications of the genotoxic susceptibility as a function of pesticide exposure was measured by MN tests and DNA damage induction in the peripheral leukocytes of the vineyard workers. Our study demonstrated that the polymorphisms in the BER pathway could modulate the susceptibility to DNA damage caused by the pesticides. Since this repair pathway is the major cellular defense against oxidative DNA damage, our results corroborate existing evidence, which suggests an involvement of oxidative damage in the pesticide-induced genotoxic effects. Our study also reinforces the importance of considering combined effects of metabolism and repair-variable genotypes on the individual susceptibility towards DNA damage. It seems feasible to conclude that these two processes act cooperatively in determining the final response to pesticide exposure.

Brazil is a major producer of soybeans, which are planted in several federal states, but especially in Rio Grande do Sul (RS). The increasing agricultural use of the land is hereby concomitant with an increased use of pesticides. Soybean workers in this region are increasingly exposed to a wide variety of herbicides and insecticides (especially OPs). A study originating from our research group investigated a total of 127 individuals, of whom 81 were exposed and 46 were not exposed to pesticides (Benedetti et al., 2013). Both groups consisted of residents from the city of Espumoso (RS-Brazil), whose main economic income relies on soy crops. We evaluated comet assays of the peripheral leukocytes and buccal micronucleus cytome assays (BMCyt; micronuclei and nuclear buds) in exfoliated buccal cells. We observed significant increases in DNA damage in the pesticide-exposed group relative to control group. We also found the gene PON1 to express the enzyme paraoxonase, which is believed to be involved in the protection against oxidative stress in the OP metabolism. The metabolizing genes PON1, GSTM1, GSTT1 and GSTP1 were evaluated in order to analyze the influence of individual susceptibility in response to exposure. The genetic polymorphisms obtained from the exposure biomarkers showed no influence of the genotype on the DNA damage in the farmers' cells. The exposure to pesticides increased DNA damage and did not change the evaluated metabolizing genes.

Occupational risks for tobacco farmers involve the exposure to very large amounts of pesticides, which are applied to the crop fields. Contact with the pesticides is normally established via the contact to green leaves during the tobacco harvest and through the additional exposure to nicotine. Nicotine poisoning could also lead to "Green Tobacco Sickness" (GTS), which occurs, when workers absorb nicotine via the skin as they come in contact with the leaves of the mature tobacco plant. GTS is characterized by nausea, vomiting, headache, muscle weakness, and dizziness. Our group examined the occupational risk of tobacco farmers, involving 167 individuals, of whom 111 were exposed, and 56 were not exposed (Alves, 2008; Da Silva et al., 2012a; Da Silva et al., 2012b). Subjects were recruited from Venâncio Aires and Santa Cruz do Sul (RS-Brazil) between July and February in the years 2008-2010. Blood and buccal cells were collected twice during the tobacco crop cycle of every year. Once during the distribution period of the pesticides and again during the harvest period. Blood and buccal cells were also collected from a non-exposed control group (office workers, who were living in the same region as the exposed individuals). Our study evaluated exposure biomarkers indicative of early biological effects and susceptibility. Genotoxicity and mutagenicity in the tobacco farmers were investigated by comet assays and micronucleus tests of buccal cells and binucleated lymphocytes, respectively. In order to detect a potential impact of these chemicals on the farmers, superoxide dismutase (SOD), catalase (CAT) and plasma cholinesterase activities, as well as levels of thiobarbituric acid reactive substances (TBARS) were evaluated. Total contents of chemical elements in the blood were examined by particle-induced X-ray emission (PIXE) and cotinine levels were analyzed in plasma samples. In order to establish a possible correlation between a potential genetic predisposition of the metabolism of xenobiotics / repair of DNA damage and individual susceptibility towards genotoxic effects of pesticides and nicotine, farmers were genotyped for several genes. The evaluation of the DNA damage also considered the following secondary parameters: use of protective measures, time after exposure, age, and gender. As tobacco farmers were exposed to complex mixtures of pesticides during the application period, significantly higher levels of DNA damage were found in the exposed group relative to the control group. For the exposed group, the damage to the DNA was three times higher during the application period and four times higher during the harvest period relative to the control group. However, no significant difference in the activity of serum cholinesterase was observed between exposed and control group. Prior studies, examining pesticide workers, were unable to identify any correlation between chronic exposure to OPs and BChE inhibition. During the exposure period, all individuals showed symptoms related to pesticide poisoning and GTS, e.g. headache, abdominal pain, nausea, and vomiting. We observed in our study that the serum cotinine levels among the non-smoking section of the exposed individuals during the harvest period were significantly increased, suggesting absorption of nicotine through skin contact with tobacco leaves. Nuclear anomalies in the buccal mucosa cells of exposed tobacco farmers (both during the application and harvest period) showed mixtures of genotoxic and cytotoxic substances. A minor discrepancy concerning the mutagenicity was noticed between the two different periods of the tobacco cycle. During the harvest period, higher MN values were observed in buccal cells, relative to the application period. In addition, effects on the extent of pesticide-induced DNA damage and cell death as a result of the genetic polymorphisms of PON1 and CYP2A6*9 were observed. Binucleated lymphocyte responses to genetic damage were evident from higher MN levels in the exposed group (mainly during the application period) relative to the control group. Workers employed in the production of pesticides and farmers who used pesticides showed a higher risk/level of exposure and hence, were more prone to the potential deleterious health effects of pesticides. Besides, many pesticides, which are commonly used on tobacco crops, contain inorganic elements, including Mg, Al, Cl, Zn, and Br, which are known to cause DNA damage. In our study, absolute inorganic element levels in the blood samples of the exposure group (application period) were found to be increased.

Elevated levels of DNA damage in the exposure group were also observed during the harvest period, presumably via contact with green tobacco leaves and tobacco plants during the various cultivation processes and concomitant dermal nicotine absorption. Nicotine has been implicated in the generation of free radicals in human cells, directly addressing the relationship between ROS induction and observed DNA damage. Thus, synthetic and natural pesticides may induce oxidative stress, and lead to increased generation of free radicals as well as subsequent alterations in antioxidants, free oxygen-based radicals, lipid peroxidation, and the quenching of enzyme systems. In the exposure group (application period), only the antioxidant enzyme SOD showed increased activity, relative to harvest and control groups. In the harvest group, levels of body-defending antioxidant mechanisms (SOD and CAT) were increased, in order to overcome the induction of oxidative stress. These results indicate that the level of lipid peroxidation was significantly different in the harvest group relative to application and control groups. It is feasible to assume, that the internal antioxidant stimulation in the body were insufficient to scavenge all the free radicals and thus compensate for the increased levels of lipid peroxidation. Age and personal protective equipment (PPE) also showed an influence onto the results obtained from the MN tests. An increase of MN levels corresponding to age was observed for both groups (exposed and control). Interestingly, significant differences were observed between exposed individuals (application period) with complete PPE, relative to those without.

The effect of individual genotypes of metabolism genes on the level of the different biomarkers (comet assay and MN tests in binucleated lymphocytes) was examined in the exposed group. Increased damage of GSTM1 in the application group and an increased damage of CYP2A6*1/*1 in the harvest group were observed. The individual genotype of DNA repair genes in the exposure groups did not show any influence on the different biomarkers analyzed in this study. Our study demonstrates once more the importance of occupational training for farm workers, regarding safe working practices and safe working environments. Developing countries should use such data to establish occupational safety rules when using pesticides (especially in the context of tobacco crops) in order to minimize occupational risks for the workers involved.

We also evaluated the influence of micronutrient intake (vitamins B_{12} and B_6, folates) and MTHFR C677T polymorphism on DNA damage in the exposed individuals (Fernandes, 2012). We examined 110 individuals of both genders (average age: 42.3 ± 13.3 years), living and working in the city of Venâncio Aires (RS-Brazil). The examined exposure time was 30.3 ± 15.6 years. The results showed increased levels of MN in lymphocytes and modified consumption of folates and B_{12} ($p = 0.030$ and $p = 0.014$, respectively). No significant correlation between DNA damage (MN frequency, comet assay) and age, gender, smoking, years of exposure or BMI could be observed. Similar results were obtained for the genetic polymorphism of *MTHFR C677T*. A diet with appropriate folate and vitamin B_{12} supplements was able to facilitate adequate DNA repair.

Another study originating from our group followed, for over 30 years, a group of public health workers, concerned with endemic diseases. During their work, these individuals have been exposed to considerable amounts of genotoxic and mutagenic pesticides, which are used in vector control programs. Our study with this group therefore aimed at the evaluation of the mutagenicity

(MN tests in buccal cells) caused by the occupational exposure to pesticides in the "Território Entre Rios" (Piauí-Brazil) (Fianco, 2013). The study included 129 individuals, of whom 66 were public health workers (exposed group), and 63 individuals without occupational exposure to pesticides (control group). Mutagenic events were manifested through the presence of significant increased numbers of MN (14.7 ± 2.7), binucleated cells (5.9 ± 1.1) and nuclear buds (10.2 ± 1.8) in the exfoliated oral mucosa cells of the workers, relative to the control group (4.2 ± 0.9, 2.8 ± 0.8, and 3.9 ± 0.9, respectively; Mann-Whitney test). However, age, gender, exposure time, smoking, drinking, or diet did not influence the DNA damage parameters examined. According to these results, the occupational exposure of public health workers to pesticides induces mutagenic damage. Even though public health workers should be aware of the risks they are exposed to, the proper use of personal protective equipment could still be improved.

CONCLUSIONS

Our findings show in general that agricultural workers exhibit higher levels of DNA damage in somatic cells, suggesting that pesticide exposure is a potential health risk for these workers. In addition, it was possible to correlate these results to specific genetic susceptibility, to the absence or inappropriate use of PPE and to dietary habits. It became evident that continuous education is very important for exposed workers, in order to minimize the deleterious effects of the occupational exposure and the risk of contracting work-related diseases. Chronically exposed individuals were more susceptible to the clastogenic effects of pesticides. Significant differences in the cytogenetic damage were detected in individuals with symptoms of chronic intoxication (Zeljezic and Garaj-Vrhovac, 2001). Furthermore, others studies observing agricultural workers demonstrated an increase in chromosomal damage during the spraying/application season, when pesticides were used intensively (mainly in workers not using PPE). The use of PPE seems to be beneficial for the workers, which is evident from reduced cytogenetic effects (Shaham et al., 2001). The DNA damage (CA, MN and SCE) could be correlated with the exposure duration in many of these investigations (Bolognesi et al., 1993; Joksic et al., 1997; Shaham et al., 2001; Bolognesi et al., 2002; Bolognesi, 2011), and moreover seem the clastogenic effects to be cumulative for a continuous exposure to pesticide mixtures (Bolognesi, 2003; Bolognesi, 2011).

REFERENCES

1. Abass, K; Reponen, P; Mattila, S; Pelkonen, O. (2009). Metabolism of carbosulfan. Species differences in the in vitro biotransformation by

mammalian hepatic microsomes including human. Chemico-Biological Interactions. 181; 210–219.

2. Abass, K; Reponen, P; Mattila, S; Pelkonen, O. (2010). Metabolism of carbosulfan II. Human interindividual variability in its in vitro hepatic biotransformation and the identification of the cytochrome P450 isoforms involved. Chemico biological interactions. 185; 163-173.

3. Ames, B.N. (1999). Micronutrient deficiencies. A major cause of DNA damage. Annals of the New York Academy of Sciences. 889; 87-106.

4. Ames, B.N. (2001). DNA damage from micronutrient deficiencies is likely to be a major cause of cancer. Mutation research. 475; 7-20.

5. Alves, J. (2008). Avaliação da genotoxicidade e estresse oxidativo em em agricultores que trabalham na fumicultura. Canoas: ULBRA/PPGGTA. Dissertação de Mestrado. 60 pp.

6. Aprea, C. M. (2012). Mini review: Environmental and biological monitoring in the estimation of absorbed doses of pesticides. Toxicology letters. 210; 110-118.

7. Arbuckle, T. E; Lin, Z; Mery, L. S. (2001). An exploratory analysis of the effect of pesticide exposure on the risk of spontaneous abortion in an Ontario farm population. Environmental health perspectives. 109; 851-857.

8. Au, W; Serra-Torres, C. H; Cajas-Salazar, N; Shipp, B. K; Legator, M. S. (1999). Cytogenetic effects from exposure to mixed pesticides and the influence from genetic susceptibility. Environmental health perspectives. 107; 501-505.

9. Au, W. W; Navasumrit, P; Ruchirawat, M. (2004). Use of bio-markers to characterize functions of polymorphic DNA repair genotypes. International journal of hygiene and environmental health. 207; 301-313.

10. Balser, J. R. (1999). Structure and function of the cardiac sodium channels. Cardiovascular research. 42; 327-328.

11. Beetstra, S; Thomas, P; Salisbury, C; Turner, J; Fenech, M. (2005). Folic acid deficiency increases chromosomal instability, chromosome 21 aneuploidy and sensitivity to radiation-induced micronuclei. Mutation research. 578; 317-326.

12. Benedetti, D; Nunes, E; Sarmento, M. S; Porto, C; Santos, C. E. I; Dias, J. F; Da Silva, J. (2013). Genetic damage in soybean workers exposed to pesticides: Evaluation with the comet and buccal micronucleus cytome assays. Mutation research. Genetic Toxicology and Environmental Mutagenesis. 752; 28-33.

13. Bergh, A. F; Strobel, H. W. (1992). Reconstitution of the brain mixed function oxidase system: Purification of NADPH-cytochrome P450 reductase and partial purification of cytochrome P450 from whole rat brain. Journal of neurochemistry. 59; 575-581.
14. Blair, A; Grauman, D.J; Lubin, J. H; Fraumeni, J. F. (1983). Lung Cancer and Other Causes of Death Among Licensed Pesticide Applicators. Journal of the national cancer institute. 1; 31-37.
15. Bolognesi, C; Parrini, M; Bonassi, S; Ianello, G; Salanitto, A. (1993). Cytogenetic analysis of a human population occupationally exposed to pesticides. Mutation research. 285; 239-249.
16. Bolognesi, C; Perrone, E; Landini, E. (2002). Micronucleus monitoring of a floriculturist population from western Liguria, Italy. Mutagenesis, 17; 391-397.
17. Bolognesi, C. (2003). Genotoxicity of pesticides: a review of human biomonitoring studies. Mutation research. 543; 251-272.
18. Bolognesi, C; Creus, A; Ostrosky-Wegman, P; Marcos, R. (2011). Review: Micronuclei and pesticide exposure. Mutagenesis. 26; 19-26.
19. Bond, J.A; Medinsky, M.A. (1995). Health risk assessment of chemical mixtures from a research perspective. Toxicology letters. 82/83; 521-525.
20. Bonner, M. R; Williams, B. A; Rusiecki, J. A; Blair, A; Beane-Freeman, L. E; Hoppin, J. A; Dosemeci, M; Lubin, J; Sandler, D. P; Alavanja, M. C. (2010). Occupational exposure to terbufos and the incidence of cancer in the agricultural health study. Cancer causes control. 21; 871-877.
21. Calvert, G. M; Alarcon W, A; Chelminski, A; Crowley, M. S; Barrett, R; Correa, A; Higgins, S; Leon, H. L; Correia, J; Becker, A; Allen, R. H; Evans, E. (2007). Case Report: Three farmworkers who gave birth to infants with birth defects closely grouped in time and place Florida and North Carolina, 2004-2005. Environmental health perspectives. 115; 787-791.
22. Chambers, E. J; Russell, L. C; Boone, S; Chambers, H. W. (2001). The Metabolism of Organophosphorus Insecticides. In Handbook of Pesticide Toxicology, v. 2, chapter 45, p. 919-927. Mississippi State University.
23. Clapper, M. L. (2000). Genetic polymorphism and cancer risk. Current oncology reports.2; 251-256.
24. Costa, L. G. (2006). Current issues in organophosphate toxicology. Clinica chimica acta. 366; 1-13.

25. Costa, L. G; Giordano, G; Cole, T. B; Marsillach, J; Furlong, C. E. (2012). Paraoxonase 1 (PON1) as a genetic determinant of susceptibility to organophosphate toxicity. Toxicology. 307; 115-122.
26. Dangour, A. D; Whithouse, P. J; Rafferty, K; Mitchell, S. A; Smith, L; Hawkesworth, S; Vellas, B. (2010). B-vitamins and fatty acids in the prevention and treatment of Alzheimer's disease and dementia: a systematic review. Journal of Alzheimer's disease. 22; 205-224.
27. Da Silva, F. R; Da Silva, J; Nunes, E; Benedetti, D; Kahl, V; Rohr, P; Abreu, M. A; Thiesen, F. V; Kvitko, K. (2012a) Application of the buccal micronucleus cytome assay and analysis of PON1Gln192Arg and CYP2A6*9(−48T>G) polymorphisms in tobacco farmers. Environmental and Molecular Mutagenesis. 53; 525-534.
28. Da Silva, F. R; Da Silva, J; Allgayer, M. Da. C; Simon, C. F; Dias, J. F; Dos Santos, C. E. I; Salvador, M; Branco, C; Schneider, N. B; Kahl, V; Rohr, P; Kvitko, K. (2012b) Genotoxic biomonitoring of tobacco farmers: Biomarkers of exposure, of early biological effects and of susceptibility. Journal of Hazardous Materials. 225/226; 81-90.
29. De Bortoli, G. M; De Azevedo, M. B; Da Silva, L. B. (2009). Cytogenetic biomonitoring of Brazilian workers exposed to pesticides: Micronucleus analysis in buccal epithelial cells of soybean growers. Mutation research. Genetic Toxicology and Environmental Mutagenesis. 675; 1-4.
30. De La Cerda, E; Navarro-Polanco, R. A; Sánchez-Chapula, J. A. (2002). Modulation of cardiac action potential and underlying ionic currents by the pyrethroid insecticide deltamethrin. Archives of medical research. 33; 448-454.
31. DiMattina, M; Maronian, N; Ashby, H; Loriaux, D. L; Albertson, B. D.(1988). Ketoconazole inhibits multiple steroidogenic enzymes involved in androgen biosynthesis in the human ovary. Fertility and sterility. 1; 62-65.
32. Dyk, V. S. J; Pletschke, B. (2011). Review on the use of enzymes for the detection of organochlorine, organophosphate and carbamate pesticides in the environment. Chemosphere. 82; 291-307.
33. Ekstrom, A. M; Eriksson, M; Hansson, L; Lindgren, A; Signorello, L. B; Nyren, O; Hardell, L. (1999). Occupational exposures and risk of gastric cancer in a population-based case-control study. Cancer research. 59; 5932–5937.
34. Eleršek, T; Filipič, M. Organophosphorus Pesticides: Mechanisms of their toxicity. National Institute of Biology Slovenia Pesticides - The

Impacts of Pesticides Exposure, Cap. 12 (Ed) ISBN: 978-953-307-531-0, InTech (2011).

35. Eriksson, C; Brittebo, E. B. (1991). Metabolic activation of the herbicide dichlobenil in the olfactory mucosa of mice and rats. Chemico biological interactions. 79; 165-177.

36. Falck, G. C; Hirvonen, A; Scarpato, R; Saarikoski, S. T; Migliore, L; Norppa, H. (1999). Micronuclei in blood lymphocytes and genetic polymorphism for GSTM1, GSTT1 and NAT2 in pesticide-exposed greenhouse workers. Mutation research. 441; 225-237.

37. Fairbairn, D. W; Olive, P. L; O'Neill, K. L. (1995). The comet assay: A comprehensive review. Mutation research. 339 ; 37-59.

38. Fenech, M; Bonassi, S. (2011). The effect of age, gender, diet and lifestyle on DNA damage measured using micronucleus frequency in human peripheral blood lymphocytes. Mutagenesis. 26; 43-49.

39. Fernandes, S. P. (2012). Relação do hábito alimentar e polimorfismos da mthfr c677t com a instabilidade genômica em fumicultores gaúchos. Porto Alegre: UFRGS/ PPGBM. Dissertação de Mestrado. 67 pp.

40. Fianco, M. C. (2013). Avaliação do risco ocupacional do uso dos praguicidas na saúde dos agentes de combate às endemias do estado do Piauí. Canoas: ULBRA/PPGGTA-MP. Dissertação de Mestrado. 99 pp.

41. Franco, R; Sánchez-Olea, R; Reyes-Reyes, E.M; Panayiotidis, I. (2009). Minireview: Environmental toxicity, oxidative stress and apoptosis: Ménage à Trois. Mutation research. 674; 3-22.

42. Gammon, D.W; Liu, L; Becker, J.M. (2012). Carbofuran occupational dermal toxicity, exposure and risk assessment. Pest management science. 68; 362–370.

43. Gaspari, R. J; Paydarfar, D. (2007). Pathophysiology of respiratory failure following acute dichlorvos poisoning in a rodent model. Neurotoxicology. 28; 664-671.

44. Garry, V. F; Harkins, M. E; Erickson, L; Long-Simpson, L. K; Holland, S. E; Burroughs, B. L. (2002). Birth defects, season of conception, and sex of children born to pesticide applicators living in the Red River Valley of Minnesota, USA. Endocrine disruptors. 110; 441-448.

45. Gauthier, E; Fortier, I; Courchesne, F; Pepin, P; Mortimer, J; Gauvreau, D. (2001). Environmental pesticide exposure as a risk factor for Alzheimer's disease: A case-control study. Environmental research. 86; 37-45.

46. George, J; Prassad, S; Mahmood, Z; Shukla, Y. (2010). Studies on glyphosate-induced carcinogenicity in mouse skin: a proteomic approach. Proteomics. 5; 951-964.

47. George, J; Shukla, Y. (2011). Review: Pesticides and cancer: Insights into toxicoproteomic-based findings. Journal of proteomics. 74; 2713-2722.
48. Goode, E. L; Ulrich, C. M; Potter, J. D (2002). Polymorphisms in DNA repair genes and associations with cancer risk. Cancer epidemiology biomarkers & prevention. 11;1513-1530.
49. Grover, P; Danadevi, K; Mahboob, M; Rozati, R; Banu, B. S; Rahman, M. F. (2003). Evaluation of genetic damage in workers employed in pesticide production utilizing the Comet assay. Mutagenesis. 18; 201-205.
50. Guengerich, F. P; Shimada, T. (1991). Oxidation of toxic and carcinogenic chemicals by human cytochrome P-450 enzymes. Chemical research in toxicology. 4; 391-407.
51. Guengerich, F.P. (2001). Uncommon P450-catalyzed reactions. Current drug metabolism. 2; 93-115.
52. Gomes, J; Dawodu, A. H; Lloyd, O; Revitt, D. M; Anilal, S. V. (1999). Hepatic injury and disturbed amino acid metabolism in mice following prolonged exposure to organophosphorus pesticides. Human experimental toxicology. 18; 33-37.
53. Gupta, R. C. (2006). Toxicology of Organophosphate & Carbamate Compound. Elsevier Academic Press, p. 271-291.
54. Hagmar, L; Brogger, A; Hansteen, I. L; Heim, S; Hogstedt, B; Knudsen, L; Lambert, B; Linnainmaa, K; Mitelman, F; Nordenson, I; Reuterwall, C; Salomaa SI; Skerfving, S; Sorsa, M. (1994). Cancer risk in human predicted by increased levels of chromosomal aberrations in lymphocytes: Nordic Study Group on the Health Risk of Chromosome Damage. Cancer research. 54; 2919-2922.
55. Hagmar, L; Bonassi, S; Stromberg, U; Brogger, A; Knudsen, L; Norppa, H. Reuterwall, C. (1998). Chromosomal aberrations in lymphocytes predict human cancer. A report from the European Study Group on Cytogenetic Biomarkers and Health (ESCH). Cancer research. 58; 4117-4121.
56. Hardell, L; Eriksson, M. (1999). A case-control study of non-Hodgkin lymphoma and exposure to pesticides. Cancer. 6; 1353-1360.
57. Hao, B; Wang, H; Zhou, K; Li, Y; Chen, X; Zhou, G; Zhu, Y; Miao, X; Tan, W; Wei, Q; et al. (2004). Identification of genetic variants in base excision repair pathway and their associations with risk of esophageal squamous cell carcinoma. Cancer research. 64; 4378-4384.

58. Hjelle, J; Hazelton, G; Klaassen, C; Hjelle, J. (1986). Glucuronidation and sulfation in rabbit kidney. The Journal of pharmacology and experimental therapeutics. 236; 150-156.

59. Hodgson, E; Goldstein, J. A. (2001). Metabolism of toxicants: phase I reactions and pharmacogenetics, In: Introduction to Biochemical Toxicology, Hodgson, E. & Smart, R.C., (Ed.), (67-113), Wiely, New York.

60. Hoppin, J. A;. Umbach, D. M; London, S. J; Alavanja, M. C. R; Sandler, D. P. (2002). Chemical predictors of wheeze among farmer pesticide applicators in the agricultural health study. American journal of respiratory and critical care medicine. 165; 683-689.

61. Jayasinghe, S. S; Pathirana, K. D. (2012). Autonomic function following acute organophosphorus poisoning: a cohort study. PlosOne. 7; 1-8.

62. Ji, B. T; Silverman, D. T; Stewart, P. A; Blair, A; Swanson, G. M; Baris, D; Greenberg, R. S; Hayes, R. B; Brown, L. M; Lillemoe, K. D; et al. (2001). Occupational exposure to pesticides and pancreatic cancer. American journal of industrial medicine. 39; 92-99.

63. Joksic, G; Vidakovic, A; Spasojevic-Tisma, V. (1997). Cytogenetic monitoring of pesticide sprayers. Environmental research. 75; 113-118.

64. Jonnalagadda, P.R.; Jahan, P.; Venkatasubramanian, S.; Khan, I.A.; Prasad, A.Y.E.; Reddy, K.A.; Rao, M.V.; Venkaiah, K; Hasan, Q. (2012). Genotoxicity in agricultural farmers from Guntur district of South India-A case study. Human and Experimental Toxicology. 31; 741-747.

65. Kamanyire, R; Karalliedde, L. (2004). Organophosphate toxicity and occupational exposure. Occupational medicine. 54; 69-77.

66. Khan, W. A; Park, S. S; Gelboin, H. V; Bickers, D. R; Mukhtar, H. (1989). Monoclonal antibodies directed characterization of epidermal and hepatic cytochrome P-450 isozymes induced by skin application of therapeutic crude coal. Journal of investigative dermatology. 93; 40-45.

67. Kirkhorn, S. R; Garry, V. F. (2000). Agricultural lung diseases. Environmental health perspectives. 108; 705-712.

68. Kogevinas, M; Kauppinen, T; Winkelmann, R; et al. (1995). Soft tissue sarcoma and non-Hodgkin's lymphoma in workers exposed to phenoxy herbicides, chlorophenols, and dioxins: two nested case control studies. Epidemiology. 6; 396-402.

69. Kortenkamp, A; Faust, M; Scholze, M; Backhaus, T. (2007). Low-Level Exposure to Multiple Chemicals: Reason for Human Health Concerns. Environmental health perspectives. 115; 106-113.

70. Khuder, S. A; Mutgi, A. B. (1997). Meta-analyses of multiple myeloma and farming. American Journal of industrial medicine. 5; 510-516.
71. Kym, Y.I. (2007). Folate and colorectal cancer: an evidence-based critical review. Molecular nutrition & food research. 51; 267-92.
72. Lang, D. H; Rettie, A. E; Bocker, R. H. (1997). Identification of enzymes involved in the metabolism of atrazine, terbuthylazine, ametryne, and terbutryne in human liver microsomes. Chemical research in toxicology. 9; 1037-1044.
73. Lawton, M; Gasser, R; Tynes, R; Hodgson, E; Philpot, R. (1990). The flavin-containing monooxygenase enzymes expressed in rabbit liver and lung are products of related but distinctly different genes. Journal of biological chemistry. 265; 5855-5861.
74. LeBlanc, G. A; Bain, L. J; Wilson, V. S. (1997). Pesticides: multiple mechanisms of demasculinization. Molecular and cellular endocrinology. 126; 1-5.
75. Leoni, C; Balduzzi, M; Burattia, F.M; Testai, E. (2012). The contribution of human small intestine to chlorpyrifos biotransformation. Toxicology letters 215; 42-48.
76. Li, Q; Hirata, Y; Piao, S; Minari, M. (2000). The products generated during sarin shynthesis in the Tokyo sarin disaster induced inhibition of natural killer and cytotoxic T lymphocyte activity. Toxicology. 146; 209-220.
77. Li, Q. (2007). New mechanism of Organophosphorus pesticide-induced immunotoxicity. Journal of Nippon Medical School. 74; 92-105.
78. Li, Q; Kobayashi, M; Kawada, T. (2009).Chlorpyrifos induces apoptosis in human T cells. Toxicology, 255; 53-57.
79. Limon-Pacheco, J; Gonsebatt, M.E. (2009). Mini review: The role of antioxidants and antioxidant-related enzymes in protective responses to environmentally induced oxidative stress. Mutation research. 674; 3-22.
80. Manna, S; Bhattacharyya, D; Mandal, T.K; Dey, S. (2006). Neuropharmacological effects of deltamethrin in rats. Journal of veterinary science. 2; 133-136.
81. Mansour, S. A. (2004). Pesticide exposure-Egyptian scene. Toxicology. 198; 91–115.
82. Maroni, M; Colosio, C; Ferioli, A; Fait, A. (2000). Toxicology: review. Toxicology. 143; 5-91.
83. Maurizio, C; Gian, M. T; Roberto, C; Cinzia, L. R; Francesca, M; Francesco, R. et al. (2008). Pesticides and fertility: An epidemiological

study in Northeast Italy and review of the literature. Reproductive toxicology. 26; 13-18.

84. Mostafalou, S; Abdollahi, M. (2013). Pesticides and human chronic diseases: Evidences, mechanisms, and perspectives. Toxicology and applied pharmacology. 268; 157–177.

85. Moller, P; Knudsen, L. E; Loft, S; Wallin, H. (2000). The comet assay as a rapid test in biomonitoring occupational exposure to DNA-damaging agents and effect of confounding factors. Cancer epidemiology biomarkers & prevention. 9; 1005-1015.

86. Muniz, J. F; McCauley, L; Scherer, J; Lasarev, M; Koshy, M; Kow, Y.W; Nazar-Stewart, V; Kisby, G. E. (2008). Biomarkers of oxidative stress and DNA damage in agricultural workers: A pilot study. Toxicology and applied pharmacology. 227; 97-107.

87. Muntaz, M.M. (1995). Risk assessment of chemical mixtures from a public health perspective. Toxicology letter. 82183; 527-532.

88. Namba, T., Hiraki, K. (1958) PAM (pyridine-2-aldoxime methiodide) therapy for alkyl-phosphate poisoning. The journal of the american medical association. 166; 1834–1839.

89. Narahashi, T. (1996). Neuronal ion channel as the target sites of insecticides. Pharmacology & toxicology. 79; 1-14.

90. Nasuti, C; Cantalamessa, F; Falcioni, G; Gabbianelli, R. (2003). Different effects of type I and type II pyrethroids on erythrocyte plasma membrane properties and enzimatic activity in rats. Toxicology. 191; 233-244.

91. O'Leary, K; Parameswaran, N; Johnston, C. L; McIntosh, J. M; Di Monte, A. D; Quik, M. (2008). Paraquat exposure reduces nicotinic receptor-evoked dopamine release in monkey striatum. The journal of pharmacology and experimental therapeutics. 327; 124-129.

92. Ortiz, R.H; Bouchard, M. (2012). Toxicokinetic modeling of captan fungicide and its tetrahydrophthalimide biomarker of exposure in humans. Toxicology letter. 1; 27-34.

93. Pastor, S; Creus, A; Parron, T; Cebulska-Wasilewska, A; Siffel, C; Piperakis, S; Marcos R. (2003). Biomonitoring of four European populations occupationally exposed to pesticides: Use of micronuclei as biomarkers. Mutagenesis. 18; 249-258.

94. Parkinson, A. (2001). Biotransformation of xenobiotics, In: Casarett and Doull›s toxicology: the basic science of poisons, Klaassen, C.D., (Ed.), (113–186), McGraw-Hill Medical Pub. Division, ISBN: 0071124535: 44.99; 0071347216 (U.S.), New York ; London.

95. Paro, R; Tiboni, G. M; Buccione, R; Rossi, G; Cellini, V; Canipari, R; Cecconi, S.(2012). The fungicide mancozeb induces toxic effects on mammalian granulosa cells. Toxicology and applied pharmacology. 260; 155–161.
96. Patil, J. A; Patil, A; Govindwar, S. P. (2003). Biochemical effects of various pesticides on sprayers of grape gardens. Indian journal of clinical biochemistry. 2; 16-22.
97. Peters, W. H. M; Kremers, P. G. (1989). Cytochromes P-450 in the intestinal mucosa of man. Biochemical pharmacology. 38; 1535-1538.
98. Pullman, P; Valdemoro, C. (1960). Electronic structure and activity of organophosphorus inhibitors of esterases. Biochimica et biophysica acta. 43; 548-55.
99. Rahimi, R; Abdollahi, M. (2007). A review on the mechanisms involved in hyperglycemia induced by organophosphorus pesticides. Pesticide biochemistry and physiology. 88; 115–121.
100. Ranjbar, A; Pasalar, P; Sedighi, A; Abdollahi, M. (2002). Induction of oxidative stress in paraquat formulating workers. Toxicology letters. 131; 191-194.
101. Rastogi, S. K; Satyanarayan, P. V. V; Ravishankar, D; Tripathi, S. (2009). A study on oxidative stress and antioxidant status of agricultural workers exposed to organophosphorus insecticides during spraying. Indian journal of occupational and environmental medicine. 13; 131-134.
102. Ray, D; Johnson, M; Marrs, T; Coggon, D; Edwards, P; Levy, L. (1998) Organophosphorus esters: An evaluation of chronic neurotoxic effects. MRC Institute for Environment and Health, p. 1-64.
103. Ray, D; Richards, P. G (2001). The potential for toxic effects of chronic, low-dose exposure to organophosphates. Toxicology letters. 120; 343–351.
104. Reffstrup, T.K; Larsen, J.C; Meyer, O. (2010). Risk assessment of mixtures of pesticides: Current approaches and future strategies. Regulatory toxicology and pharmacology. 56; 174–192.
105. Rezg, R; Mornagui, B; El-Fazaa, S; Gharbi, N. (2010). Organophosphorus pesticides as food chain contaminants and type 2 diabetes: a review. Trends in food science & technology. 21; 345-357.
106. Rohr, P. Da Silva, J., Erdtmann, B., Saffi, J., Guecheva, T.N., Henriques, A.P., Kvitko, K. (2011). Ber gene polymorphisms (OGG1 Ser326Cys and XRCC1 Arg194Trp) and modulation of DNA damage due to pesticides exposure. Environmental and Molecular Mutagenesis. 52; 20-27.

107. Saadeh, A. M; Farsakh, N. A; Al-Ali, M. K. (1997). Cardiac manifestations of acute carbamate and organophosphate poisoning. Heart. 77; 461-464.
108. Samai, M; Boccuti, S; Samai, H. H; Gard, P. R; Chatterjee, P. K. (2010). Modulation of antioxidant enzyme expression and activity by paraquat in renal epithelial NRK-52E cells. Sierra leone journal of biomedical research. 2; 103-114.
109. Sanborn, M; Kerr, K.J; Sanin, L.H; Cole, D.C; Bassil, K.L; Vakil, C. (2007). Non-cancer health effects of pesticides. Systematic review and implications for family doctors. Canadian family physician. 53; 712-720.
110. Sarhan, O. M. M; Al-Sahhaf, Z. Y. (2011). Histological and Biochemical Effects of Diazinon on Liver and Kidney of Rabbits. Life science journal. 4; 1183-1189.
111. Shaham, J; Kaufman, Z; Gurvich, R; Levi, Z. (2001). Frequency of sister-chromatid exchange among greenhouse farmers exposed to pesticides. Mutation research. 491; 71-80.
112. Seema, S. S; Tirpude, B. H. (2008). Pattern of histo pathological changes of liver in poisoning. Journal of indian academy of forensic medicine. 30; 63-68.
113. Shadnia S, Azizi E, Hosseini R, Khoei S, Fouladdel S, Pajoumand A, et al. (2007). Evaluation of oxidative stress and genotoxicity in organophosphorus insecticide formulators. Human & experimental toxicology. 24; 439-445.
114. Shukla, Y; Arora, A. (2001). Transplacental carcinogenic potential of the carbamate fungicide mancozeb. Journal of pathology, toxicology and oncology. 20; 127-131.
115. Simpson, W. M; Schuman, S. H. (2002). Recognition and Management of Acute Pesticide Poisoning. American family physician. 65; 1599-1604.
116. Spencer, C. I; Yuill, K. H; Borg, J. J; Hancox, J. C; Kozlowski, R. Z. (2001). Actions of pyrethroid insecticides on sodium currents, action potentials, and contractile rhythm in isolated mammalian ventricular myocytes and perfused hearts. The Journal pharmacology experimental therapeutics. 298; 1067-1082.
117. Stallones, L; Beseler, C. (2002). Pesticide Poisoning and Depressive Symptoms among Farm Residents. Annals of epidemiology. 12; 389-394.
118. Suiter, D.R; Scharf, M.E. (2012). Insecticide basics for the pest management professional. Cooperative Extension, the University of Georgia College of Agricultural and environmental sciences, bulletin. 1352; 1-28.

119. Sulbatos, L. G. (1994). Mammalian toxicology of organophosphorous pesticides. Journal toxicology environmental health. 43; 271-289.
120. Swanson, D.A; Liu, M.J; Baker, P.J; Garrett, L; Stitzel, M; Wu, J; Harris, M; Banerjee, R; Shane, B. (2001). Brody LC targeted disruption of the methionine synthase gene in mice. Molecular and cellular biology. 21; 1058–1065.
121. Zeljezic, D; Garaj-Vrhovac, V. (2001). Chromosomal aberration and single cell gel electrophoresis (Comet) assay in the longitudinal risk assessment of occupational exposure to pesticides. Mutagenesis. 16; 359-363.
122. Zeljezic, D; Garaj-Vrhovac, V. (2002). Sister chromatid exchange and proliferative rate index in the longitudinal risk assessment of occupational exposure to pesticides. Chemosphere. 46; 295-303.
123. Wald, D.S; Kasturiratne, A; Simmonds, M. (2010). Effect of folic acid, with or without other B vitamins, on cognitive disorders: meta-analysis of randomized trials. The american journal of medicine. 123; 522-527.
124. Wild, D. (1975). Mutagenicity studies on organophosphorus inseticides. Mutation research. 32; 133-150.

Chapter 4

BIOSENSORS FOR PESTICIDE DETECTION: NEW TRENDS

Audrey Sassolas[1], Beatriz Prieto-Simón[2], and Jean-Louis Marty[1]

[1]Laboratoire IMAGES EA 4218, Université de Perpignan via Domitia, Perpignan, France
[2]Nanobioengineering Group, Institute for Bioengineering of Catalonia, Barcelone, Spain

ABSTRACT

Due to the large amounts of pesticides commonly used and their impact on health, prompt and accurate pesticide analysis is important. This review gives an overview of recent advances and new trends in biosensors for pesticide detection. Optical, electrochemical and piezoelectric biosensors have been reported based on the detection method. In this review biosensors have been classified according to the immobilized biorecognition element: enzymes, cells, antibodies and, more rarely, DNA. The use of tailor-designed biomolecules, such as aptamers and molecularly imprinted polymers, is reviewed. Artificial Neural Networks, that allow the analysis of pesticide mixtures are also presented. Recent advances in the field of nanomaterials merit special mention. The incorporation of nanomaterials provides highly sensitive sensing devices allowing the efficient detection of pesticides.

INTRODUCTION

In agriculture, farmers use numerous pesticides to protect crops and seeds before and after harvesting. Pesticide is a term used in broad sense for organic toxic compounds used to control insects, bacteria, weeds, nematodes, rodents and other pests. The pesticide residues may enter into the food chain through air, water and soil. They affect ecosystems and cause several health problems to animals and humans. Pesticides can be carcinogenic and cytotoxic. They can produce bone marrow and nerve disorders, infertility, and immunological and respiratory diseases.

Detection of pesticides at the levels established by the Environmental Protection Agency (EPA) remains a challenge. Chromatographic methods coupled to selective detectors have been traditionally used for pesticide analysis due to their sensitivity, reliability and efficiency. Nevertheless, they are time-consuming and laborious, and require expensive equipments and highly-trained technicians. Over the past decade, considerable attention has been given to the development of biosensors for the detection of pesticides as a promising alternative. A biosensor is a self-contained device that integrates an immobilized biological element (e.g. enzyme, DNA probe, antibody) that recognizes the analyte (e.g. enzyme substrate, complementary DNA, antigen) and a transduction element used to convert the (bio)chemical signal resulting from the interaction of the analyte with the bioreceptor into an electronic one. According to the signal transduction technique, biosensors are classified into electrochemical, optical, piezoelectric and mechanical biosensors. Electrochemical transducers have been widely used in biosensors for pesticides detection due to their high sensitivity [1-3]. Additionally, their low cost, simple design and small size, make them excellent candidates for the development of portable biosensors [4-8]. According to the biorecognition element, enzymatic, whole cell, immunochemical, and DNA biosensors have been developed for pesticides detection.

This review presents a state-of-the-art update in pesticide biosensors. To clearly report the last advances, biosensors have been classified according to the immobilized recognition element. New trends in the field of pesticide analysis are also reviewed. Aptamers are shown as good candidates to replace the conventional antibodies and, thus, to be the biorecognition elements in more robust and stable biosensors for pesticide detection. Due to exceptional characteristics, molecular imprinted polymers (MIPs) are innovative affinity-based recognition elements that are exploited for the development of environmental sensors. The use of Artificial Neural Networks (ANNs) coupled with a sensor array could substantially improve the selectivity and allow exact identification of pesticides present in a sample. Recent reports on the properties of nanomaterials show nanoparticles and nanotubes as promising tools to improve the efficiency of biosensors for the detection of pesticides.

ENZYME BIOSENSORS

Enzyme biosensors for pesticide detection are based on measurements of enzyme inhibition or on direct measurements of compounds involved in the enzymatic reaction.

Inhibition-Based Biosensors

Cholinesterase-Based Biosensors

Enzymatic detection of pesticides is mainly based on cholinesterase (ChE) inhibition [6-10]. Organophosphate and carbamate insecticides are the main ChE inhibitors (**Table 1**). Other compounds, such as heavy metals, fluoride, nerve gas or nicotine, can also inhibit ChE enzyme. Although this lack of selectivity, ChE-based biosensors are shown as powerful tools when a rapid toxicity screening is required.

Mono-Enzymatic Biosensors

Two types of natural ChE enzymes are known: acetylcholinesterase (AChE) and butyrylcholine-sterase (BChE). These enzymes have different substrates: AChE preferentially hydrolyzes acetyl esters, such as acetylcholine (Equation (1)), whereas BChE hydrolyzes butyrylcholine (Equation (2)):

$$\text{Acetylcholine} + H_2O \xrightarrow{\text{AChE}} \text{Choline} + \text{Acetic acid} \quad (1)$$

$$\text{Butyrylcholine} + H_2O \xrightarrow{\text{BChE}} \text{Choline} + \text{Butyric acid} \quad (2)$$

The pH variation produced by the acid formation can be measured using electrochemical methods, such as potentiometry [11]. This pH change can also be measured using pH-sensitive spectrophotometric indicators [12,13] or pH sensitive fluorescence indicators [14].

Artificial substrates, acetylthiocholine for AChE and butyrylthiocholine for BChE, have been also used. The enzymatic hydrolysis of these substrates produces electroactive thiocholine (Equations (3) and (4)).

$$\text{Acetylthiocholine} + H_2O \xrightarrow{\text{AChE}} \text{Thiocholine} + \text{Acetic acid} \quad (3)$$

$$\text{Butyrylthiocholine} + H_2O \xrightarrow{\text{AChE}} \text{Thiocholine} + \text{Butyric acid} \quad (4)$$

$$2\,\text{Thiocholine} + H_2O \xrightarrow[\text{oxidation}]{\text{Anodic}} \text{Dithiobischoline} + 2H^+ + 2e^- \quad (5)$$

This system has two advantages over the bi-enzymatic ChE/ChOD biosensors. First, it has a simpler design. Secondly, the detection potential is lower than the one used for the oxidation of H_2O_2.

La Rosa et al. proposed the use of 4-aminophenyl acetate as alternative ChE substrate [15,16]. They oxidize the enzymatic product 4-aminophenol at

+250 mV vs SCE. Electrochemical biosensors for pesticide detection based on the use of this substrate avoid interferences from the oxidation of other electroactive compounds [15-18]. However, 4-aminophenyl acetate is not commercially available and its use involves a laborious and time-consuming synthesis. Moreover, this substrate is unstable and requires special storage conditions (nitrogen atmosphere, below 0°C).

Bi-Enzymatic Biosensors

In most cases, ChE is coupled to choline oxidase (ChOD) [6]. AChE hydrolyzes its natural substrate to choline and acetic acid (Equation (1)). Since choline is not electrochemically active, ChOD is used to produce H_2O_2, which can be oxidized onto the platinum electrode at around + 0.7 V vs Ag/AgCl (Equations (7) and (8)). However, an over-potential is required, favouring the oxidation of interfering electroactive species present in real samples. To overcome this drawback, different approaches have been proposed, such as the use of nanomaterial-modified electrodes. A biosensor for the detection of pesticides and nerve agents was developed by immobilizing AChE and ChOD onto Au-Pt bimetallic NPs [19]. The synergistic effect of these nanoparticles increased the surface area and facilitated the electron transfer process, reducing the applied potential for the detection of H_2O_2. Alternatively to H_2O_2 oxidation, ChE inhibition can be followed using a Clark electrode able to measure the oxygen consumed by the ChOD catalyzed reaction (Equations (6)-(8)) [20].

$$\text{Acetylcholine} + H_2O \xrightarrow{\text{AChE}} \text{Choline} + \text{Acetic acid} \quad (1)$$

$$\text{Choline} + O_2 \xrightarrow{\text{ChOD}} \text{Betaine} + H_2O \quad (6)$$

$$H_2O \xrightarrow[\text{vs Ag/AgCl}]{+0.7\text{ V}} O_2 + 2H^+ + 2e^- \quad (7)$$

$$O_2 + 4H^+ + 4e^- \xrightarrow[\text{vs Ag/AgCl}]{-0.6\text{ V}} H_2O \quad (8)$$

AChE was also coupled to tyrosinase [18]. In this case, AChE enzymatic hydrolysis of phenyl acetate produces phenol compounds, characterized by a high oxidation potential. For this reason, tyrosinase enzyme was used to convert the phenol to quinone, compound that can be electrochemically reduced to catechol at −150 mV vs Ag/AgCl.

Tri-Enzymatic Biosensors

Peroxidase may be added to the bi-enzyme system to develop a tri-enzymatic biosensor. Karousos et al. used a Quartz Crystal Microbalance (QCM) sensor based on three enzymes for the determination of organophosphorus and carbamate pesticides [21].

Table 1. Characteristics of electrochemical cholinesterase-based biosensors for pesticide detection.

Target analyte	Detection technique	Enzyme immobilization technique	Electroactive materials	Linearity range (M)	Detection limit (M)	References
Organophosphorus insecticides						
Chlorpyrifos	CV	Covalent binding	Exfoliated graphite nanoplatelets	ND	1.58×10^{-10}	[170]
Chlorpyrifos	SWV	Cross-linking	SWCNT	$10^{-11} - 10^{-6}$	10^{-12}	[171]
Chlorpyrifos	Amperometry	Covalence	ZnS NPs	$1.5 \times 10^{-9} - 4 \times 10^{-8}$	ND	[172]
Chlorpyrifos oxon	CV	Entrapment	PEDOT:PSS	ND	4×10^{-9}	[173]
Chlorpyrifos oxon	Amperometry	Entrapment	7,7,8,8-tetracyano quinodimethane	$6 \times 10^{-9} - 2.4 \times 10^{-9}$	6×10^{-9}	[174,175]
Paraoxon	Amperometry	Affinity	MWCNT	$3.6 \times 10^{-14} - 3.6 \times 10^{-11}$	5×10^{-15}	[176]
Paraoxon	Fluorescence	Adsorption	CdTe QDs			
Paraoxon	Amperometry	Entrapment	-	$1.3 \times 10^{-7} - 5 \times 10^{-6}$ M.	3.5×10^{-2}	[177]
Paraoxon	Amperometry	Adsorption	AuNPs, graphene oxide nanosheets	ND	10^{-13}	[178]
Paraoxon	Amperometry	Cross-linking	CoPc-Prussian blue	$7.3 \times 10^{-9} - 1.8 \times 10^{-8}$	7.3×10^{-9}	[179]
Methylparaoxon	Amperometry	Entrapment	CoPc	$2 \times 10^{-9} - 4 \times 10^{-6}$	2.6×10^{-9}	[180]
Methylparaoxon	Amperometry	Affinity	MWCNT	$3.8 \times 10^{-14} - 3.8 \times 10^{-11}$	5.3×10^{-15}	[176]
Triazophos	Amperometry	Adsorption	MWCNT	$3 \times 10^{-8} - 7.8 \times 10^{-6}$	10^{-8}	[181]
Dichlorvos	Amperometry	Adsorption	-	ND	10^{-10}	[182]
Dichlorvos	Amperometry	Entrapment	CoPc	ND	7×10^{-12}	[23]
Dichlorvos	Amperometry	Entrapment	CoPc	$2 \times 10^{-10} - 10^{-8}$	9.6×10^{-11}	[180]
Dichlorvos	Amperometry	Adsorption	-	Up to 10^{-16}	10^{-17}	[183]
Dichlorvos				$4.52 \times 10^{-11} - 4.52 \times 10^{-8}$	1.13×10^{-11}	
Omethoate	Amperometry	Cross-linking	Prussian blue	$2.34 \times 10^{-10} - 4.69 \times 10^{-8}$	7.04×10^{-11}	[184]
Trichlorfon				$1.16 \times 10^{-10} - 1.94 \times 10^{-8}$	1.94×10^{-11}	
Phoxim				$1.68 \times 10^{-10} - 3.35 \times 10^{-8}$	3.35×10^{-11}	
Trichlorfon	Amperometry	Adsorption	TiO$_2$ and PbO$_2$ particles	$10^{-8} - 2 \times 10^{-5}$	10^{-10}	[185]
Monocrotophos	Amperometry	Adsorption	AuNPs	$4.5 \times 10^{-9} - 4.5 \times 10^{-6}$	2.7×10^{-9}	[186]
Monocrotophos	Amperometry	Covalent binding	AuNPs-QDs	$4.5 \times 10^{-9} - 4.5 \times 10^{-6}$	1.3×10^{-9}	[155]
Acephate	FET	Affinity	CNT	ND	5.45×10^{-14}	[187]
Dimethoate	Amperometry	Adsorption	CNTs, zirconia NPs, Au colloid coated Fe$_3$O$_4$ magnetic NPs, prussian blue	$4.4 \times 10^{-6} - 4.4 \times 10^{-2}$	2.4×10^{-6}	[188]
Carbamate insecticides						
Aldicarb	Amperometry	Cross-linking	CoPc-Prussian blue	$6.3 \times 10^{-8} - 1.6 \times 10^{-7}$	1.3×10^{-2}	[179]
Carbaryl	Amperometry	Adsorption	-	$2.5 \times 10^{-8} - 5 \times 10^{-7}$	1.5×10^{-8}	[189]
Carbaryl	Amperometry	Covalent binding	QDs	$5 \times 10^{-9} - 2.5 \times 10^{-7}$	3×10^{-9}	[155]
Carbaryl	Amperometry	Cross-linking	CoPc-Prussian blue	$1.2 \times 10^{-7} - 4.9 \times 10^{-7}$	1.2×10^{-7}	[179]
Carbaryl	Amperometry	Adsorption	MWCNT	$5 \times 10^{-13} - 5 \times 10^{-10}$	5×10^{-15}	[190]
Carbaryl	Amperometry	Entrapment	CoPc	$9 \times 10^{-8} - 4 \times 10^{-6}$	1.6×10^{-7}	[180]
Carbofuran	DPV	Adsorption	CNTs-AuNPs	$4.8 \times 10^{-9} - 9 \times 10^{-8}$	4×10^{-8}	[191]
Carbofuran	Amperometry	Cross-linking	CoPc	$10^{-10} - 10^{-7}$	4.9×10^{-10}	[192]
Carbofuran	Amperometry	Entrapment	CoPc	$4 \times 10^{-9} - 8 \times 10^{-8}$	4.5×10^{-9}	[180]

Acetylcholine was converted to choline by AChE and then, choline was converted to hydrogen peroxide by choline oxidase. In the presence of HRP, H_2O_2 oxidized 3,3'-diaminobenzidine to an insoluble product that precipitated out and adsorbed on the crystal surface causing a decrease in the resonant frequency of the crystal. AChE inhibition caused by pesticides reduced the amount of QCM-detectable precipitate produced. This QCM-enzyme sensor system allowed detecting carbaryl and dicholorvos concentrations down to 1 ppm.

ChE Sources

The enzyme source has an important effect on the biosensor performance. Several AChE enzymes are available from different sources, such as Electric eel, Bovine or Human erythrocytes, Horse serum and Human blood. Generally, ChE enzymes isolated from insects are more sensitive than those extracted from other sources. The use of recombinant ChE enzymes also allows improvements on the sensitivity of biosensors [22]. As an example, Valdes-Ramirez and co-workers compared the use of three AChEs in biosensors for the detection of dichlorovos in a sample of apple skin [23]. The use of genetically modified AChE decreased four orders of magnitude the detection limit found for the use of AChE from wild type Drosophila melanogaster and Electriceel.

Tyrosinase-Based Biosensors

Tyrosinase oxidizes monophenols in two consecutive steps: first, the enzyme catalyzes the o-hydroxylation of monophenol to o-diphenol (cresolate activity, Equation (9)) which, in a second step, is oxidized to its corresponding o-quinone (catecholase activity, Equation (10)):

$$\text{Monophenol} + O_2 \xrightarrow{\text{Cresolate activity}} \text{Catechol} \qquad (9)$$

$$\text{Catechol} + O_2 \xrightleftharpoons[-0.2 \text{ V vs Ag/AgCl}]{\text{Catecholase activity}} \text{O-quinone} \qquad (10)$$

Tyrosinase is inhibited by different compounds, such as carbamate pesticides and atrazine. Numerous electrochemical biosensors based on the inhibition of tyrosinase activity have been reported [24-29] (**Table 2**).

Tyrosinase biosensors suffer from poor specificity since many substrates and inhibitors can interfere. The enzyme is inherently unstable, reducing the lifetime of the tyrosinase-based biosensors. However, tyrosinase can stand high temperatures and the organic solvents used to dissolve the pesticides.

Alkaline Phosphatase (ALP)-Based Biosensors

Alkaline phosphatase catalyses the following reaction:

$$\text{Phosphate monoester} + H_2O \rightarrow \text{alcohol} + \text{phosphate} \qquad (11)$$

ALP is inhibited by different compounds. Several ALPbased biosensors for the detection of pesticides have been developed using different enzyme substrates depending on the transduction method.

Table 2. Characteristics of electrochemical inhibition-based biosensors using tyrosinase for pesticide detection.

Target analyte	Detection technique	Enzyme immobilization technique	Electroactive materials	Linearity range (M)	Detection limit (M)	References
Organophosphorus insecticides						
Methyl parathion	Amperometry	Cross-linking	CoPc	$2.28 \times 10^{-8} - 3.8 \times 10^{-7}$	ND	[26]
Diazinon	Amperometry	Cross-linking	CoPc	$6.24 \times 10^{-8} - 1.64 \times 10^{-7}$	ND	[26]
Dichlorvos	Amperometry	Cross-linking + entrapment	1,2-naphthoquinone-4-sulfonate (NQS)	Up to 8×10^{-6}	6×10^{-8}	[25]
Dimethoate				$2 \times 10^{-6} - 2 \times 10^{-1}$	10^{-6}	
Pirimicarb				$2 \times 10^{-5} - 5 \times 10^{-3}$	10^{-5}	
Paraoxon	Amperometry	Adsorption	-	$10^{-5} - 10^{-2}$	5×10^{-6}	[29]
Malathion				$10^{-5} - 10^{-2}$	5×10^{-6}	
Paraoxon	Amperometry	Cross-linking	Prussian blue	$10^{-7} - 10^{-6}$	10^{-7}	[193]
Carbamate insecticides						
Carbofuran	Amperometry	Cross-linking	CoPc	$2.26 \times 10^{-8} - 4.07 \times 10^{-7}$	ND	[26]
Carbofuran				$10^{-5} - 10^{-2}$	5×10^{-6}	
Aldicarb	Amperometry	Adsorption	-	$10^{-5} - 10^{-2}$	5×10^{-6}	[29]
Carbaryl				$10^{-5} - 10^{-2}$	5×10^{-6}	
Carbaryl	Amperometry	Cross-linking	CoPc	$4.97 \times 10^{-8} - 2.48 \times 10^{-7}$	ND	[26]
Thiodicarb	Square Wave Voltammetry (SWV)	Entrapment	-	$3.75 \times 10^{-7} - 2.23 \times 10^{-6}$	1.58×10^{-7}	[24]

Ayyagari et al. described a chemiluminescent ALPbased biosensor for the detection of paraoxon [30]. The biosensor was based on the measurement of the intensity of the light generated by ALP-catalyzed dephosphorylation of a chemiluminescent substrate, chloro 3-(4-methoxy spiro [1,2-dioxetane-3-2'-tricyclo-[3.3.1.1]-decan]-4- yl) phenyl phosphate.

A fluorescent ALP-based biosensor for the detection of organochlorine, pesticides (carbamate and fenitrothion), heavy metals and CN^- was also described [31]. ALP enzyme catalyzed the hydrolysis of 1-naphthyl phosphate to fluorescent 1-naphthol.

Mazzei and co-workers developed electrochemical ALP-based biosensors for the detection of malathion and 2,4-dichlorophenoxyacetic acid (2,4-D) by using 3-indoxyl phosphate, phenyl phosphate or ascorbate-2- phosphate as enzyme substrates [32]. Another electrochemical ALP-based biosensor was also described for the screening of several environmental pollutants. The biosensor was based on the entrapment of ALP in a hybrid sol-gel/chitosan film, deposited on the surface of a screen-printed electrode [33]. The substrate ascorbic acid 2-phosphate was catalyzed by the enyme to produce ascorbic acid, which was monitored by amperometry.

Peroxidase-Based Biosensors

Peroxidase molecules can be first oxidized by H_2O_2 and then reduced by phenolic compounds. This process involves two enzyme intermediates: compounds I and II (**Figure 1**). Phenolic compounds are thus oxidized to quinones or free radical products, able to be electrochemically reduced on the electrode surface. Several organic and inorganic compounds have been reported to inhibit the enzyme activity of peroxidase by coordinating compound I. A biosensor based on the inhibition of peroxidase was described for the detection of thiodicarb, a carbamate pesticide [34]. HRP was covalently bound on a gold electrode. In the presence of hydrogen peroxide, hydroquinone was oxidized by peroxidase to p-benzoquinone which could be electrochemically reduced to hydroquinone at a potential of –0.072 V vs Ag/AgCl. The presence of inhibitor compounds induced a decrease of the biosensor current response.

Acid Phosphatase

Acid phosphatase (AP) is reversibly inhibited by some pesticides. AP has been used with glucose oxidase (GOD) to develop a bienzymatic biosensor for the electrochemical detection of Malathion, methyl parathion and paraoxon [37]. Both enzymes were coupled on a commercial H_2O_2 sensing electrode. This system is based on the following reactions:

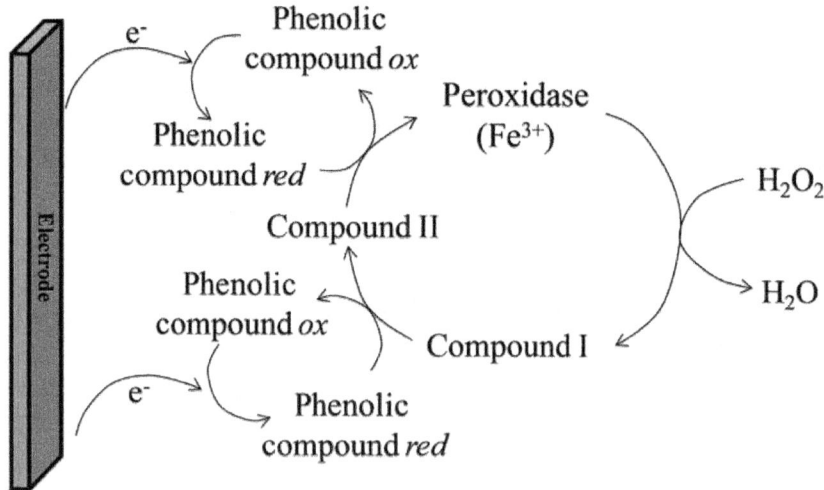

Figure 1. Scheme of the reactions occurring at the surface of a peroxidase-modified electrode. ox: oxidized form, red: reduced form [35,36].

$$\text{Glucose} - 6 - \text{phosphate} + H_2O \xrightarrow{AP} \text{Glucose} + \text{inorganic phosphate}$$

$$\text{Glucose} + O_2 \xrightarrow{GOD} \text{Gluconolactone} + H_2O$$

Catalytic Biosensors

Organophosphorus Hydrolase (OPH)

OPH is an enzyme that hydrolyzes organophosphorus pesticides [38], such as parathion, methyl parathion [39] or paraoxon [40,41]. This enzyme hydrolyzes P-O, P-S and P-CN bonds generating two protons, able to be electrochemically detected, and an alcohol, which in many cases is chromophoric and/or electroactive.

However, these biosensors show lower sensitivity values and higher detection limits than cholinesterase-based biosensors. Moreover, they can only detect some organophosphorus (OP) compounds.

Table 3 summarizes the performances of some OPHbased biosensors reported in the literature.

Glutathion-S-Transferase

Glutathion-S-transferase (GST) was used to develop a fiber-optic biosensor for the detection of atrazine [42]. The enzyme was immobilized by cross-linking on a membrane that was supported on an inner glass disk by means of an intermediate binder sol-gel layer. Bromcresol green was incorparated in the sol-gel as pH indicator. GST catalyzed the nucleophile attack of GSH on atrazine, releasing H^+. This pH variation was optically measured by colour changes of bromcresol green.

WHOLE CELL BIOSENSORS

Microbial Biosensors

To develop a microbial biosensor, microorganisms have to be immobilized onto a transducer using different chemical (e.g. cross-linking) or physical techniques (e.g. entrapment) [43].

Table 3. Characteristics of hydrolase-based biosensors.

Target analyte	Detection technique	Enzyme immobilization technique	Electroactive materials	Linearity range (M)	Detection limit (M)	References
Paraoxon	Amperometry	Covalent binding	SWCNT	$5 \times 10^{-7} - 8.5 \times 10^{-6}$	10^{-8}	[41]
Paraoxon	Amperometry	Entrapment	Mesoporous carbon	$2 \times 10^{-7} - 8 \times 10^{-6}$	1.2×10^{-7}	[40]
Paraoxon	Amperometry	Entrapment	MCNT	Up to 4×10^{-6}	15×10^{-8}	[194]
Paraoxon	Amperometry	Cross-linking	MWNT	$5 \times 10^{-7} - 2 \times 10^{-6}$	0.314×10^{-6}	[195]
Paraoxon	Optic	Cross-linking	-	$10^{-6} - 4.8 \times 10^{-4}$	2×10^{-6}	[196]
Ethyl Parathion	Amperometry	Covalent binding	-	ND	$<3.4 \times 10^{-9}$	[197]
Methyl parathion	Amperometry	Covalent binding	AuNPs – MWCN-QDs	$1.9 \times 10^{-8} - 7.6 \times 10^{-7}$	3.8×10^{-9}	[39]
Methyl parathion	Amperometry	Entrapment	MCNT	Up to 2×10^{-6}	8×10^{-7}	[194]
Parathion	Optic	Cross-linking	-	$10^{-5} - 2 \times 10^{-4}$	2×10^{-6}	[196]
Paraoxon	Fluorescence	Affinity	-	$10^{-6} - 8 \times 10^{-4}$	ND	[198]
Diisopropyl phosphorofluoridate (DFP)	Fluorescence	Affinity	-	$2 \times 10^{-6} - 4 \times 10^{-4}$	ND	[198]

Microorganisms have several advantages as sensing elements in the development of biosensors. They are able to metabolise a wide range of chemical compounds. The use of whole cells, as a source of intracellular enzymes, avoids expensive protocols of enzyme purification. The enzyme is maintained in its natural environment improving its stability and activity. The main limitation of the use of whole cells is the diffusion of substrate and products through the cell wall resulting in a slow response as compared to enzyme-based biosensors. To overcome this drawback, cells can be permeabilised [44].

Electrochemical Microbial Biosensors

Amperometric Detection

Amperometric microbial biosensors have been widely developed for the determination of biochemical oxygen demand (BOD) in order to measure biodegradable organic pollutants in aqueous samples. Most of BOD biosensors consist of a microbial film sandwiched between a porous cellulose membrane and a gas-permeable membrane. Organic substrates, present in wastewater samples, diffuse through the dialysis membrane and are assimilated by the immobilized microbial population, increasing the bacterial respiration rate. Therefore, less dissolved oxygen diffuses through the gas-permeable Teflon membrane to be detected by a Clark oxygen electrode [45]. Different microbial strains were used as biosensing element such as Arxula adeninivorans [46], Bacillus subtilis [47], Serratia marcescens [48] or yeast [49]. Single microorganisms metabolize a limited range of organic pollutants, which may result in an inaccurate estimation of BOD values. To overcome this problem, mixed cultures (e.g. Bacillus subtilis and Trichosporon cutaneum [50]) or activated sludges [51] were used.

Mulchandani's group developed amperometric microbial biosensors for the determination of organophosphate pesticides with p-nitrophenyl substituent (e.g. paraoxon, methyl parathion, parathion, fenitrothion and ethyl pnitrophenol thiobenzenephosphonate (EPN)) [52]. These biosensors were based on the co-immobilization of microorganisms and OPH (free or expressed on the cell surface of other microorganisms). OPH hydrolyzes the pesticide and releases p-nitrophenol. Released p-nitrophenol can be oxidized by some microorganisms, such as or Pseudomonas putida JS444. Two detection strategies were used:

- OPH hydrolyzes the organophosphorus compounds to produce p-nitrophenol. Released p-nitrophenol was degraded by some bacteria, such as Pseudomonas putida JS444. This degradation resulted in electroactive compounds, amperometrically detected [53,54].
- The degradation of p-nitrophenol by some microbes, such as Arthrobacter sp. JS443, consumes oxygen. A Clark oxygen electrode was used to measure oxygen concentration changes [55-57].

Potentiometric Detection

Conventional potentiometric microbial biosensors have been developed using ion-selective electrodes (e.g. pH, ammonium) or gas sensing electrodes (e.g. pCO_2) coated with an immobilized microbial layer. Assimilation of substrates by microbes causes changes in potential due to ion accumulation or depletion [43].

A potentiometric biosensor for the direct detection of paraoxon was based on the immobilization of recombinant E. Coli on a glass pH electrode. Bacteria was engineered to contain the opd gene that encodes the OPH enzyme [58]. Entrapped OPH-active bacteria hydrolyzed OP compounds producing two protons. The quantity of released H^+ was correlated to the concentration of hydrolyzed paraoxon.

Optical Microbial Biosensors

Optical microbial biosensors allowing the detection of pollutants such as phenols and heavy metals have been developed [52,59]. However, only few optical microbial biosensors allowing the detection of pesticides have been reported. A disposable colorimetric microbial biosensor for the detection of methyl parathion pesticide was described [60]. Whole cells of Flavobacterium sp. were immobilized on a glass fiber filter paper. The OPH activity of Flavobacterium sp. hydrolyzed methyl parathion into p-nitrophenol that can be detected at 410 nm.

Plant Tissue and Photosynthesis-Based Biosensors

Plant Tissue-Based Biosensors

The use of plant tissue is an attractive alternative to enzymatic biosensors. Tissue that acts as enzyme source presents many advantages [61]:

- High stability and activity resulting from the maintenance of the enzyme in its natural environ-ment;
- Long lifetime of biosensors;
- High reproducibility of the experimental results;
- Availability and low price of a wide range of plant tissues;
- Avoidance of tedious and time-consuming enzyme extraction and purification steps;
- Presence of the required cofactors in the used tissue.

Planktonic algae have been widely used to develop biosensors for pollutants present in the aquatic ecosystems. Biosensors based on immobilized Chlorella vulgaris microalgae were reported [62-64]. Those biosensors were based on the inhibition of enzymes located on the external membrane, such as alkaline phosphatases and esterases, by heavy metals and pesticides.

Photosynthesis-Based Biosensors

Different types of photosynthetic materials were used as recognition element for the development of biosensors: whole cells (e.g. microalgae), chloroplasts or thylakoids and photosystem II (PS II) [65]. PS II is a supramolecular pigment-protein complex located in the thylakoid membrane. It catalyzes the light-induced transfer of electrons from water to plastoquinone in a process that evolved oxygen. The activity of PS II can be inhibited by several groups of herbicides and heavy metals [66].

The measurement of oxygen evolution using a Clarktype electrode is a standard procedure for the determination of the photosynthetic activity [61,67,68]. The incorporation of several types of photosystem II specific artificial electron acceptors as electroactive mediators allows to maximize the photosynthetic activity. Other biosensors are not based on the use of a Clark-type electrode. In these cases, alga were immobilized on the surface of ITO electrode [69] or SPE [70].

Optical photosynthesis-based biosensors have also been described based on the fluorescence induced by chlorophyll a. The light absorbed by chlorophyll molecules of PSII may be assimilated into the light reactions of the photosynthesis or may be released as fluorescence or thermal energy. Herbicides inhibit photosynthetic electron flow by blocking the PSII quinone binding site causing an increase in the chlorophyll fluorescence emission. Based on this principle, herbicide biosensors based on the measurement of the algal chlorophyll fluorescence at 682 nm (under 469 nm excitation light) were developed [71-73]. Recently, three microalgae species (Dictyosphaerium chlorelloides, Scenedesmus intermedius and Scenedesmus sp.) were entrapped within a silica matrix and the increase in the amount of chlorophyll fluorescence signal was used to quantify simazine [74].

IMMUNOSENSORS

Immunosensors are characterized by the highly selective affinity interactions between immobilized antibodies (Ab) or antigens (Ag), on the transducer surface, and their specific analytes, Ag or Ab respectively [75-77]. Unlike enzyme-based biosensors, able to evaluate total toxicity, immuno-sensors are specific for a molecule.

Several immunosensors for pesticides detection have been described, based on electrochemical, optical, piezoelectric and mechanical transduction methods.

Electrochemical Immunosensors

Table 4 presents performance characteristics of some electrochemical immunosensors.

Amperometric Detection

Numerous amperometric immunosensors for the detection of pesticides have been reported [78-81]. A simple amperometric immunosensor was developed for the analysis of 2,4-D in the presence of organic solvents, required to solubilize it from soil [81]. An amperometric immunosensor based on a carbon paste SPE incorporateing a conducting polyaniline (PANI)/poly(vinylsulphonic acid) (PVS) copolymer was developed for the detection of atrazine [82]. Free and HRP-labeled atrazine competed for their binding to the Ab previously immobilized onto the PANI-PVS-modified electrode surface. The addition of HRP substrate, hydrogen peroxide, enabled the catalytic reaction inducing a flow of electrons from the electrode surface through the molecular wires of the PANI/PVS copolymer.

Table 4. Characteristics of some electrochemical immunosensors.

Target analyte	Immobilization technique	Electroactive materials	Linearity range (M)	Detection limit (M)	References
Amperometry					
Phenanthrene (PAH)	Adsorption of BSA-phenanthrene conjugate	-	$2.8 \times 10^{-9} - 2.5 \times 10^{-7}$	4.5×10^{-9}	[199]
Paraoxon	Adsorption + Nafion film	AuNPs	$8.7 \times 10^{-8} - 6.9 \times 10^{-6}$	4.4×10^{-8}	[200]
Atrazine	Affinity	-	$7 \times 10^{-10} - 1.35 \times 10^{-8}$	1.7×10^{-10}	[201]
Atrazine	Adsorption	PANI-PVS copolymer	$5.5 \times 10^{-10} - 2.3 \times 10^{-8}$	4.6×10^{-10}	[82]
17-β estradiol	Adsorption of BSA-estradiol conjugate	-	$2.2 \times 10^{-12} - 3.6 \times 10^{-5}$	9.2×10^{-13}	[202]
Pichloram	Adsorption of BSA-pichloram conjugate	Gold nanoclusters	$3.6 \times 10^{-9} - 3.6 \times 10^{-5}$	1.8×10^{-9}	[203]
Diuron	Adsoprtion of hapten-BSA conjugate	Prussian blue, AuNPs	$4.3 \times 10^{-12} - 4.3 \times 10^{-5}$	4.3×10^{-12}	[156]
Naphthalene	Adsorption	AuNPs	$3.9 \times 10^{-9} - 7.8 \times 10^{-7}$	6.2×10^{-10}	[204]
Impedance spectroscopy					
Atrazine	Affinity	-	$4.6 \times 10^{-8} - 1.4 \times 10^{-6}$	9.3×10^{-8}	[97]
Atrazine	Adsorption of BSA-atrazine conjugate	-	ND	2.7×10^{-8}	[90]
Atrazine	Adsorption of BSA-atrazine conjugate	-	ND	3.9×10^{-8}	[91]
Atrazine	Covalent immobilization of BSA-atrazine conjugate	-	ND	1.9×10^{-10}	[92]
Atrazine	Affinity	-	$4.6 \times 10^{-10} - 4.6 \times 10^{-6}$	4.6×10^{-11}	[93]
Atrazine	Affinity/entrapment	-	$4.6 \times 10^{-10} - 9.3 \times 10^{-7}$	4.6×10^{-10}	[94]
2,4-D	Covalent binding	-	$2 \times 10^{-10} - 2 \times 10^{-6}$	ND	[205]

Conductometric Detection

To our knowledge, only a few conductometric immunosensors for environmental analysis have been described, probably due to their low specificity.

Valera and co-workers used this method for atrazine detection [83-85]. Atrazine was covalently immobilized on interdigitated µ-electrodes (IDµE) [85]. Detection of free atrazin was achieved through a competitive reaction with immobilized atrazine for the antibody added in solution. The detection method was based on the use of antibodies labeled with gold nanoparticles. Their presence amplified the conductive signal. This biosensor is adapted for the detection of atrazine in red wines since none matrix effect related to red wine samples were observed.

Potentiometric Detection

A few potentiometric immunosensors have been described for environmental analysis [86-88]. A potentiometric biosensor was developed for terbuthylazine (TBA), a herbicide widely used in agriculture [88]. Free TBA and immobilized TBA-BSA conjugate competed for their binding to urease-labeled specific Ab. The addition of urease substrate, urea, enabled the potentiometric measurement of the ammonia produced in an inversely proportional amount of the TBA present in the sample.

Electrochemical Impedance Spectroscopy

Several impedimetric immunosensors for the detection of pesticides have been reported [89-97]. The electron transfer resistance at the interface between the electrode and the solution changes slightly when the immobilized biomolecule binds the analyte. Impedance spectroscopy allows a label-free detection with many potential advantages, such as higher signal-to noise ratio, ease of detection, lower assay cost, faster assays and shorter analysis times. However, regeneration of the sensing surface is typically time-consuming and not reproducible [4].

Table 4 presents characteristics of some impedimetric immunosensors for pesticide analysis.

Valera and co-workers developed biosensors for atrazine detection using a competitive immuno-assay. AtrazineBSA conjugate was immobilized on the IDμE surface, either by adsorption [90,91] or by covalent binding [92]. Best performances were obtained when the conjugate was covalently bound to the IDμE area previously activated with (3-glycidoxypropyl)trimethoxysilane. The detection limit was 1.9×10^{-10} M. Recently, an impedimetric immunosensor for atrazine detection was developed by immobilizing anti-atrazine antibody modified with histidine-tag onto a polypyrrole (PPy) film N-substituted by nitrilotriacetic acid (NTA) electrogenerated on a gold electrode [93]. After coordination of Cu^{2+} to poly-NTA, anchoring of histidine-tagged atrazine was achieved by affinity interactions between histidine groups and the chelated Cu^{2+} centers. In the presence of atrazine, the interaction of the analyte with the immobilized antibody triggered an increase of the charge transfer resistance proportional to the pesticide concentration. The detection limit was 4.6×10^{-11} M.

Optical Immunosensors

Optical immunosensors are based on the measurement of changes in the optical characteristics induced by the formation of Ab-Ag complexes.

SPR

SPR-based biosensors have been reported for the detection of different pesticides (**Table 5**). SPR allows realtime monitoring and does not require labeled molecules [98].

Mauriz and co-workers used a commercial SPR (SENSIA) for the on-line monitoring of pesticides in real water samples [99-104]. Pesticides-BSA conjugates were immobilized through a self-assembled monolayer (SAM) onto a gold electrode to obtain a reusable sensing surface. The same sensing surface was used for more than 200 assays, showing good reproducibility. Several conjugates were immobilized on the sensing surface of one individual flow cell to simultaneously detect DTT, chlorpyrifos and carbaryl.

Miura and co-workers developed a SPR-based immunosensor for the competitive detection of 2,4-D [105- 108]. A 2,4-D-ovalbumine conjugate was immobilized onto the Au surface of the sensor chip to compete with free 2,4-D for their selective binding to monoclonal anti- 2,4-D. Amplification through avidin-biotin interactions was described to enhance the sensitivity of the sensor. The amplification was based on successive incubations with a biotinylated secondary antibody against anti-2,4-D, avidin and finally biotin-BSA molecules. The sensor signal was amplified by a factor of 10 [108].

Fluorescence Polarisation

Immunospecies have to be conjugated to fluorescent labels (e.g. cyanine) to develop immuno-sensors. However, fluorescent organic dyes suffer from a photo-bleaching problem. This drawback can be overcome by the use of nanoparticles as fluorescent reporters. For this purpose, europium chelate-dyed polystyrene NPs have been used to develop immunosensors for the detection of atrazine [109].

Total Internal Reflection Fluorescence (TIRF)

TIRF was used to develop immunosensors for water pollution control [75,110-115]. Barzen et al. described a prototype of a portable TIRF-based immunosensor combined with a flow injection system to monitor surface water quality [111].

Table 5. Characteristics of some SPR-based immunosensors.

Target	Immobilization technique	Linearity range (M)	Detection limit (M)	References
DTT	Covalent binding	$1.7 \times 10^{-10} - 5 \times 10^{-9}$	5.64×10^{-11}	[104]
2,4-D	Adsorption of 2,4-D—BSA conjugate	$2.3 \times 10^{-9} - 4.5 \times 10^{-6}$	2.3×10^{-9}	[107]
2,4-D	Adsorption of 2,4-D—ovalbumin conjugate	$4.5 \times 10^{-10} - 1.4 \times 10^{-6}$	4.5×10^{-10}	[106]
2,4-D	Covalent immobilization of 2,4-D—BSA conjugate	ND	3.6×10^{-11}	[108]
2,4-D	Covalent immobilization of 2,4-D—BSA conjugate	4.5×10^{-10}	4.5×10^{-10}	[105]
Chorpyrifos	Covalent binding	$6.6 \times 10^{-10} - 1.4 \times 10^{-7}$	1.4×10^{-10}	[104]
Carbaryl	Covalent binding	$8.2 \times 10^{-9} - 7.3 \times 10^{-8}$	4.5×10^{-9}	[104]
Isoproturon	Covalent binding	$6.3 \times 10^{-9} - 7.9 \times 10^{-8}$	4.8×10^{-10}	[206]
Atrazine	Covalent immobilization of atrazine—BSA conjugate	$1.3 \times 10^{-10} - 3.7 \times 10^{-9}$	9.3×10^{-11}	[99]
TNT	Adsorption of TNP—BSA conjugate	$2.6 \times 10^{-11} - 1.3 \times 10^{-6}$	2.6×10^{-11}	[207]
TNT	Covalent immobilization on AuNPs	ND	4.4×10^{-11}	[208]
TNT	Covalent binding	$3.5 \times 10^{-11} - 1.3 \times 10^{-7}$	ND	[209]
TNT	Covalent binding	Up to 3.1×10^{-8}	4.8×10^{-10}	[210]
2,4-dinitrotoluene	Covalent binding	$4.4 \times 10^{-9} - 4.4 \times 10^{-7}$	8.8×10^{-11}	[211]
2,4-dichlorophenol	Affinity (protein G)	ND	1.2×10^{-7}	[212]

Aminodextran-analyte conjugates were adsorbed onto the transducer surface. Specific antibodies labeled with a fluorophore were incubated with the target analytes present in the sample. Then, free labeled-antibodies bound immobilized analyte derivatives. A collimated laser beam (635 nm) coupled to the transducer was guided by total internal reflection causing excitation of the bound fluorescent antibody in the evanescent field (**Figure 2**). This TIRF-based immunosensor, called the RIver ANAlyser (RIANA), was used to monitor the levels of atrazine, simazine and alachlor in the Ebre Delta and the estuarine area of Portugal [77]. Studies were focused on the evaluation of matrix effects, interferences due to the presence of crossreactant substances and on the validation of the sensor. Results show the efficiency of the RIANA immunosensor for monitoring natural waters in compliance with the Drinking Water Directive of the European Union.

Automated Water Analyser Computer Supported System (AWACSS) was developed using the same principle as RIANA immunosensor. However, improvements were made in three critical areas: 1) expanded multi-analyte analysis capability allowing for simultaneous detection of up to 30 analytes; 2) novel design approaches to the optical detection and fluidics including miniaturized integrated optics and micro-fluidics and 3) intelligent remote surveillance and control for unattended continuous monitoring [116].

Polarisation-Modulation InfraRed Reflection-Absorption Spectroscopy (PM-IRRAS)

Reflection-absorption IR spectroscopy combined with polarization modulation was proposed as a novel optical transduction mode. Recently, PM-IRRAS was used to develop immunosensors allowing the detection of environmental pollutants [117-119].

This principle was applied by Pradier and co-workers to determine atrazine using an indirect competitive format [119]. First, sensor chips were coated with ovalbumine-atrazine derivatives. The surfaces were analyzed by PM-IRRAS and the integrated area of the peptide bands was measured. Then, the different steps of the conventional ELISA test were performed. Successive binding of anti-atrazine antibody and secondary anti-rabbit immunoglobulin G antibody resulted in a change of the IR absorption properties of the organic film at the sensor surface. Detection of atrazine was based on the analysis of the amide I and II bands. When the concentration of free pesticide increased, the intensity of amide bands decreased.

Piezoelectric Immunosensors

Piezoelectric immunosensors are devices based on materials such as quartz crystals with Ab or Ag immobilized on their surface. A QCM sensor is a mass-sensitive device able to measure very small mass changes. Several QCM immunosensors for environmental monitoring have been described [120], but only few of them were developed for the detection of pesticides. A QCM immunosensor was developed for the analysis of carbaryl, and 3,5,6-trichloro-2-pyridinol (TCP), the main metabolite of the insecticide chlorpyrifos and of the herbicide triclopyr [121]. The biosensor was based on the immobilization of hapten conjugates onto the gold electrode via SAM. This covalent immobilization allowed the reusability of the modified electrode surface for at least 150 assays without significant loss of sensitivity.

Mechanical Immunosensors

Microcantilever sensors are based on a response due to either surface stress variation or mass loading. Interaction between an immobilized ligand (e.g. antibody) and an analyte (e.g. an antigen) causes a surface stress change of the cantilever and can be detected as changes in the cantilever deflection. Microcanlitever sensors offer many advantages: label-free detection, high precision, reliability, reduced size and easy manufacture of multielement sensor arrays [122].

Some microcantilever immunosensors have been developed for the detection of pesticides such as atrazine[123,124], DDT [125] and 2,4-D [123].

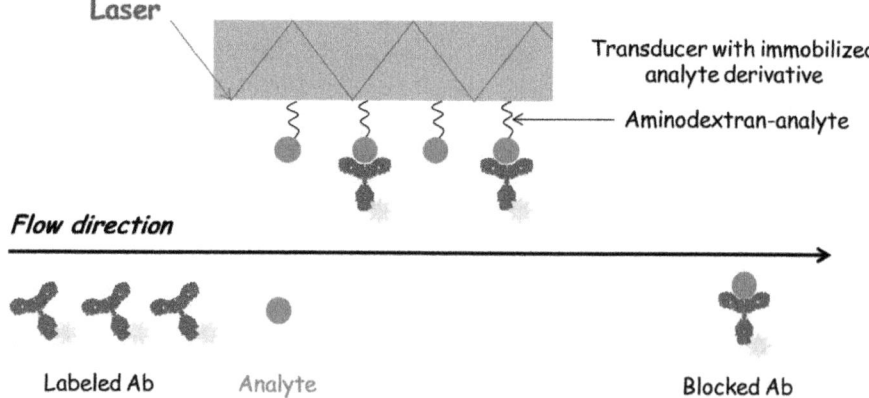

Figure 2. Principle of a TIRF-based immunosensor [111].

Suri et al. developed an ultrasensitive cantilever-based immunosensor for the detection of atrazine. A thiolated atrazine antibody was immobilized on a gold coated cantilever. The binding of atrazine to the immobilized Ab changes the surface stress causing bending of the cantilever. The detection limit was 1 ppt (4.65 pM). Recently, a cantilever-based competitive immunosensor was developed for the detection of 2,6 dichlorobenzamide (BAM), which is the most frequent found pesticide residue in European groundwater [126].

DNA BIOSENSORS

DNA biosensors exploit the preferential binding of complementary single-stranded nucleic acid sequences. They usually rely on the immobilization of a single-stranded DNA probe onto a surface able to recognize its complementary DNA target sequence by hybridization [122].

Recently, an electrochemical DNA biosensor was developed to study DNA damage caused by several pesticides, such as atrazine, 2,4-D, glufosinate ammonium, carbofuran, paraoxon-ethyl and difluorobenzuron [127]. A biotinylated DNA probe was immobilized on a streptavidin-modified electrode surface. This DNA probe was hybridized with biotinylated complementary DNA target analyte. Streptavidin labeled with ferrocene was further attached to the hybridized biotinylated DNA. The close proximity of ferrocene to the electrode surface induced a current signal. The presence of pesticides caused an unwinding of the DNA and thus a decrease of the ferrocene oxidation current observed in voltammetric experiments. Paraoxon-ethyl and atrazine caused the fastest and most severe damage to DNA.

NEW TRENDS

Aptamers

In 1990, Ellington's group [128], Gold's group and Robertson's group independently reported the development of an in vitro selection technique which allowed the discovery of specific nucleic acid sequences that bind non-nucleic acid targets with high affinity and specificity. The technique was called SELEX (Selection Evolution of Ligands by EXponential enrichment) and the resulting DNA or RNA oligonucleotides are referred to as aptamers [129-131].

Aptamers show high affinity towards a wide range of target analytes, including proteins, metal ions and pathogenic microorganisms. Aptamers possess several competitive advantages over antibodies, such as their accurate and reproducible chemical production [132]. Immunization and animals hosts are not necessary to produce aptamers. The selected nucleic acids bind their targets with affinity and specificity comparable to those of antibodies. Aptamers are more stable than antibodies. They can be selected in extreme conditions whereas antibodies are only stable in physiological conditions. Aptamers can also undergo reversible denaturation and they can be easily modified with new functional groups without affecting their activity. Due to its many advantages, numerous aptamer-based biosensors have been developed for the detection of a wide range of targets [133-137].

To our knowledge, few aptamers for the detection of pesticides have been selected. Recently, a DNA aptamer specific for acetamiprid was described [138]. The potential of aptamers for the pesticide detection has not still been exploited but aptamer-based biosensors could be an alternative to the conventional methods of pesticide analysis.

Molecularly Imprinted Polymers (MIPs)

Molecular imprinting, which allows the formation of specific recognition sites in polymers, is used to develop MIP-based sensors in the areas of environmental, food and pharmaceutical analysis [139].

The overall principle of molecular imprinting is presented in **Figure 3**. The template interacts with functional monomers either by the formation of covalent bonds or by self-association. Then, these monomers are polymerized around the template with the help of a crosslinker in the presence of a porogenic solvent. Template molecules are removed by extensive washing steps to disrupt the interactions between the template and the monomers. This process allows to obtain synthetic polymers possessing specific cavities complementary to the template in size, shape and position of the functional group [140,141]. The choice of the chemical reagents making up the MIP must be judicious in order to create highly specific cavities designed for the template molecule.

MIPs have been used as artificial recognition elements of biosensors for pesticide detection. An optical sensor for the detection of pesticides (chloropyrifos, diazinon and glyphosate) was developed by forming MIP onto optical fibers [142]. In this case, a luminescent lanthanide (europium), used as spectroscopic probe, was incorporated into the polymer. Detection of the analyte was based upon the changes that occur in the lanthanide spectrum when pesticide was incorporated to Eu^{3+}. The increase of the pesticide concentration induced an increase in the luminescence intensity of the spectra. Several electrochemical MIP-based sensors have been described [143-145]. An electrochemical sensor for 2,4-D was developed by electropolymerization of polypyrrole on a glassy carbon electrode in the presence of template 2,4-D molecules.

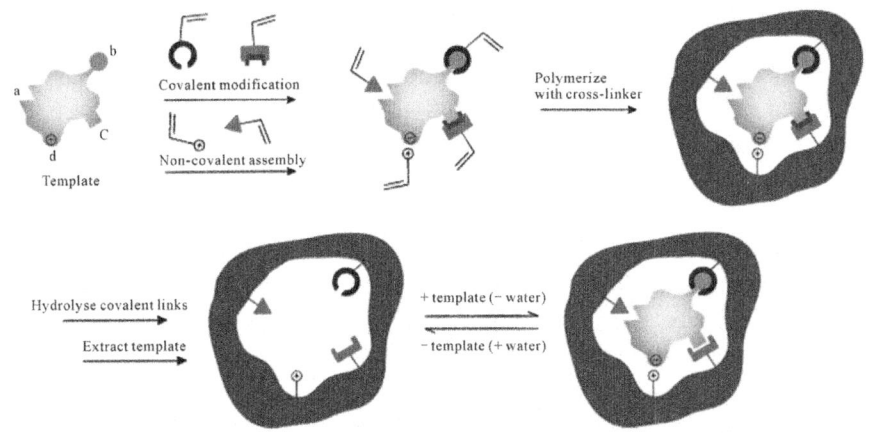

Figure 3. Schematic representation of the molecular imprinting [140].

During the electropolymerization step, 2,4-D molecules were embedded in the imprinted polypyrrole by hydrogen bond and electrostatic interacttions. Then, the template molecules were removed from the polymer by overoxidized process at +1.3 V in 0.2 M Na_2HPO_4 solution for 10 min.

Analysis of Pesticide Mixtures: The Artificial Neural Networks

Development of detection systems able to detect several analytes simultaneously represents a promising tool in environmental monitoring and screening. As it has been previously mentioned, numerous organophosphorus and carbamate insecticides can inhibit cholinesterase activity. One limitation of the enzymatic inhibition tests is the difficulty in discriminating between different inhibitors. To solve this problem, a sensor array can be coupled with an Artificial Neural Network (ANN) in order to precisely identify the inhibitors present in the sample. An ANN is a systematic procedure of data processing inspired by the nervous system function in animals. It combines the response of different enzymes to find a pattern that relates inhibitor concentrations with the inhibition percentages observed. Several intelligent biosensors for the analysis of pesticide mixtures have been developed based on the principle of the AChE inhibition and chemometric data analysis using ANNs [146]. Marty's group constructed ANNs to model the combined response of two pesticides (chlorpyrifos oxon and chlorfenvinfos) using sensors incorporating wild-type electric eel AChE and drosophila mutant AChE, associated or not with PTE [147]. These two types of AChEs were selected according to their different sensitivities to OP insecticides. The developed system was applied to the determination of pesticides in real water samples.

Nanomaterials

The emerging synergy between nanotechnology and sensors has been exploited over the past few years [148- 150]. Intensive research efforts have been performed for the design of efficient nanomaterial-based biosensors that exhibit high sensitivity and stability. The immobilization of nanomaterials onto sensing devices generates novel interfaces that enable the sensitive optical or electrochemical detection of analytes. Recently, some nanomaterials have been used for the design of electrochemical enzyme biosensors. Their high conductivity properties have been reported to enhance the electron transfer between the enzyme redox center and the electrode surface [151]. The electrocatalytic action of nanomaterials decreases the overpotential associated to electroactive compounds, minimizing the interferences present in the sample. In few cases, nanomaterials have been used as labels to amplify the signal measured.

Nanoparticle-Based Electrochemical Biosensors

Enzyme Biosensors

Various kinds of nanoparticles, such as QDs and AuNPs, have been used for the development of electrochemical enzyme biosensors. An enzyme biosensor was developed for the amperometric detection of trichlorfon using poly (N-vinyl-2-pyrrolidone) (PVP)-capped CdS QDs [152]. The formation of PVP-QD nanostructures on the electrode surface provided a favourable microenvironment and led to a highly sensitive and stable electrochemical detection of the enzymatically generated thiocholine product. The detection limit was 4.8×10^{-8} M. Another kind of nanoparticles are AuNPs. Their unique property to provide a suitable microenvironment for immobilization of biomolecules retaining their bioactivity is a major advantage for the preparation of biosensors. Moreover, AuNPs facilitate direct electron transfer between immobilized redox proteins and the electrode surface [153]. An electrochemical biosensor based on colloidal AuNP modified sol-gel interface was developed for the detection of monocrotophos, carbaryl and methyl parathion [154]. The assembled AuNPs on a sol-gel derived silicate network provided a conductive pathway to electron transfer and favored the interface enzymatic hydrolysis reaction, increasing the sensitivity of the amperometric response. This biosensor presented good stability, retaining 90% of its initial current response after a 30-day storage period. Recently, an efficient biosensor for the detection of monocrotophos was developed by combining the unique properties of AuNPS with those of QDs. This new electrochemical system based on CdTe QDs-AuNPs electrode was more sensitive than those based on QDs or AuNPs alone [155].

Immunosensors

Recently, an electrochemical immunosensor was developed for rapid screening of diuron, a substituted phenyl urea herbicide [156]. Low cost ablated electrodes fabriccated on polystyrene substrate were modified with Prussian Blue (PB)-AuNP film. The electrodeposition of PBAuNP film enhanced electron transfer in the vicinity of the gold electrode increasing the sensitivity of the system as compared to unmodified gold electrodes. A conductimetric immunosensor for the detection of atrazine was also developed using antibodies labelled with nanoparticles [85]. The authors showed that AuNPs amplify the conductive signal and hence allow the detection of atrazine by means of DC measurements.

Nanoparticle-Based Optical Biosensors

Nanoparticles have also been used for the development of efficient optical biosensors [157]. QDs are candidates to replace conventional fluorescent markers. These semiconductor particles are more photostable than an organic fluorophore. Moreover, QDs exhibit higher fluorescence quantum yields than conventional organic fluorophores, allowing higher sensitivity. Recently, an optical biosensor was developed for the detection of monocrotophos using CdTe as fluorescence probe [158]. Using positively charged chitosan, CdTe and acetylcholinesterase were assembled onto a quartz surface by a layer-by-layer technique. In the absence of pesticide, acetylcholine was biocatalytically hydrolyzed inducing the production of choline and acetic acid. The released acid resulted in pH decrease that was sensed by the immobilized pH indicator (CdTe). The presence of monocrotophos induced a change of the fluorescence intensity that was related to the pesticide concentration.

Optic properties of AuNPs have been exploited for the development of localized SPR (LSPR) sensor [159,160]. The absorption band of AuNPs results when the incident photon frequency is resonant with the collective oscillation of the conduction electrons and is known as the localized surface plasmon resonance (LSPR). The resonance frequency of the LSPR is highly dependent upon the local environment of the nanoparticle and more specifically upon the binding events that occur to the functionalized NPs. The LSPR was used to develop a biosensor for the detection of paraoxon by immobilizing AChE onto AuNPs layer using a self-assembling technique [157]. In the presence of pesticides, the enzymatic activity was inhibited causing a change of the light attenuation. The detection limit with optimal conditions was 0.2 ppb. The biosensor retained 94% of its original activity after 6 cycles of inhibition with 500 ppb paraoxon followed by reactivation of AChE with 0.5 mM 2-pyriding-adoxime methoiodide. In addition, the sensor retained its activity after 2 months storage in dry state at 4°C.

Nanotube-Based Electrochemical Biosensors

Carbon nanotubes (CNTs) consist of cylindrical graphene sheets with nanometer diameters. They present unique mechanical, physical and chemical properties [148]. CNTs include both single-walled and multiwalled structures. Since their discovery, CNTs have been used in nanoelectronics, biomedical engineering, biosensing and bioanalysis.

Electrochemical biosensors, particularly enzyme electrodes, have benefited from the ability of CNT-modified transducers to promote the electron transfer reactions of enzymatically generated species [161]. Recently, CNTs have been used for the development of biosensors based on the inhibition of AChE activity [162-165]. An amperometric biosensor based on layer-by-layer assembly of single walled CNT-poly (diallyldimethylammonium chloride) and AChE was developed for the analysis of carbaryl [164]. The biosensor showed good sensitivity and stability towards the monitoring of pesticides in water. The detection limit was 4.9×10^{-15} M.

In some cases, the authors developed efficient biosensors for the detection of pesticides by associating the properties of CNTs with those of nanoparticles [39,166].

CONCLUSIONS

Biosensors are good candidates for the environmental monitoring. They exploit the remarkable specificity of recognition elements to design efficient analytical tools that can detect the presence of pesticides in complex samples. The biological elements used for environmental sensing are classically enzymes, antibodies and whole cells. Immunosensors, based on the highly selective and sensitive Ab-Ag reaction, allow the identification of a particular pesticide. Their high specificity can be sometimes a disadvantage. Enzyme-based biosensors allow the detection of broad families of pollutants. Thus, they often offer a general toxicity "index" [167], without providing specific information about a particular pesticide. Genetically engineered AChEs, showing better performance than native enzymes, have been widely exploited in enzyme inhibition-based biosensors for the detection of pesticides. Some genetically engineered microorganisms have also been used to develop microbial biosensors for pesticide detection [150]. In the last decade, aptamers have been used as new molecular recognition elements to develop biosensors

[133,134]. However, unique properties of aptamers have not been yet exploited for the development of pesticide analysis. We believe that these recognition elements could be used, in the near future, for the development of efficient aptasensors allowing the detection of pesticides. Although biological receptors have specific molecular affinity and have been widely used in biosensing technology, they are often produced via complex protocols with a high cost and require specific handling conditions because of their poor stability [168]. The high specificity and stability of MIPs render them as promising alternatives to enzymes and Abs [169]. Additional advantages include their compatibility with microfabrication technology and their cost-effectiveness compared to conventional biological receptors.

Nanotechnology is playing an important role in the development of efficient biosensors for the pesticide detection [148-150]. Different types of nanomaterials (e.g. nanoparticles and nanotubes) with different properties have been used. They offer exciting new opportunities to improve the performance of biosensors for the detection of pesticides.

The use of biosensors in environmental field is still limited in comparison to medical applications. Most commercial biosensors are for medical applications, whereas only few are adapted for the environmental monitoring. Thus, there is still a challenge to develop improved and more reliable devices allowing the detection of pesticides.

The use of biosensors in environmental field is still limited in comparison to medical applications. Most commercial biosensors are for medical applications, whereas only few are adapted for the environmental monitoring. Thus, there is still a challenge to develop improved and more reliable devices allowing the detection of pesticides.

REFERENCES

1. S. Andreescu and J. L. Marty, "Twenty Years Research in Cholinesterase Biosensors: From Basic Research to Practical Applications," Biomolecular Engineering, Vol. 23, No. 1, 2006, pp. 1-15. doi:10.1016/j.bioeng.2006.01.001

2. M. D. Luque de Castro and M. C. Herrera, "Enzyme_ Inhibition-Based Biosensors and Biosensing Systems: Questionable Analytical Devices," Biosensors and Bioelectronics, Vol. 18, No. 2-3, 2003, pp. 279-294. doi:10.1016/S0956-5663(02)00175-6

3. N. Jaffrezic-Renault, "New Trends in Biosensors for Organophosphorus Pesticides," Sensors, Vol. 1, No. 2, 2001, pp. 60-74. doi:10.3390/s10100060

4. D. Grieshaber, R. MacKenzie, J. Vöros and E. Reimhult, "Electrochemical Biosensors—Sensor Principles and Architectures," Sensors, Vol. 8, No. 3, 2008, pp. 1400-1458. doi:10.3390/s8031400
5. R. S. Freire, C. A. Pessao, L. D. Mello and L. T. Kubota, "Direct Electron Transfer: An Approach for Electrochemical Biosensors with Higher Selectivity and Sensitivity," Journal of Brazilian Chemical Society, Vol. 14, No. 2, 2003, pp. 230-243. doi:10.1590/S0103-50532003000200008
6. D. R. Thévenot, K. Toth, R. A. Durst and G. S. Wilson, "Electrochemical Biosensors: Recommended Definitions and Classification," Pure Applied Chemistry, Vol. 71, No. 12, 1999, pp. 2333-2348. doi:10.1016/S0956-5663(01)00115-4
7. N. J. Ronkainen, H. B. Halsall and W. R. Heineman, "Electrochemical Biosensors," Chemical Society Reviews, Vol. 39, No. 11, 2010, pp. 1747-1763. doi:10.1039/b714449k
8. U. Yogeswaran and S.-M. Chen, "A Review on the Electrochemical Sensors and Biosensors Composed of Nanowires as Sensing Material," Sensors, Vol. 8, No. 1, 2008, pp. 290-313. doi:10.3390/s8010290
9. S. Liu, L. Yuan, X. Yue, Z. Zheng and Z. Tang, "Recent Advances in Nanosensors for Organophosphates Pesticide Detection," Advanced Powder Technology, Vol. 19, No. 5, 2008, pp. 419-441. doi:10.1016/S0921-8831(08)60910-3
10. B. Prieto-Simon, M. Campas, S. Andreescu and J. L. Marty, "Trends in Flow-Based Biosensing for Pesticide Assessment," Sensors, Vol. 6, No. 10, 2006, pp. 1161- 1186. doi:10.3390/s6101161
11. B. Liu, Y.-H. Yang, Z.-Y. Wu, H. Wang, G.-L. Shen and R.-Q. Yu, "A Potentiometric Acetylcholinesterase Biosensor Based on Plasma-Polymerized Film," Sensors and Actuators B, Vol. 104, No. 2, 2005, pp. 186-190. doi:10.1016/j.snb.2004.04.093
12. V. G. Andreou and Y. D. Clonis, "A Portable Fiber-Optic Pesticide Biosensor Based On Immobilized Cholinesterase and Sol-Gel Entrapped Bromocresol Purple for In-Field Use," Biosensors and Bioelectronics, Vol. 17, No. 1-2, 2002, pp. 61-69. doi:10.1016/S0956-5663(01)00261-5
13. F. C. Wong, M. Ahmad, L. Y. Heng and L. B. Peng, "An Optical Biosensor for Dichlovos Using Stacked Sol-Gel Films Containing Acetylcholinesterase and a Lipophilic Chromoionophore," Talanta, Vol. 69, No. 4, 2006, pp. 888-893. doi:10.1016/j.talanta.2005.11.034
14. H. C. Tsai and R. A. Doong, "Simultaneous Determination of Ph, Urea, Acetylcholine and Heavy Metals Using Array-Based Enzymatic Optical

Biosensor," Biosensors and Bioelectronics, Vol. 20, No. 9, 2005, pp. 1796-1804. doi:10.1016/j.bios.2004.07.008

15. C. La Rosa, F. Pariente, L. Hernandez and E. Lorenzo, "Amperometric Flow-Through Biosensor for the Determination of Pesticides," Analytica Chimica Acta, Vol. 308, No. 1-3, 1995, pp. 129-136. doi:10.1016/0003-2670(94)00529-U

16. C. La Rosa, F. Pariente, L. Hernandez and E. Lorenzo, "Determination of Organophosphorus and Carbamic Pesticides with an Acetylcholinesterase Amperometric Biosensor Using 4-Aminophenyl Acetate as Substrate," Analytica Chimica Acta, Vol. 295, No. 3, 1994, pp. 273-282. doi:10.1016/0003-2670(94)80232-7

17. S. Andreescu, T. Noguer, V. Mageau and J. L. Marty, "Screen-Printed Electrode Based on AChE for the Detection of Pesticides in Presence of Organic Solvents," Talanta, Vol. 57, No. 1, 2002, pp. 169-176. doi:10.1016/S0039-9140(02)00017-6

18. S. Andreescu, A. Avramescu, C. Bala, V. Mageau and J. L. Marty, "Detection of Organophosphorus Insecticides with Immobilized Acetylcholinesterase—Comparative Study of Two Enzyme Sensors," Analytical and Bioanalytical Chemistry, Vol. 374, No. 1, 2002, pp. 39-45. doi:10.1007/s00216-002-1442-4

19. S. U padhyay, G. R. Rao, M. K. Sharma, B. K. Bhattacharya, V. K. Rao and R. Vijayaraghavan, "Immobilization of Acetylcholineesterase-Choline Oxidase on a GoldPlatinum Bimetallic Nanoparticles Modified Glassy Carbon Electrode for the Sensitive Detection of Organophosphate Pesticides, Carbamates and Nerves Agents," Biosensors and Bioelectronics, Vol. 25, No. 4, 2009, pp. 832-838. doi:10.1016/j.bios.2009.08.036

20. L. Campanella, M. Achilli, M. P. Sammartino and Tomassetti, "Butyrylcholine Enzyme Sensor for Determining Organophosphorus Inhibitors," Journal of Electroanalytical Chemistry, Vol. 321, No. 2, 1991, pp. 237-249. doi:10.1016/0022-0728(91)85599-K

21. N. G. Karousos, S. Aouabdi, A. S. Way and S. M. Reddy, "Quartz Crystal Microbalance Determination of Organophosphorus and Carbamate Pesticides," Analytica Chimica Acta, Vol. 469, No. 2, 2002, pp. 189-196. doi:10.1016/S0003-2670(02)00668-2

22. M. Campas, B. Prieto-Simon and J. L. Marty, "A Review of the Use of Genetically Engineered Enzymes in Electrochemical Biosensors," Seminars in Cell and Developmental Biology, Vol. 20, No. 1, 2009, pp. 3-9. doi:10.1016/j.semcdb.2009.01.009

23. G. Valdes-Ramirez, D. Fournier, M. T. Ramirez-Silva and J. L. Marty, "Sensitive Amperometric Biosensor for Dichlorovos Quantification: Application to Detection of Residues on Apple Skin," Talanta, Vol. 74, No. 4, 2008, pp. 741-746.doi:10.1016/j.talanta.2007.07.004
24. F. De Lima, B. Lucca, A. M. J. Barbosa, V. S. Ferreira, S. K. Moccelini, A. C. Franzoi and I. C. Vieira, "Biosensor Based on Pequi Polyphenol Oxidase Immobilized on Chitosan Crosslinked with Cyanuric Chloride for Thiodicarb Determination," Enzyme and Microbial Technology, Vol. 47, No. 4, 2010, pp. 153-158. doi:10.1016/j.enzmictec.2010.05.006
25. J. C. Vidal, S. Esteban, J. Gil and J. R. Castillo, "A Comparative Study of Immobilization Methods of a Tyrosinase Enzyme on Electrodes and Their Application to the Detection of Dichlorvos Organophosphorus Insecticide," Talanta, Vol. 68, No. 3, 2006, pp. 791-799.doi:10.1016/j.talanta.2005.06.038
26. Y. D. de Albuquerque and L. F. Ferreira, "Amperometric Biosensing of Carbamate and Organophosphate Pesticides Utilizing Screen-Printed Tyrosinase-Modified Electrodes," Analytica Chimica Acta, Vol. 596, No. 2, 2007, pp. 210-221.doi:10.1016/j.aca.2007.06.013
27. G.-Y. Kim, M.-S. Kang and S.-H. Moon, "Sub-strateBound Tyrosinase Electrode Using Gold Nanoparticles Anchored to Pyrroloquinoline Quinone for a Pesticide Biosensor," Sensors and Actuators B, Vol. 133, No. 1, 2008, pp. 1-4. doi:10.1016/j.snb.2008.01.055
28. G. Y. Kim, J. Shim, M. S. Kang and S. H. Moon, "Optimized Coverage of Gold Nanoparticles at Tyrosinase Electrode for Measurement of a Pesticide in Various Water Samples," Journal of Hazardous Materials, Vol. 156, No. 1-3, 2008, pp. 141-147.doi:10.1016/j.jhazmat.2007.12.007
29. L. Campanella, D. Lelo, E. Martini and M. Tomassetti, "Organophosphorus and Carbamate Pesticide Analysis Using an Inhibition Tyrosinase Organic Phase Enzyme Sensor; Comparison by Butyrylcholinesterase + Choline Oxidase Opee and Application to Natural Waters," Analytica Chimica Acta, Vol. 587, No. 1, 2007, pp. 22-32.doi:10.1016/j.aca.2007.01.023
30. M. S. Ayyagari, S. Kamtekar, R. Pande, K. A. Marx, J. Kumar, S. K. Tripathy and D. L. Kaplan, "Biosensors for Pesticide Detection Based on Alkaline Phosphatase-Catalyzed Chemiluminescence," Materials Science and Engineering: C, Vol. 2, No. 4, 1995, pp. 191-196. doi:10.1016/0928-4931(95)00077-1
31. F. Garcia Sanchez, A. Navas Diaz, M. C. Ramos Peinado and C. Belledone, "Free and Sol-Gel Immobilized Alkaline Phosphatase-Based Biosensor for the Determination of Pesticides and Inorganic Compounds,"

Analytica Chimica Acta, Vol. 484, No. 1, 2003, pp. 45-51.doi:10.1016/S0003-2670(03)00310-6

32. F. Mazzei, F. Botrè, S. Montilla, R. Pilloton, E. Podesta and C. Botrè, "Alkaline Phosphatase Inhibition Based Electrochemical Sensors for the Detection of Pesticides," Journal of Electroanalytical Chemistry, Vol. 574, No. 1, 2004, pp. 95-100.doi:10.1016/j.jelechem.2004.08.004

33. L. K. Shyuan, L. Y. Heng, M. Ahmad, S. A. Aziz and Z. Ishak, "Evaluation of Pesticide and Heavy Mtal Toxicity Using Immobilised Enzyme Alkaline Phosphatase with an Electrochemical Biosensor," Asian Journal of Biochemistry, Vol. 3, No. 6, 2008, pp. 359-365. doi:10.3923/ajb.2008.359.365

34. S. K. Moccelini, I. C. Vieira, F. De Lima, B. Lucca, A. M. J. Barbosa and V. S. Ferreira, "Determination of Thiodicarb Using a Biosensor Based on Alfalfa Sprout Peroxidase Immobilized in Self-Assembled Monolayers," Talanta, Vol. 82, No. 1, 2010, pp. 164-170.doi:10.1016/j.talanta.2010.04.015

35. S. Yang, Y. Li, X. Jiang, Z. Chen and X. Lin, "Horseradish Peroxidase Biosensor Based on Layer-by-Layer Technique for the Determination of Phenolic Compounds," Sensors and Actuators B, Vol. 114, No. 2, 2006, pp. 774-780. doi:10.1016/j.snb.2005.07.035

36. S. Yang, Z. Chen, X. Jin and X. Lin, "HRP Biosensor Based on Sugar-Lectin Biospecific Interactions for the Determination of Phenolic Compounds," Electrochimica Acta, Vol. 52, No. 1, 2006, pp. 200-205. doi:10.1016/j.electacta.2006.04.059

37. F. Mazzei, F. Botrè and C. Botrè, "Acid Phosphatase/Glucose Oxidase-Based Biosensors for the Determination of Pesticides," Analytica Chimica Acta, Vol. 336, No. 1, 1996, pp. 67-75. doi:10.1016/S0003-2670(96)00378-9

38. A. Mulchandani, W. Chen, P. Mulchandani, J. Wang and K. R. Rogers, "Biosensors for Direct Determination of Organophosphate Pesticides," Biosensors and Bioelectronics, Vol. 16, No. 4, 2001, pp. 225-230. doi:10.1016/S0956-5663(01)00126-9

39. D. Du, W. Chen, W. Zhang, D. Liu, H. Li and Y. Lin, "Covalent Coupling of Organophosphorus Hydrolase Loaded Quantum Dots to Carbon Nanotube. Au Nanocomposite for Enhanced Detection of Methyl Parathion," Biosensors and Bioelectronics, Vol. 25, No. 6, 2010, pp. 1370-1375. doi:10.1016/j.bios.2009.10.032

40. J. H. Lee, J. Y. Park, K. Min, H. J. Cha, S. S. Choi and Y. J. Yoo, "A Novel Organophosphorus Hydrolase-Based Biosensor Using Mesoporous

Carbons and Carbon Black for the Detection of Organophosphate Nerve Agents," Biosensors and Bioelectronics, Vol. 25, No. 7, 2010, pp. 1566-1570. doi:10.1016/j.bios.2009.10.013

41. V. A. Pedrosa, S. Paliwal, S. Balasubramanian, D. Nepal, V. Davis, J. Wild, E. Ramanculov and A. Simonian, "Enhanced Stability of Enzyme Organophosphate Hydrolase Interfaced on the Carbon Nanotubes," Colloids and Surfaces B Biointerfaces, Vol. 77, No. 1, 2010, pp. 69-74. doi:10.1016/j.colsurfb.2010.01.009

42. V. G. Andreou and Y. D. Clonis, "Novel Fiber-Optic Biosensor Based on Immobilized Glutathione S-Transferase and Sol-Gel Entrapped Bromcresol Green for the Determination of Atrazine," Analytica Chimica Acta, Vol. 460, No. 2, 2002, pp. 151-161.doi:10.1016/S0003-2670(02)00250-7

43. Y. Lei, W. Chen and A. Mulchandani, "Microbial Biosensors," Analytica Chimica Acta, Vol. 568, No. 1-2, 2006, pp. 200-210. doi:10.1016/j.aca.2005.11.065

44. S. F. D'Souza, "Microbial Biosensors," Biosensors and Bioelectronics, Vol. 16, 2001, pp. 337-353. doi:10.1016/S0956-5663(01)00125-7

45. J. Liu and B. Mattiasson, "Microbial BOD Sensors for Wastewater Analysis," Water Research, Vol. 36, No. 15, 2002, pp. 3786-3802. doi:10.1016/S0043-1354(02)00101-X

46. C. Chan, M. Lehmann, K. Chan, P. Chan, C. Chan, B. Gruendig, G. Kunze and R. Renneberg, "Designing an Amperometric Thick-Film Microbial BOD Sensor," Biosensors and Bioelectronics, Vol. 15, No. 7-8, 2000, pp. 343-353. doi:10.1016/S0043-1354(02)00101-X

47. K. Riedel, R. Renneberg, M. Kühn and F. Scheller, "A Fast Estimation of Biochemical Oxygen Demand Using Microbial Sensors," Applied Microbiology and Biotechnology, Vol. 28, No. 3, 1988, pp. 316-318. doi:10.1007/BF00250463

48. M. N. Kim and H. S. Kwon, "Biochemical Oxygen Demand Sensor Using Serratia marcescens LSY 4," Biosensors and Bioelectronics, Vol. 14, No. 1, 1999, pp. 1-7.doi:10.1016/S0956-5663(98)00107-9

49. M. Hikuma, H. Suzuki, T. Yasuda, I. Karube and S. Suzuki, "Amperometric Estimation of BOD by Using Living Immobilized Yeast," European Journal of Applied Microbiology and Biotechnology, Vol. 8, No. 4, 1979, pp. 289-297. doi:10.1007/BF00508793

50. J. Jia, M. Tang, X. Chen, L. Qi and S. Dong, "Co-Immobilized Microbial Biosensor for Bod Estimation Based on Sol-Gel Derived Composite

Material," Biosensors and Bioelectronics, Vol. 18, No. 8, 2003, pp. 1023-1029. doi:10.1016/S0956-5663(02)00225-7

51. J. Liu, G. Olsson and B. Mattiasson, "Short-Term BOD (BOD_{st}) as a Parameter for On-Line Monitoring of Biological Treatment Process Part I. A Novel Design of BOD Biosensor for Easy Renewal of Bio-Receptor," Biosensors and Bioelectronics, Vol. 20, No. 3, 2004, pp. 562-570. doi:10.1016/j.bios.2004.03.008

52. L. Su, W. Jia, C. Hou and Y. Lei, "Microbial Biosensors: A Review," Biosensors and Bioelectronics, Vol. 26, No. 5, 2011, pp. 1788-1799. doi:10.1016/j.bios.2010.09.005

53. P. Mulchandani, C. M. Hangarter, Y. Lei, W. Chen and A. Mulchandani, "Amperometric Microbial Biosensor for p- Nitrophenol Using Moraxella sp.-Modified Carbon Paste Electrode," Biosensors and Bioelectronics, Vol. 21, No. 12, 2005, pp. 523-527.doi:10.1016/j.bios.2004.11.011

54. Y. Lei, P. Mulchandani, W. Chen, J. Wang and A. Mulchandani, "Whole Cell-Enzyme Hybrid Amperometric Biosensor for Direct Determination of Organophosphorous Nerve Agents with p-Nitrophenyl Substituent," Biotechnology and Bioengineering, Vol. 85, No. 7, 2004, pp. 706-712. doi:10.1002/bit.20022

55. Y. Lei, P. Mulchandani, W. Chen and A. Mulchandani, "Direct Determination of p-Nitrophenyl Substituent Organophosphorus Nerve Agents Using a Recombinant Pseudomonas putida JS444-Modified Clark Oxygen Electrode," Journal of Agricultural and Food Chemistry, Vol. 53, 2005, pp. 524-527. doi:10.1021/jf048943t

56. Y. Lei, P. Mulchandani, W. Chen and A. Mulchandani, "Biosensor for Direct Determination of Fenitrothion and EPN Using Recombinant Pseudomonas putida JS444 with Surface Expressed Organophosphorus Hydrolase. 1. Modified Clark Oxygen Electrode," Sensors, Vol. 6, No. 4, 2006, pp. 466-472. doi:10.3390/s6040466

57. P. Mulchandani, W. Chen and A. Mulchandani, "Microbial Biosensor for Direct Determination of NitrophenylSubstituted Organophosphate Nerve Agents Using Genetically Engineered Moraxella sp.," Analytica Chimica Acta, Vol. 568, No. 1-2, 2006, pp. 217-221. doi:10.1016/j.aca.2005.11.063

58. E. I. Rainina, E. N. Efremenco, S. D. Varfolomeyev, A. L. Simonian and J. R. Wild, "The Development Of A New Biosensor Based on Recombinant E. coli for the Direct Detection of Organophosphorus Neurotoxins," Biosensors and Bioelectronics, Vol. 11, No. 10, 1996, pp. 991- 1000. doi:10.1016/0956-5663(96)87658-5

59. F. Lagarde and N. Jaffrezic-Renault, "Cell-Based Electrochemical Biosensors for Water Quality Assessment," Analytical and Bioanalytical Chemistry, Vol. 400, No. 4, 2011, pp. 947-964. doi:10.1007/s00216-011-4816-7
60. J. Kumar, S. K. Jha and S. F. D'Souza, "Optical Microbial Biosensor for Detection of Methyl Parathion Pesticide Using Flavobacterium sp. Whole Cells Adsorbed on Glass Fiber Filters as Disposable Biocomponent," Biosensors and Bioelectronics, Vol. 21, No. 11, 2006, pp. 2100-2105. doi:10.1016/j.bios.2005.10.012
61. M. Campas, R. Carpentier and R. Rouillon, "Plant-Tissue-and Photosynthesis-Based Biosensors," Biotechology Advances, Vol. 26, No. 4, 2008, pp. 370-378.doi:10.1016/j.biotechadv.2008.04.001
62. C. Chouteau, S. V. Dzyadevych, J.-M. Chovelon and C. Durrieu, "Development of Novel Conductometric Biosensors Based on Immobilised Whole Cell Chlorella vulgaris Microalgae," Biosensors and Bioelectronics, Vol. 19, No. 9, 2004, pp. 1089-1096.doi:10.1016/j.bios.2003.10.012
63. C. Chouteau, S. V. Dzyadevych, C. Durrieu and J.-M. Chovelon, "A Bi-Enzymatic Whole Cell Conductometric Biosensor for Heavy Metal Ions and Pesticides Detection in Water Samples," Biosensors and Bioelectronics, Vol. 21, No. 2, 2005, pp. 273-281.doi:10.1016/j.bios.2004.09.032
64. H. Guedri and C. Durrieu, "A Self-Assembled Monolayers Based Conductimetric Algal Whole Cell Biosensor for Water Monitoring," Microchimica Acta, Vol. 163, No. 3-4, 2008, pp. 179-184. doi:10.1007/s00604-008-0017-2
65. R. Rouillon, S. A. Piletsky, E. V. Piletska, P. Euzet and R. Carpentier, "Comparison of the Immobilization Techniques for Photosystem II," In: M. T. Giardi and E. V. Piletska, Eds., Biotechnological Applications of Photosynthesis Proteins: Biochips, Biosensors and Biodevices, Landes Bioscience, Georgetown, 2006, pp. 73-83.
66. M. T. Giardi and E. Pace, "Photosystem II-Based Biosensors for the Detection of Photosynthetic Herbicides," In: M. Giardi and E. V. Piletska, Eds., Biotechnological Applications of Photosynthetic Proteins; Biochips, Biosensors and Biodevices, Landes Bioscience, Georgetown, 2006, pp. 147-154. doi:10.1007/978-0-387-36672-2_13
67. R. Rouillon, J.-J. Mestres and J. L. Marty, "Entrapment of Chloroplasts and Thylakoids in Polyvinylalcohol-SbQ. Optimization of Membrane

Preparation and Storage Conditions," Analytica Chimica Acta, Vol. 311, No. 3, 1995, pp. 437-442. doi:10.1016/0003-2670(95)00031-T

68. M. Koblizek, J. Masojidek, J. Komenda, T. Kucera, R. Pilloton, A. K. Mattoo and M. T. Giardi, "A Sensitive Photosystem II-Based Biosensor for Detection of a Class of Herbicides," Biotechnology Bioengeering, Vol. 60, No. 6, 1998, pp. 664-669.doi:10.1002/(SICI)1097-0290

69. I. Shitanda, K. Takada, T. Sakai and T. Tatsuma, "Compact Amperometric Algal Biosensors for the Evaluation of Water Toxicity," Analytica Chimica Acta, Vol. 530, No. 2, 2005, pp. 191-197. doi:10.1016/j.aca.2004.09.073

70. I. Shitanda, S. Takamatsu, K. Watanabe and M. Itagaki, "Amperometric Screen-Printed Algal Biosensor with Flow Injection Analysis System for Detection of Environmental Toxic Compounds," Electrochimica Acta, Vol. 54, No. 21, 2009, pp. 4933-4936.doi:10.1016/j.electacta.2009.04.005

71. H. Nguyen-Ngoc and C. Tran-Minh, "Fluorescent Biosensor Using Whole Cells in an Inorganic Translucent Matrix," Analytica Chimica Acta, Vol. 583, No. 1, 2007, pp. 161-165. doi:10.1016/j.aca.2006.10.005

72. C. Védrine, J. C. Leclerc, C. Durrieu and C. Tran-Minh, "Optical Whole-Cell Biosensor Using Chlorella Vulgaris Designed for Monitoring Herbicides," Biosensors and Bioelectronics, Vol. 18, No. 4, 2003, pp. 457-463. doi:10.1016/S0956-5663(02)00157-4

73. A. Ventrella, L. Catucci and A. Agostiano, "Herbicides Affect Fluorescence and Electron Transfer Activity of Spinach Chloroplasts, Thylakoid Membranes and Isolated Photosystem II," Bioelectrochemistry, Vol. 79, No. 1, 2010, pp. 43-49.doi:10.1016/j.bioelechem.2009.10.008

74. E. Pena-Vazquez, E. Maneiro, C. Perez-Conde, M. C. Moreno-Bondi and E. Costas, "Microalgae Fiber Optic Biosensors for Herbicide Monitoring Using Sol-Gel Technology," Biosensors and Bioelectronics, Vol. 24, No.12, 2009, pp. 3538-3543.doi:10.1016/j.bios.2009.05.013

75. X. Jiang, D. Li, X. Xu, Y. Ying, Y. Li, Z. Ye and J. Wang, "Immunosensors for Detection of Pesticides Residues," Biosensors and Bioelectronics, Vol. 23, No. 11, 2008, pp. 1577-1587.doi:10.1016/j.bios.2008.01.035

76. C.R. Suri, R. Boro, Y. Nangia, S. Gandhi, P. Sharma, N. Wangoo, K. Rajesh and G. S. Shekhawat, "Immunoanalytical Techniques for Analyzing Pesticides in the Environment," Trends in Analytical Chemistry, Vol. 28, No. 1, 2009, pp. 29-39.doi:10.1016/j.trac.2008.09.017

77. E. Mallat, D. Barcelo, C. Barzen, G. Gauglitz and R. Abuknesha, "Immunosensors for Pesticide Determination in Natural Waters," Trends in Analytical Chemistry, Vol. 20, No. 3, 2001, pp. 124-132. doi:10.1016/S0165-9936(00)00082-0

78. F. F. Bier, E. Ehrentreich-Förster, R. Dölling, A. V. Eremenko and F. W. Scheller, "A Redox Label Immunosensor on Basis of a Bi-Enzyme Electrode," Analytica Chimica Acta, Vol. 344, No. 1-2, 1997, pp. 119-124. doi:10.1016/S0003-2670(97)00050-0
79. T. Kalab and P. Skadal, "A Disposable Amperometric Immunosensor for 2,4-Dichlorophenoxyacetic Acid," Analytica Chimica Acta, Vol. 304, No. 3, 1995, pp. 361-368.doi:10.1016/0003-2670(94)00641-X
80. P. Skadal and T. Kalab, "A Multichannel Immunochemical Sensor for the Determination of 2,4-Dichlorophenoxyacetic Acid," Analytica Chimica Acta, Vol. 316, No. 1, 1995, pp. 73-78. doi:10.1016/0003-2670(95)00342-W
81. S. Kröger, S. J. Setford and A. P. F. Turner, "Immunosensor for 2,4-Dichlorophenoxyacetic Acid in Aqueous/Organic Solvents Soil Extracts," Analytical Chemistry, Vol. 70, No. 23, 1998, pp. 5047-5053. doi:10.1021/ac9805100
82. K. Grennan, G. Strachan, A. J. Porter, A. Killard and M. R. Smyth, "Atrazine Analysis Using an Amperometric Immunosensor Based on Single-Chain Antibody Fragments and Regeneration-Free Multi-Calibrant Measurement," Analytica Chimica Acta, Vol. 500, No. 1-2, 2003, pp. 287-298. doi:10.1016/S0003-2670(03)00942-5
83. E. Valera, D. Muniz and A. Rodriguez, "Fabrication of Flexible Interdigitated μ-Electrodes (FIDμEs) for the Development of a Conductimetric Immunosensor for Atrazine Detection Based on Antibodies Labelled with Gold Nanoparticles," Microelectronic Engineering, Vol. 87, No. 2, 2010, pp. 167-173. doi:10.1016/j.mee.2009.07.001
84. E. Valera, J. Ramon-Azcon, A. Barranco, B. Alfaro, F. Sanchez-Baeza, M. P. Marco and A. Rodriguez, "Determination of Atrazine Residues in Red Wine Samples. A Conductimetric Solution," Food Chemistry, Vol. 122, No. 3, 2010, pp. 888-894.doi:10.1016/j.foodchem.2010.03.030
85. E. Valera, J. Ramon-Azcon, F. J. Sanchez, M. P. Marco and A. Rodriguez, "Conductimetric Immunosensor for Atrazine Detection Based on Antibodies Labelled with Gold Nanoparticles," Sensors and Actuators B, Vol. 134, No. 1, 2008, pp. 95-103.doi:10.1016/j.snb.2008.04.023
86. Y. V. Plekhanova, A. N. Reshetilov, E. V. Yazynina, A. V. Zherdev and B. B. Dzantiev, "A New Assay Format for Electrochemical Immunosensors: PolyelectrolyteBased Separation on Membrane Carriers Combined with Detection of Peroxidase Activity by pH-Sensitive Field-Effect Transistor," Biosensors and Bioelectronics, Vol. 19, No. 2, 2003, pp. 109-114.doi:10.1016/S0956-5663(03)00176-3

87. M. F. Yulaev, R. A. Sidtikov, N. M. Dmitrieva, E. V. Yazynima, A. V. Zherdev and B. B. Dzantiev, "Development of a Potentiometric Immunosensor for Herbicide Simazine and Its Application for Food Testing," Sensors and Actuators B, Vol. 75, No. 1-2, 2001, pp. 129-135. doi:10.1016/S0925-4005(01)00551-2
88. L. Mosiello, C. Laconi, M. Del Gallo, C. Ercole and A. Lepidi, "Development of a Monoclonal Antibody Based Potentiometric Biosensor for Terbuthylazine Detection," Sensors and Actuators B, Vol. 95, No. 1-3, 2003, pp. 315- 320. doi:10.1016/S0925-4005(03)00431-3
89. H. B. Fredj, S. Helali, Z. Sassi, N. Jaffrezic-Renault and A. Abdelghani, "Polyaniline Based Immunosensor for Atrazine Sensing," Sensor Letters, Vol. 7, No. 5, 2009, pp. 661- 666.doi:10.1166/sl.2009.1126
90. A. Rodriguez, E. Valera, J. Ramon-Azcon, F. J. Sanchez, M. P. Marco and L. M. Castaner, "Single Frequency Impedimetric Immunosensor for Atrazine Detection," Sensors and Actuators B, Vol. 129, No. 2, 2008, pp. 921-928. doi:10.1016/j.snb.2007.10.003
91. E. Valera, J. Ramon-Azcon, A. Rodriguez, L. M. Castaner, F. J. Sanchez and M. P. Marco, "Impedimetric immunosensor for Atrazine Detection Using Interdigitated µ-Electrode (IDµE's)," Sensors and Actuators B, Vol. 125, No. 2, 2007, pp. 526-537.doi:10.1016/j.snb.2007.02.048
92. J. Ramon-Azcon, E. Valera, A. Rodriguez, A. Barranco, B. Alfaro, F. Sanchez-Baeza and M. P. Marco, "An Impedimetric Immunosensor Based on Interdigitated Microelectrodes (IDµE) for the Determination of Atrazine Residues in Food Samples," Biosensors and Bioelectronics, Vol. 23, No. 9, 2008, pp. 1367-1373. doi:10.1016/j.bios.2007.12.010
93. R. E. Ionescu, C. Gondran, L. Bouffier, N. Jaffrezic-Renault, C. Martelet and S. Cosnier, "Label-Free Impedimetric Immunosensor for Sensitive Detection of Atrazine," Electrochimica Acta, Vol. 55, No. 21, 2010, pp. 6228-6232.doi:10.1016/j.electacta.2009.11.029
94. C. Esseghaier, S. Helali, H. Fredj, A. Tlili and A. Abdelghani, "Polypyrrole-Neutravidin Layer for Impedimetric Biosensor," Sensors and Actuators B, Vol. 131, No. 2, 2008, pp. 584-589. doi:10.1016/j.snb.2007.12.043
95. S. Helali, C. Martelet, A. Abdelghani, M. A. Maaref and N. Jaffrezic-Renault, "A Disposable Immunomagnetic Electrochemical Sensor Based on Functionalised Magnetic Beads on Gold Surface for the Detection of Atrazine," Electrochimica Acta, Vol. 51, No. 24, 2006, pp. 5182- 5186. doi:10.1016/j.electacta.2006.03.086

96. B. Corry, J. Uilk and C. Crawley, "Probing Direct Binding Affinity in Electrochemical Antibody-Based Sensors," Analytica Chimica Acta, Vol. 496, No. 1-2, 2003, pp. 103-116.doi:10.1016/j.aca.2003.01.001
97. S. Hleli, C. Martelet, A. Abdelghani, N. Burais and N. Jaffrezic-Renault, "Atrazine Analysis Using an Impedimetric Immunosensor Based on Mixed Biotinylated SelfAssembled Monolayer," Sensors and Actuators B, Vol. 113, No. 2, 2006, pp. 711-717.doi:10.1016/j.snb.2005.07.023
98. J. Homola, S. S. Yee and G. Gauglitz, "Surface Plasmon Resonance Sensors: Review," Sensors and Actuators B, Vol. 54, No. 1-2, 1999, pp. 3-15. doi:10.1016/S0925-4005(98)00321-9
99. M. Farre, E. Martinez, J. Ramon, A. Navarro, J. Radjenovic, E. Mauriz, L. Lechuga, M. P. Marco and D. Barcelo, "Part per Trillion Determination of Atrazine in Natural Water Samples by a Surface Plasmon Resonance Immunosensor," Analytical and Bioanalytical Chemistry, Vol. 388, No. 1, 2007, pp. 207-214. doi:10.1007/s00216-007-1214-2
100. E. Mauriz, A. Calle, J. J. Manclus, A. Montoya and L. M. Lechuga, "Multi-Analyte SPR Immunoassays for Environmental Biosensing of Pesticides," Analytical and Bioanalytical Chemistry, Vol. 387, No. 4, 2006, pp. 1449- 1458. doi:10.1007/s00216-006-0800-z
101. E. Mauriz, A. Calle, A. Abad, A. Montoya, A. Hildebrandt, D. Barcelo and L. M. Lechuga, "Determination of Car0 baryl in Natural Water Samples by a Surface Plasmon Resonance Flow-Through Immunosensor," Biosensors and Bioelectronics, Vol. 21, No. 11, 2006, pp. 2129-2136. doi:10.1016/j.bios.2005.10.013
102. E. Mauriz, A. Calle, J. J. Manclus, A. Montoya, A. M. Escuela, J. R. Sendra and L. M. Lechuga, "Single and Multi-Analyte Surface Plasmon Resonance Assays for Simultaneous Detection Of Cholinesterase Inhibiting Pesticides," Sensors and Actuators B, Vol. 118, No. 1-2, 2006, pp. 399-407. doi:10.1016/j.snb.2006.04.085
103. E. Mauriz, A. Calle, L. M. Lechuga, J. Quintana, A. Montoya and J. J. Manclus, "Real-Time Detection of Chlorpyrifos at Part Per Trillion Levels in Ground, Surface and Drinking Water Samples by a Portable Surface Plasmon Resonance Immunosensor," Analytica Chimica Acta, Vol. 561, No. 1-2, 2006, pp. 40-47. doi:10.1016/j.aca.2005.12.069
104. E. Mauriz, A. Calle, A. Montoya and L. M. Lechuga, "Determination of Environmental Organic Pollutants with a Portable Optical Immunosensor," Talanta, Vol. 69, No. 2, 2006, pp. 359-364. doi:10.1016/j.talanta.2005.09.049

105. S. J. Kim, K. V. Gobi, H. Tanaka, Y. Shoyama and N. Miura, "A Simple and Versatile Self-Assembled Monolayer Based Surface Plasmon Resonance Immunosensor for Highly Sensitive Detection of 2,4-D from Natural Water Resources," Sensors and Actuators B, Vol. 130, No. 1, 2008, pp. 281-289. doi:10.1016/j.snb.2007.08.023

106. K. V. Gobi, S. J. Kim, H. Tanaka, Y. Shoyama and N. Miura, "Novel Surface Plasmon Resonance (SPR) Immunosensor Based on Monomolecular Layer of Physically-Adsorbed Ovalbumin Conjugate for Detection of 2,4-Dichlorophenoxyacetic Acid and Atomic Force Microscopy Study," Sensors and Actuators B, Vol. 123, No. 1, 2007, pp. 583-593.doi:10.1016/j.snb.2006.09.056

107. K. V. Gobi, H. Tanaka, Y. Shoyama and N. Miura, "Highly Sensitive Regenerable Immunosensor for LabelFree Detection of 2,4-Dichlorophenoxyacetic Acid at ppb Levels by Using Surface Plasmon Resonance Imaging," Sensors and Actuators B, Vol. 111-112, 2005, pp. 562- 571. doi:10.1016/j.snb.2005.03.118

108. S. J. Kim, K. V. Gobi, H. Iwasaka, H. Tanaka and N. Miura, "Novel Miniature SPR Immunosensor Equipped with All-in-One Multichannel Sensor Chip for Detecting Low-Molecular-Weight Analytes," Biosensors and Bioelectronics, Vol. 23, No. 5, 2007, pp. 701-707. doi:10.1016/j.bios.2007.08.010

109. C. M. Cummins, M. E. Koivunen, A. Stephanian, S. J. Gee, B. D. Hammock and I. M. Kennedy, "Application of Europium(III) Chelate-Dyed Nanoparticle Labels in a Competitive Atrazine Fluoroimmunoassay on an ITO Waveguide," Biosensors and Bioelectronics, Vol. 21, No. 7, 2003, pp. 1077-1085.

110. A. Klotz, A. Brecht, C. Barzen, G. Gauglitz, R. D. Harris, G. R. Quigley, J. S. Wilkinson and R. Abuknesha, "Immunofluorescence Sensor for Water Analysis," Sensors and Actuators B, Vol. 51, No. 1-3, 1998, pp. 181-187. doi:10.1016/S0925-4005(98)00187-7

111. C. Barzen, A. Brecht and G. Gauglitz, "Optical MultipleAnalyte Immunosensor for Water Pollution Control," Biosensors and Bioelectronics, Vol. 17, No. 4, 2002, pp. 289-295.doi:10.1016/S0956-5663(01)00297-4,

112. F. Long, H. C. Shi, M. He and A. N. Zhu, "Sensitive and Rapid Detection of 2,4-Dicholorophenoxyacetic Acid in Water Samples by Using Evanescent Wave All-Fiber Immunosensor," Biosensors and Bioelectronics, Vol. 23, No. 9, 2008, pp. 1361-1366.doi:10.1016/j.bios.2007.12.004

113. F. Long, M. He, H. C. Shi and A. N. Zhu, "Development of Evanescent Wave All-Fiber Immunosensor for Environmental Water Analysis," Biosensors and Bioelectronics, Vol. 23, No. 7, 2008, pp. 952-958. doi:10.1016/j.bios.2007.09.013

114. S. Rodriguez-Mozaz, M. L. de Alda and D. Barcelo, "Analysis of Bisphenol A in Natural Waters by Means of an Optical Immunosensor," Water Research, Vol. 39, No. 20, 2005, pp. 5071-5079. doi:10.1016/j.watres.2005.09.023

115. J. Tschmelak, N. Kappel and G. Gauglitz, "TIRF-Based Biosensor for Sensitive Detection of Progesterone in Milk Based on Ultra-Sensitive Progesterone Detection in Water," Analytical and Bioanalytical Chemistry, Vol. 382, No. 8, 2005, pp. 1895-1903. doi:10.1007/s00216-005-3261-x

116. J. Tschmelak, G. Proll, J. Riedt, J. Kaiser, P. Kraemmer, L. Barzaga, J. S. Wilkinson, P. Hua, J. P. Hole, R. Nudd, M. Jackson, R. Abuknesha, D. Barcelo, S. RodriguezMozaz, M. J. de Alda, F. Sacher, J. Stien, J. Slobodnik, P. Oswald, H. Kozmenko, E. Korenkova, L. Tothova, Z. Krascsenits and G. Gauglitz, "Automated Water Analyser Computer Supported System (AWACSS) Part I: Project Objectives, Basic Technology, Immunoassay Development, Software Design and Networking," Biosensors and Bioelectronics, Vol. 20, No. 8, 2005, pp. 1499-1508. doi:10.1016/j.bios.2004.07.032

117. S. Boujday, S. Nasri, M. Salmain and C.-M. Pradier, "Surface IR Immunosensors for Label-Free Detection of Benzo[a]pyrene," Biosensors and Bioelectronics, Vol. 26, No. 4, 2010, pp. 1750-1754.

118. S. Boujday, C. Gu, M. Girardot, M. Salmain and C. M. Pradier, "Surface IR Applied to Rapid and Direct Immunosensing of Environmental Pollutants," Talanta, Vol. 78, No. 1, 2009, pp. 165-170. doi:10.1016/j.talanta.2008.10.064

119. M. Salmain, N. Fischer-Durand and C. M. Pradier, "Infrared optical immunosensors: Application to the MeasUrement of the Herbicide Atrazine," Analytical Biochemistry, Vol. 373, No. 1, 2008, pp. 61-70. doi:10.1016/j.ab.2007.10.031

120. S. Kurosawa, J.-W. Park, H. Aiwaza, S.-I. Wakida, H. Tao and K. Ishihara, "Quartz Crystal Microbalance Immunosensor for Environmental Monitoring," Biosensors and Bioelectronics, Vol. 22, No. 4, 2006, pp. 473-481. doi:10.1016/j.bios.2006.06.030

121. C. March, J. J. Manclus, Y. Jimenez, A. Arnau and A. Montoya, "A Piezoelectric Immunosensor for the Determination of Pesticide Residues

and Metabolites in Fruit Juices," Talanta, Vol. 78, No. 3, 2009, pp. 827-833. doi:10.1016/j.talanta.2008.12.058
122. A. Sassolas, B. D. Leca-Bouvier and L. J. Blum, "DNA Biosensors and Microarrays," Chemical Reviews, Vol. 108, No. 1, 2008, pp. 109-139. doi:10.1021/cr0684467
123. J. Kaur, K. V. Singh, A. H. Schmid, G. C. Varshney, C. R. Suri and M. Raje, "Atomic Force Spectroscopy-Based Study of Antibody Pesticide Interactions for Characterization of Immunosensor Surface," Biosensors and Bioelectronics, Vol. 20, No. 2, 2004, pp. 284-293. doi:10.1016/j.bios.2004.01.012
124. C. R. Suri, J. Kaur, S. Gandhi and G. S. Shekhawat, "Label-free Ultra-Sensitive Detection of Atrazine Based on Nanomechanics," Nanotechnology, Vol. 19, No. 23, 2008, p. 235502. doi:10.1088/0957-4484/19/23/235502
125. M. Alvarez, A. Calle, J. Tamayo, L. Lechuga, A. Abad and A. Montoya, "Development of Nanomechanical Biosensors for Detection of the Pesticide DDT," Biosensors and Bioelectronics, Vol. 18, No. 5-6, 2003, pp. 649-653. doi:10.1016/S0956-5663(03)00035-6
126. M. Bache, R. Taboryski, S. Schmid, J. Aamand and M. H. Jakobsen, "Investigations on Antibody Binding to a Microcantilever Coated with a BAM Pesticide Residue," Nanoscale Research Letters, Vol. 6, 2011, p. 386.
127. A. M. Nowicka, A. Kowalczyk, Z. Stojek and M. Hepel, "Nanogravimetric and Voltammetric DNA-Hybridization Biosensors for Studies of DNA Damage by Common Toxicants and Pollutants," Biophysical Chemistry, Vol. 146, No. 1, 2010, pp. 42-53. doi:10.1016/j.bpc.2009.10.003
128. A. D. Ellington and J. W. Szostak, "In Vitro Selection of RNA Molecules that Bind Specific Ligands," Nature, Vol. 346, No. 6287, 1990, pp. 818-822. doi:10.1038/346818a0
129. C. L. A. Hamula, J. W. Guthrie, H. Zhang, X.-F. Li and X. C. Le, "Selection and Analytical Applications of Aptamers," Trends in Analytical Chemistry, Vol. 30, No. 10, 2011, pp. 1587-1597. doi:10.1016/j.trac.2011.08.006
130. R. Stoltenburg, C. Reinemann and B. Strehlitz, "SELEX- A(r)evolutionary Method to Generate High-Affinity Nucleic Acid Ligands," Biomolecular Engineering, Vol. 24, No. 4, 2007, pp. 381-403. doi:10.1016/j.bioeng.2007.06.001
131. A. Famulok and J. W. Szostak, "In Vitro Selection of Specific Ligand Binding Nucleic Acids," Angewandte Chemie-International Edition, Vol. 31, No. 8, 1992, pp. 979-988. doi:10.1002/anie.199209791

132. S. D. Jayasena, "Aptamers: An Emerging Class of Molecules That Rival Antibodies in Diagnostics," Clinical Chemistry, Vol. 45, No. 9, 1999, pp. 1628-1650.
133. A. Sassolas, L. J. Blum and B. D. Leca-Bouvier, "Optical Detection Systems Using Immobilized Aptamers," Biosensors and Bioelectronics, Vol. 26, No. 9, 2011, pp. 3725- 3736. doi:10.1016/j.bios.2011.02.031
134. A. Sassolas, L. J. Blum and B. D. Leca-Bouvier, "Electrochemical Aptasensors," Electroanalysis, Vol. 21, 2009, pp. 1237-1250. doi:10.1002/elan.200804554
135. T. Yuan, Z.-Y. Liu, L.-Z. Hu and G.-B. Xu, "Electrochemical and Electrochemiluminescent Aptasensors," Chinese Journal of Analytical Chemistry, Vol. 39, No. 7, 2011, pp. 972-977.doi:10.1016/S1872-2040(10)60451-3
136. A.-E. Radi, "Electrochemical Aptamer-Based Biosensors: Recent Advances and Perspectives," International Journal of Electrochemistry, 2011, Article ID 863196.
137. Y. Xu, G. Cheng, P. He and Y. Fang, "A Review: Electrochemical Aptasensors with Various Detection Strategies," Electroanalysis, Vol. 21, No. 11, 2009, pp. 1251- 1259.doi:10.1002/elan.200804561
138. J. He, Y. Liu, M. Fan and X. Liu, "Isolation and Identification of the Dna Aptamer Target to Acetamiprid," Journal of Agricultural and Food Chemistry, Vol. 59, No. 5, 2011, pp. 1582-1586. doi:10.1021/jf104189g
139. A. L. Hillberg, K. R. Brain and C. J. Allender, "Molecular imprinted Polymer Sensors: Implications for Therapeutics," Advanced Drug Delivery Reviews, Vol. 57, No. 12, 2005, pp. 1875-1889. doi:10.1016/j.addr.2005.07.016
140. A. G. Mayes and M. J. Whitcombe, "Synthetic Strategies for the Generation of Molecularly Imprinted Organic Polymers," Advanced Drug Delivery Reviews, Vol. 57, No. 12, 2005, pp. 1742-1778. doi:10.1016/j.addr.2005.07.011
141. V. Pichon and F. Chapuis-Hugon, "Role of Molecularly Imprinted Polymers for Selective Determination of Environmental Pollutants—A Review," Analytica Chimica Acta, Vol. 622, No. 1-2, 2008, pp. 48-61. doi:10.1016/j.aca.2008.05.057
142. A. L. Jenkins, R. Yin and J. L. Jensen, "Molecularly Imprinted Polymer Sensors for Pesticide and Insecticide Detection in Water," Analyst, Vol. 126, 2001, pp. 798- 802.doi:10.1039/b008853f
143. C. Xie, S. Gao, Q. Guo and K. Xu, "Electrochemical Sensor for 2,4-Dichlorophenoxy Acetic Acid Using Molecularly Imprinted

Polypyrrole Membrane as Recognition Element," Microchimica Acta, Vol. 169, No. 1, 2010, pp. 145- 152. doi:10.1007/s00604-010-0303-7

144. C. Pellicer, A. Gomez-Caballero, N. Unceta, M. A. Goicolea and R. J. Barrio, "Using a Portable Device Based on Screen-Printed Sensor Modified with a Molecularly Imprinted Polymer for the Determination of the Insecticide Fenitrothion in Forest Samples," Analytical Methods, Vol. 2, No. 9, 2010, pp. 1280-1285. doi:10.1039/c0ay00329h

145. E. Pardieu, H. Cheap, C. Vedrine, M. Lazerges, Y. Lattach, F. Garnier, S. remira and C. Pernelle, "Molecularly Imprinted Conducting Polymer Based Electrochemical Sensor for Detection of Atrazine," Analytica Chimica Acta, Vol. 649, No. 2, 2009, pp. 236-245.doi:10.1016/j.aca.2009.07.029

146. M. Cortina-Puig, G. Istamboulie, T. Noguer and J. L. Marty, "Analysis of Pesticide Mixtures Using Intelligent Biosensors," In: V. S. Somerset, Ed., Intelligent and Biosensors, 2010, pp. 205-216.

147. G. Istamboulie, M. Cortina-Puig, J. L. Marty and T. Noguer, "The Use of Artificial Neural Networks for the Selective Detection of Two Organophosphate Insecticides: Chlorpyrifos and Chlorfenvinfos,"Talanta, Vol. 79, No. 2, 2009, pp. 507-511.doi:10.1016/j.talanta.2009.04.014

148. X. Zhang, Q. Guo and D. Cui, "Recent Advances in Nanotechnology Applied to Biosensors," Sensors, Vol. 9, No. 2, 2009, pp. 1033-1053. doi:10.3390/s90201033

149. R. Singh, "Prospects of Nanobiomaterials for Biosensing," International Journal of Electrochemistry, 2011, Article ID 125487.

150. A. Ballesteros-Gomez and S. Rubio, "Recent Advances in Environmental Analysis," Analytical Chemistry, Vol. 81, No. 12, 2009, pp. 4601-4622. doi:10.1021/ac200921j

151. A. Sassolas, L. J. Blum and B. D. Leca-Bouvier, "Immobilization Strategies to Develop Enzymatic Biosensors," Biotechnology Advances, in press,doi:10.1016/j.biotechadv.2011.09.003

152. X.-H. Li, Z. Xie, H. Min, C. Li, M. Liu and Y. J. Xian, "Development of Quantum Dots Modified Acetylcholinesterase Biosensor for the Detection of Trichlorfon," Electroanalysis, Vol. 18, No. 22, 2006, pp. 2163-2167. doi:10.1002/elan.200603615

153. J. Pingarron, P. Yanez-Sedeno and A. Gonzalez-Cortes, "Gold Nanoparticles-Based Electrochemical Biosensors," Electrochimica Acta, Vol. 53, No. 19, 2008, pp. 5848- 5866.doi:10.1016/j.electacta.2008.03.005

154. D. Du, S. Chen, J. Cai and A. Zhang, "Electrochemical Pesticide Sensitivity Test Using Acetylcholinesterase Biosensor Based on Colloidal

Gold Nanoparticle Modified Sol-Gel Interface," Talanta, Vol. 74, No. 4, 2008, pp. 766-772. doi:10.1016/j.talanta.2007.07.014

155. D. Du, S. Chen, D. Song, H. Li and X. Chen, "Development of Acetylcholinesterase Biosensor Based on CdTe Quantum Dots/Gold Nanoparticles Modified Chitosan Microspheres Interface," Biosensors and Bioelectronics, Vol. 24, No. 3, 2008, pp. 475-479.doi:10.1016/j.bios.2008.05.005

156. P. Sharma, K. Sablok, V. Bhalla and C. R. Suri, "A Novel Disposable Electrochemical Immunosensor for Phenyl Urea Herbicide Diuron," Biosensors and Bioelectronics, Vol. 26, No. 10, 2011, pp. 4209-4212. doi:10.1016/j.bios.2011.03.019

157. T.-J. Lin, K.-T. Huang and C.-Y. Liu, "Determination of Organophosphorous Pesticides by a Novel Biosensor Based on Localized Surface Plasmon Resonance," Biosensors and Bioelectronics, Vol. 22, No. 4, 2006, pp. 513-518. doi:10.1016/j.bios.2006.05.007

158. X. Sun, B. Liu and K. Xia, "A Sensitive and Regenerable Biosensor for Organophosphate Pesticide Based on Self- Assembled Multilayer Film with CdTe as Fluorescence Probe," Luminescence, Vol. 26, No. 6, 2011, pp. 616-621. doi:10.1002/bio.1284

159. J. Fu, B. PArk and Y. Zhao, "Limitation of a Localized Surface Plasmon Resonance Sensor for Salmonella Detection," Sensors and Actuators B, Vol. 141, No. 1, 2009, pp. 276-283.doi:10.1016/j.snb.2009.06.020

160. L.-K. Chau, Y.-F. Lin, S.-F. Cheng and T.-J. Lin, "Fiber-Optic Chemical and Biochemical Probes Based on Localized Surface Plasmon Resonance," Sensors and Actuators B, Vol. 113, No. 1, 2006, pp. 100-105. doi:10.1016/j.snb.2005.02.034

161. J. Wang, "Carbon-Nanotube Based Electrochemical Biosensors: A Review," Electroanalysis, Vol. 17, No. 1, 2005, pp. 7-14. doi:10.1002/elan.200403113

162. D. Du, X. Huang, J. Cai and A. Zhang, "Comparison of Pesticide Sensitivity by Electrochemical Test Based on Acetylcholinesterase Biosensor," Biosensors and Bioelectronics, Vol. 23, No. 2, 2007, pp. 285-289. doi:10.1016/j.bios.2007.05.002

163. A. C. Oliveira and L. H. Mascaro, "Evaluation of Acetylcholinesterase Biosensor Based on Carbon Nanotube Paste in the Determination of Chlorphenvinphos," International Journal of Analytical Chemistry, 2011, Article ID 974216. doi:10.1155/2011/974216

164. S. Firdoz, F. Ma, X. Yue, Z. Dai, A. Kumar and B. Jiang, "A Novel Amperometric Biosensor Based on Single Walled Carbon Nanotubes

with Acetylcholine Esterase for the Detection of Carbaryl Pestcide in Water," Talanta, Vol. 83, No. 1, 2010, pp. 269-273.doi:10.1016/j.talanta.2010.09.028

165. Y. Qu, Q. Sun, F. Xiao, G. Shi and L. Jin, "Layer- by-Layer Self-Assembled Acetylcholienesterase/ PAMAMAu on CNTs Modified Electrode for Sensing Pesticides," Bioelectrochemistry, Vol. 77, No. 2, 2010, pp. 139-144.doi:10.1016/j.bioelechem.2009.08.001

166. S. Chen, J. Huang, D. Du, J. Li, H. Tu, D. Liu and A. Zhang, "Methyl Parathion Hydrolase Based Nanocomposite Biosensors for Highly Sensitive and Selective Determination of Methyl Parathion," Biosensors and Bioelectronics, Vol. 26, No. 11, 2011, pp. 4320-4325.doi:10.1016/j.bios.2011.04.025

167. J. L. Marty, B. Leca and T. Noguer, "Biosensors for the Detection of Pesticides," Analusis Magazine, Vol. 26, 1998, pp. 144-149. doi:10.1051/analusis:199826060144

168. M. Blanco-Lopez, "Electrochemical Sensors Ased on Molecularly Imprinted Polymers," Trends in Analytical Chemistry, Vol. 23, No. 1, 2004, pp. 36-48. doi:10.1016/S0165-9936(04)00102-5

169. S. A. Piletsky and A. P. F. Turner, "Electrochemical Sensors Based on Molecularly Imprinted Polymers," Electroanalysis, Vol. 14, No. 5, 2002, pp. 317-323.doi:10.1002/1521-4109

170. A. C. Ion, I. Ion, A. Culetu, D. Gherase, C. A. Moldovan, R. Iosub and A. Dinescu, "Acetylcholinesterase Voltammetric Biosensors Based on Carbon Nanostructure-Chitosan Composite Material for Organophosphate Pesticides," Materials Science and Engineering, Vol. 30, No. 6, 2010, pp. 817-821. doi:10.1016/j.msec.2010.03.017

171. S. Viswanathan, H. Radecka and J. Radecki, "Electrochemical Biosensor for Pesticides Based on Acetylcholinesterase Immobilized on Polyaniline Deposited on Vertically Assembled Carbon Nanotubes Wrapped with ssDNA," Biosensors and Bioelectronics, Vol. 24, No. 9, 2009, pp. 2772-2777. doi:10.1016/j.bios.2009.01.044

172. N. Chauhan, J. Narang and C. S. Pundir, "Immobilization of Rat Brain Acetylcholinesterase on ZnS and Poly(Indole-5-carboxylic acid) Modified Au Electrode for Detection of Organophosphorus Insecticides," Biosensors and Bioelectronics, Vol. 29, No. 1, 2011, pp. 82-88. doi:10.1016/j.bios.2011.07.070

173. G. Istamboulie, T. Sikora, E. Jubete, E. Ochoteco, J. L. Marty and T. Noguer, "Screen-Printed Poly(3,4-Ethylenedioxythiophene) (PEDOT): A New Electrochemical Mediator for Acetylcholinesterase-Based

Biosensors," Talanta, Vol. 82, No. 3, 2010, pp. 957-961.doi:10.1016/j.talanta.2010.05.070

174. A. Hildebrandt, J. Ribas, R. Bragos, J. L. Marty, M. Tresanchez and S. Lacorte, "Development of a Portable Biosensor for Screening Neurotoxic Agents in Water Samples," Talanta, Vol. 75, No. 5, 2008, pp. 1208-1213. doi:10.1016/j.talanta.2008.01.033

175. A. Hildebrandt, R. Bragos, S. Lacorte and J. L. Marty, "Performance of a Portable Biosensor for the Analysis of Organophosphorus and Carbamate Insecticides in Water and Food," Sensors and Actuators B, Vol. 133, No. 1, 2010, pp. 195-201.doi:10.1016/j.snb.2008.02.017

176. Y. Ivanov, I. Marinov, K. Gabrovska, N. Dimcheva and T. Godjevargova, "Amperometric Biosensor Based on a SiteSpecific Immobilization of Acetylcholinesterase via Affinity Bonds on a Nanostructured Polymer Membrane With Intergrated Multiwall Carbon Nanotubes," Journal of Molecular Catalysis B: Enzymatic, Vol. 63, No. 3-4, 2010, pp. 141-148. doi:10.1016/j.molcatb.2010.01.005

177. R. Sinha, M. Ganesana, S. Andreescu and L. Stanciu, "AChE Biosensor Based on Zinc Oxide Sol-Gel for the Detection of Pesticides," Analytica Chimica Acta, Vol. 661, No. 2, 2010, pp. 195-199. doi:10.1016/j.aca.2009.12.020

178. Y. Wang, S. Zhang, D. Du, Y. Shao, Z. Li, J. Wang, M. H. Englehard, J. Li and Y. H. Lin, "Self Assembly of Acetylcholinesterase on a Gold Nanoparticles-Graphene Nanosheet Hybrid for Organophosphate Pesticide Detection Using Polyelectrolyte as a Linker," Journal of Materials Chemitsry, Vol. 21, No. 22, 2011, pp. 5319-5325.

179. F. Arduini, F. Ricci, C. S. Tuta, D. Moscone, A. Amine and G. Palleschi, "Detection of Carbamic and Organophosphorous Pesticides in Water Samples Using a Cholinesterase Biosensor Based on Prussian Blue-Modified Screen-Printed Electrode," Analytica Chimica Acta, Vol. 580, No. 2, 2006, pp. 155-162. doi:10.1016/j.aca.2006.07.052

180. G. Valdes-Ramirez, M. Cortina, M. T. Ramirez-Silva and J. L. Marty, "Acetylcholinesterase-Based Biosensors for Quantification of Carbofuran, Methylparaoxon, and Dichlorvos in 5% Acetonitrile," Analytical and Bioanalytical Chemistry, Vol. 392, No. 4, 2008, pp. 699-707.doi:10.1007/s00216-008-2290-7

181. D. Du, X. Huang, J. Cai and A. Zhang, "Amperometric Detection of Triazophos Pesticide Using Acetylcholinesterase Biosensor Based on Miltiwall Carbon Nanotube-Chitosan Matrix," Sensors and Actuators B, Vol. 127, No. 2, 2007, pp. 531-535.doi:10.1016/j.snb.2007.05.006

182. A. Vakurov, C. E. Simpson, C. L. Daly, T. D. Gibson and P. A. Millner, "Acetylcholinesterase-Based Biosensor Electrodes for Organophosphate Pesticide Detection. I. Modification of Carbon Surface for Immobilization of Acetylcholinesterase," Biosensors and Bioelectronics, Vol. 20, No. 6, 2004, pp. 1118-1125.doi:10.1016/j.bios.2004.03.039

183. S. Sotiropoulou, D. Fournier and N. A. Chaniotakis, "Genetically Engineered Acetylcholinesterase-Based Biosensor for Attomolar Detection of Dichlorvos," Biosensors and Bioelectronics, Vol. 20, No. 11, 2005, pp. 2347-2352.doi:10.1016/j.bios.2004.08.026

184. X. Sun and X. Wang, "Acetylcholinesterase Biosensor Based on Prussian Blue-Modified Electrode for Detecting Organophosphorous Pesticides," Biosensors and Bioelectronics, Vol. 25, No. 1-2, 2010, pp. 2611-2614. doi:10.1016/j.bios.2010.04.028

185. Y. Wei, Y. Li, Y. Qu, F. Xiao, G. Shi and L. Jin, "A Novel Biosensor Based on Photoelectro-Synergistic Catalysis for Flow-Injection Analysis System/Amperometric Detection of Organophosphorous Pesticides," Analytica Chimica Acta, Vol. 643, No. 1-2, 2009, pp. 13-18. doi:10.1016/j.aca.2009.03.045

186. D. Du, S. Chen, J. Cai and A. Zhang, "Immobilization of Acetylcholinesterase on Gold Nanoparticles Embedded in Sol-Gel Film for Amperometric Detection of Organophosphorous," Biosensors and Bioelectronics, Vol. 23, No. 1, 2007, pp. 130-134.doi:10.1016/j.bios.2007.03.008

187. A. Ishii, S. Takeda, S. Hattori, K. Sueoka and K. Mukasa, "Ultrasensitive Detection of Organophosphate Insecticides by Carbon Field-Effect Transistor," Colloids and Surfaces A, Vol. 313-314, 2008, pp. 456-460.

188. N. Gan, X. Yang, D. Xie, Y. Wu and W. Wen, "A Disposable Organophosphorus Pesticides Enzyme Biosensor Based On Magnetic Nano-Particles Modified Screen Printed Carbon Electrode," Sensors, Vol. 10, No. 1, 2010, pp. 665-638. doi:10.3390/s100100625

189. D. Du, J. Ding, J. Cai and A. Zhang, "Determination of Carbaryl Pesticide Using Amperometric Acetylcholinesterase Sensor Formed by Electrochemically Deposited Chitosan," Colloids and Surfaces B Biointerfaces, Vol. 58, No. 1, 2007, pp. 145-150.doi:10.1016/j.colsurfb.2007.03.006

190. S. P. Zhang, L. G. Shan, Z. R. Tian, Y. Zheng, L. Y. Shi and D. S. Zhang, "Study of Enzyme Biosensor Based on Carbon Nanotubes Modified Electrode for Detection of Pesticides Residue," Chinese Chemical Letters, Vol. 19, No. 5, 2008, pp. 592-594.doi:10.1016/j.cclet.2008.03.014

191. Y. Qu, Q. Sun, F. Xiao, G. Shi and L. Jin, "Layerby-Layer Self-Assembled Acetylcholinesterase/PAMAMAu on CNTs Modified Electrode for Sensing Pesticides," Bioelectrochemistry, Vol. 77, No. 2, 2010, pp. 139-144.doi:10.1016/j.bioelechem.2009.08.001

192. S. Laschi, D. Ogonczyk, I. Palchetti and M. Mascini, "Evaluation of Pesticide-Induced Acetylcholinesterase Inhibition by Means of Disposable Carbon-Modiifed Electrochemical Biosensors," Enzyme and Microbial Technology, Vol. 40, No. 3, 2007, pp. 485-489.doi:10.1016/j.enzmictec.2006.08.004

193. S. Sajjadi, H. Ghourchian and H. Tavakoli, "Choline Oxidase as a Selective Recognition Element for Determination of Paraoxon," Biosensors and Bioelectronics, Vol. 24, No. 8, 2009, pp. 2509-2514. doi:10.1016/j.bios.2009.01.008

194. R. P. Deo, J. Wang, I. Block, A. Mulchandani, K. A. Joshi, M. Trojanowicz, F. Scholz, W. Chen and Y. H. Lin, "Determination of Organophosphate Pesticides at a Carbon Nanotube/Organophosphorus Hydrolase Electrochemical biosensor," Analytica Chimica Acta, Vol. 530, No. 2, 2005, pp. 185-189. doi:10.1016/j.aca.2004.09.072

195. T. Laothanachareon, V. Champreda, P. Sritongkham, M. Somasundrum and W. Surareungchai, "Cross-Linked Enzyme Crystals of Organophosphate Hydrolase for Electrochemical Detection of Organophosphorus Compounds," World Journal of Microbiology and Biotechnology, Vol. 24, No. 12, 2008, pp. 3049-3055.doi:10.1007/s11274-008-9851-yOpen

196. A. Mulchandani, S. T. Pan and W. Chen, "Fiber-Optic Enzyme Biosensor for Direct Determination of Organophosphate Nerve Agents," Biotechnology Progress, Vol. 15, No. 1, 1999, pp. 130-134. doi:10.1021/bp980111q

197. V. Sacks, I. Eshkenazi, T. Neufeld, C. Dosoretz and J. Rishpon, "Immobilized Parathion Hydrolase: An Amperometric Sensor for Parathion," Analytical Chemistry, Vol. 72, No. 9, 2000, pp. 2055-2058. doi:10.1021/ac9911488

198. L. Viveros, S. Paliwal, D. McCrae, J. Wild and A. Simonian, "A Fluorescence-Based Biosensor for the Detection of Organophosphate Pesticides and Chemical Warfare Agents," Sensors and Actuators B, Vol. 115, No. 1, 2006, pp. 150-157.doi:10.1016/j.snb.2005.08.032

199. K. A. Fähnrich, M. Pacda and G. G. Guilbault, "Diposable Amperometric Immunosensor for the Detection of Polycyclic Aromatic Hydrocarbons (PAHs) Using ScreenPrinted Electrodes," Biosensors and Bioelectronics, Vol. 18, No. 1, 2003, pp. 73-82. doi:10.1016/S0956-5663(02)00112-4

200. S.-Q. Hu, J.-W. Xie, Q.-H. Xu, K. T. Rong, G.-L. Shen and R.-Q. Yu, "A Label-Free Electrochemical Immunosensor Based on Gold Nanoparticles for Detection of Paraoxon," Talanta, Vol. 61, No. 6, 2003, pp. 769-777. doi:10.1016/S0039-9140(03)00368-0

201. E. Zacco, R. Galve, M. P. Marco, S. Alegret and M. I. Pividori, "Electrochemical Biosensing of Pesticide Residues Based on Affinity Biocomposite Platforms," Biosensors and Bioelectronics, Vol. 22, No. 8, 2007, pp. 1707- 1715. doi:10.1016/j.bios.2006.07.037

202. D. Butler and G. G. Guilbault, "Disposable Amperometric Immunosensor for the Detection of 17-β Estradiol Using Screen-Printed Electrodes," Sensors and Actuators B, Vol. 113, No. 2, 2006, pp. 692-699. doi:10.1016/j.snb.2005.07.019

203. L. Chen, G. Zeng, Y. Zhang, L. Tang, D. Huang, C. Liu, Y. Pang and J. Luo, "Trace Detection of Pichloram Using an Electrochemical Immunosensor Based on 3D Au Nanoclusters," Analytical Biochemistry, Vol. 407, No. 2, 2010, pp. 172-179.doi:10.1016/j.ab.2010.08.001

204. Y. Zhang and H.-S. Zhuang, "Amperometric Immunosensor Based on Layer-by-Layer Assembly of Thiourea and Nano-Gold Particles on Gold Electrode for Determination of Naphthalene," Chinese Journal of Analytical Chemistry, Vol. 38, No. 2, 2010, pp. 153-157.doi:10.1016/S1872-2040(09)60021-9

205. I. Navratilova and P. Skladal, "The Immunosensors for Measurement of 2,4-Dichlorophenoxyacetic Acid Based on Electrochemical Impedance Spectroscopy," Bioelectrochemistry, Vol. 62, No. 1, 2004, pp. 11-18. doi:10.1016/j.bioelechem.2003.10.004

206. M. F. Gouzy, M. Kess and P. M. Kramer, "A SPR-Based Immunosensor for the Detection of Isoproturon," Biosensors and Bioelectronics, Vol. 24, No. 6, 2009, pp. 1563-1568.doi:10.1016/j.bios.2008.08.005

207. D. R. Shankaran, K. Matsumoto, K. Toko and N. Miura, "Development and Comparison of Two Immunoassays for the Detection of 2,4,6-Trinitrotoluene (TNT) Based on Surface Plasmon Resonance," Sensors and Actuators B, Vol. 114, No. 1, 2006, pp. 71-79.doi:10.1016/j.snb.2005.04.013

208. T. Kawaguchi, D. R. Shankaran, G. Y. Kim, K. Matsumoto, K. Toko and N. Miura, "Surface Plasmon Resonance Au Nanoparticle for Detection of TNT," Sensors and Actuators B, Vol. 133, No. 2, 2008, pp. 467-472. doi:10.1016/j.snb.2008.03.005

209. T. Kawaguchi, D. R. Shankaran, S. J. Kim, K. V. Gobi, K. Matsumoto, K. Toko and N. Miura, "Fabrication of a Novel Immunosensor Using

Functionalized Self-Assembled Monolayer for Trace Lavel Detection of TNT by Surface Plasmon Resonance," Talanta, Vol. 72, No. 2, 2007, pp. 554-560. doi:10.1016/j.talanta.2006.11.020

210. P. Singh, T. Onodera, Y. Mizuta, K. Matsumoto, N. Miura and K. Toko, "Dendrimer Modified Biochip for Detection of 2,4,6 Trinitrotoluene on SPR Immunosensor: Fabrication and Advantages," Sensors and Actuators B, Vol. 137, No. 2, 2009, pp. 403-409.doi:10.1016/j.snb.2008.12.027

211. K. Nagatomo, T. Kawaguchi, N. Miura, K. Toko and K. Matsumoto, "Development of a Sensitive Surface Plasmon Resonance Immunosensor for Detection of 2,4-Dinitrotoluene with a Novel Oligo (Ethylene Glycol)-Based Sensor Surface," Talanta, Vol. 79, No. 4, 2009, pp. 1142-1148. doi:10.1016/j.talanta.2009.02.018

212. N. Soh, T. Tokuda, T. Watanabe, K. Mishima, T. Imato, T. Masadome, Y. Asano, S. Okutani, O. Niwa and S. Brown, "A Surface Plasmon Resonance Immunosensor for Detecting a Dioxin Precursor Using a Gold Binding Polypeptide," Talanta, Vol. 60, No. 4, 2003, pp. 733-745. doi:10.1016/S0039-9140(03)00139-5

Chapter 5

QUANTITATIVE ESTIMATION OF PESTICIDE-LIKENESS FOR AGROCHEMICAL DISCOVERY

Sorin Avram[1], Simona Funar-Timofei[1], Ana Borota[1], Sridhar Rao Chennamaneni[2], Anil Kumar Manchala[2], and Sorel Muresan[3]

[1]Department of Computational Chemistry, Institute of Chemistry of Romanian Academy Timisoara, 24 Mihai Viteazul Avenue, 300223 Timisoara, Romania.

[2]GVK Biosciences Pvt. Ltd., S1, Phase-1, Technocrats Industrial Estate, Balanagar, Hyderabad 500 037, India.

[3]Food Control Department, Banat's University of Agricultural Sciences and Veterinary Medicine, Calea Aradului 119, 300645 Timisoara, Romania.

ABSTRACT

Background

The design of chemical libraries, an early step in agrochemical discovery programs, is frequently addressed by means of qualitative physicochemical and/or topological rule-based methods. The aim of this study is to develop quantitative estimates of herbicide- (QEH), insecticide- (QEI), fungicide- (QEF), and, finally, pesticide-likeness (QEP).

In the assessment of these definitions, we relied on the concept of desirability functions.

Results

We found a simple function, shared by the three classes of pesticides, parameterized particularly, for six, easy to compute, independent and interpretable, molecular properties: molecular weight, logP, number of hydrogen bond acceptors, number of hydrogen bond donors, number of rotatable bounds and number of aromatic rings. Subsequently, we describe the scoring of each pesticide class by the corresponding quantitative estimate. In a comparative study, we assessed the performance of the scoring functions using extensive datasets of patented pesticides.

Conclusions

The hereby-established quantitative assessment has the ability to rank compounds whether they fail well-established pesticide-likeness rules or not, and offer an efficient way to prioritize (class-specific) pesticides. These findings are valuable for the efficient estimation of pesticide-likeness of vast chemical libraries in the field of agrochemical discovery.

BACKGROUND

In the past years, the systematic identification of new lead compounds has gained increasing attention in both pharmaceutical and agrochemical industries. The progress of combinatorial chemistry (the parallel synthesis of large numbers of compounds) and high-throughput screening (the parallel testing for bioactivity of large numbers of compounds) facilitated the exploration of extensive chemical spaces for chemicals with desirable properties. In order to conduct effectively a drug/agrochemical discovery program, a screening library should contain compounds displaying reasonable properties to ease the passage to final products. Thus, in the early stages of such programs, *in silico* approaches are used to design chemical libraries [1],[2]. Oral bioavailability or membrane permeability have often been connected to simple molecular descriptors such as logP, molecular weight, or the counts of hydrogen bond acceptors and donors in a molecule [3]. Hence, over the years, simple rule-based models were derived based upon physicochemical and structural property of available datasets. These qualitative approaches (also referred to as filters) retain or reject molecules depending on a set of strict threshold values for key molecular descriptors (often combined with the presence or absence of undesirable chemical groups). This provides a rapid way to select molecules showing increased likelihood to exhibit the specific property for which the filter has been designed for [4]-[7].

In drug discovery, Lipinski's rule of five (Ro5) is considered to be the reference in defining physicochemical and structural properties profiles for optimal bioavailability of drug candidates [3]. Upper limits of five basic molecular descriptors were established based upon a set of known drugs, i.e., molecular weight ≤ 500, octanol/water partition coefficient (hydrophobicity) ≤ 5, number of hydrogen bond donors ≤ 5 and number of hydrogen bond acceptors ≤ 10. Molecules that would obey these rules should exert acceptable solubility and cell permeability properties and were defined as 'drug-like' [3]. Although Ro5 is considered predictive for oral bioavailability, 16% of oral drugs violate at least one of the criteria and 6% fail two or more [8]. Other simplified rule-based definitions of drug-likeness were established by Veber [9] and Ghose [10].

In the field of agrochemical discovery, Lipinski's Ro5 approach was quickly adopted to profile agrochemicals, i.e., herbicides, insecticides and fungicides [11],[12]. In this sense, a referential paper was published by Tice [11], who defined, using Ro5 molecular descriptors, criteria to identify herbicides and insecticides, the two major classes of pesticides (see Table 1). Clarke & Delaney added further molecular properties known to influence absorption and distribution of agrochemicals, i.e., predicted solubility, melting point, ΔlogP, charge, acidity and basicity, percentage of aromatic atoms and non-carbon atoms [12]. In a more recent work Clarke [13] established upper limits of Abraham descriptors McGowan volume, hydrogen bond acidity and the hydrogen bond basicity. Investigating the constitutive properties of a representative library of marketed pesticides, from different periods of registration, Hao et al. [14] defined simple and easy to implement rules for pesticide-likeness, by including molecular weight (MW), hydrophobicity (LogP), number of H-bond acceptors (HBA) and donors (HBD), number of rotatable bonds (RB) and number of aromatic bounds.

Table 1. Rule-based filters for drugs and pesticides

Rule	Lipinski	Tice		Hao
Class	Drugs	Herbicides	Insecticides	Pesticides
MW	≤ 500	150 – 500	150 – 500	≤ 435
MLogP(*CLogP)	≤ 5	≤ 3.5	0 - 5	≤ 6*
HBD	≤ 5	≤ 3	≤ 2	≤ 2
HBA	≤ 10	2 - 12	1 – 8	≤ 6
RB	-	< 12	< 12	≤ 9
aromatic bonds	-	-	-	≤ 17

*MLogP [15] values were computed for Lipiniski's [3] and Tice's [11] rules and CLogP [16] values for Hao's [14] according to the original publications.

To overcome the hard boundaries established by traditional filters for drug-likeness, Bickerton et al. [8] developed the so-called quantitative estimate of drug-likeness (QED) which combines the simplicity of rules-based methods and the ranking advantages of continuous models. The approach relies on a small number of relevant, accessible and quick to compute, molecular descriptors describing the distribution of a set of molecules. So-called desirability functions [17], i.e., functions that describe the distribution of the data, have been fitted for each descriptor. Hence, QED defines drug-like molecules on a continuous scale, ranging from zero (the least drug-like) to one (the most drug-like) [8].

We consider that the field of agrochemical discovery would benefit from a similar treatment of pesticide-likeness. Thus, in this study, we aim to establish quantitative estimates of pesticide-likeness. Three main classes of pesticides are considered herein, i.e., herbicides, insecticides and fungicides, and, accordingly, we describe the quantitative estimate of herbicide-likeness (QEH), of insecticide-likeness (QEI) and of fungicide-likeness (QEF). We found a simple type of function that accurately describes six physicochemical properties over the three pesticide classes. Furthermore, we compare the performance of this quantitative approach to well known rule-based methods defining pesticide-likeness using a large library of patented compounds for agrochemical applications and discuss the results. For practical reasons and for the purpose of this paper, we will denominate the ensemble of scoring functions dedicated to pesticide-likeness as QEPest-SFs.

RESULTS AND DISCUSSION

The Assessment of a Common Desirability Function for Pesticides

We applied the concept of desirability [17] to provide a quantitative metric for assessing pesticide-classes-likeness and subsequently pesticide-likeness. The desirability function approach was originally proposed by Harrington [17] and later refined by Derringer and Suich [18]. The approach consists of employing one/several functions to characterize the properties of several dependent variables, normalize (scale between zero and one) and combine the resulted terms using the geometric mean. Since we deal with molecular data sets, we followed the procedure of Bickerton's et al. [8] which derived series of desirability functions, each for a different molecular descriptor.

Here, we sought to find a type of function (as simple as possible) that would accurately fit distributions resulted from molecular properties describing herbicides, insecticides and fungicides. Firstly, we computed a number of 15 molecular descriptors (see Additional file 1: Table S1) for the 1685 marketed pesticides (see *Marketed pesticide set* section in Methods). The resulted distributions of the three pesticide-classes were fitted as described in *Curve fitting* section in Methods. We found six independent (see Additional file 1: Figure S1) molecular descriptors, closest to those enumerated in Table 1 showing adequate distribution of data and accurate fitting curves (for the three pesticide classes), i.e., MW, LogP, HBA, HBD, RB and arR (number of aromatic rings). We examined the first fifty equations ranked, increasingly, according to the lowest sum of squared absolute error, as computed by the fitting algorithm. Accordingly, we selected the function showing the smallest sum of ranks among the three classes of pesticides. Thus, a simple function f

(eq. 1) was selected, parameterized by o, a, b, c, coefficients computed for each distribution of pesticide-class and molecular descriptor (see Additional file 1: Table S2).

$$f = o + a \cdot e^{-e^{\frac{-x-b}{c}} - \frac{x-b}{c} + 1} \quad (1)$$

In order to assure reasonable desirability scores, function f was scaled between zero and one by division with maximum values (see Additional file 1: Table S2). Thus, the value of the resulted desirability function df, increases as the 'desirability' of the corresponding response increases (see Figures 1, 2 and 3). The accuracy of the fittings is reported in Additional file 1: Table S3.

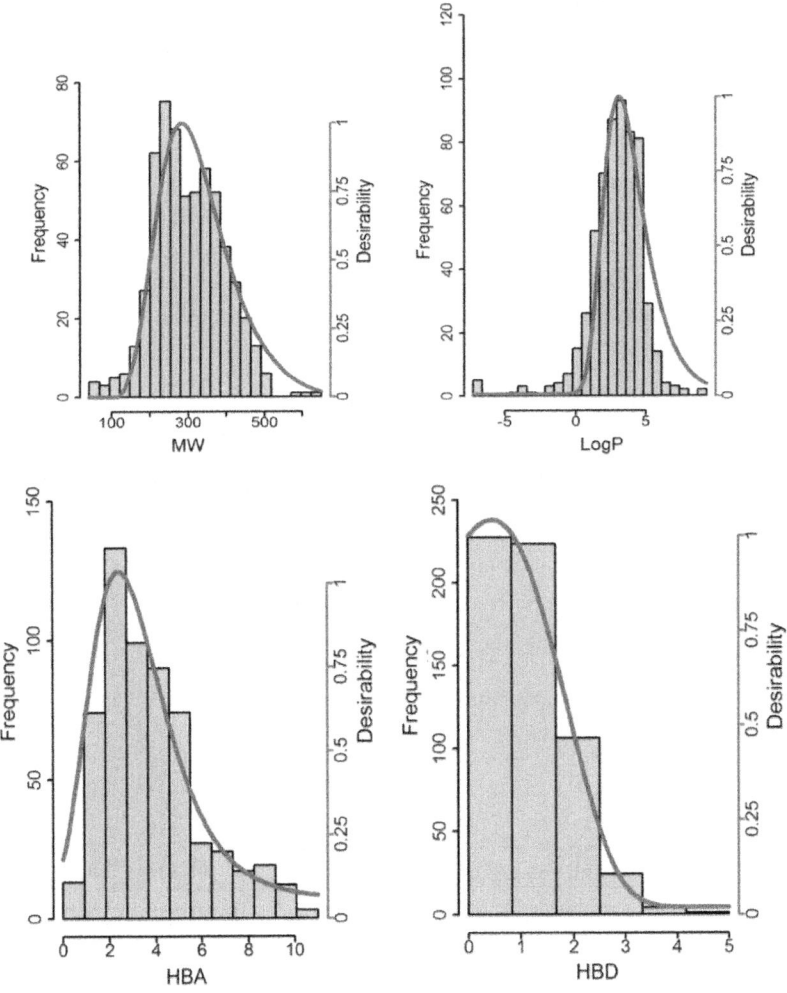

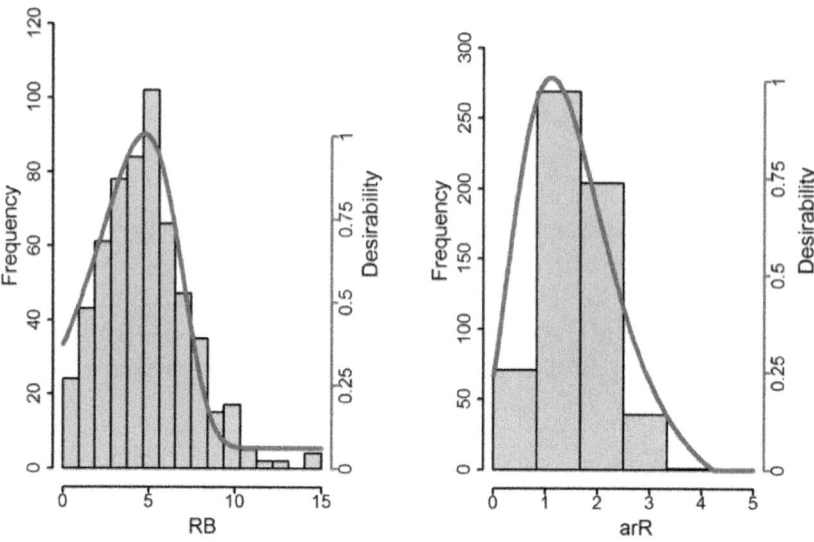

Figure 1. Frequency counts and desirability function plots of herbicides. Histograms and desirability functions (red curve, see right scale) of six molecular descriptors, i.e., MW (molecular weight), LogP (log of the octanol–water partition coefficient), HBA (number hydrogen bond acceptors), HBD (number hydrogen bond donors), RB (number of rotatable bonds), arR (number of aromatic rings) computed for the herbicides subset.

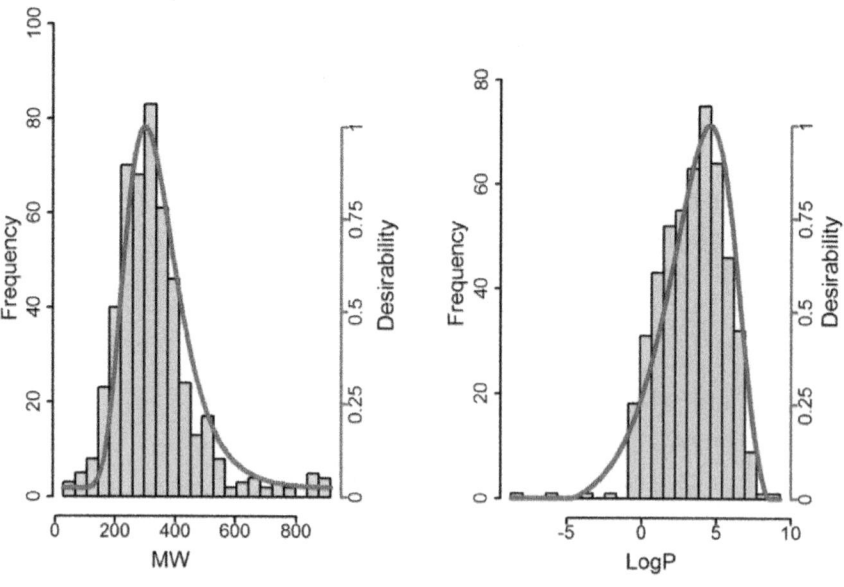

Quantitative Estimation of Pesticide-Likeness for Agrochemical Discovery

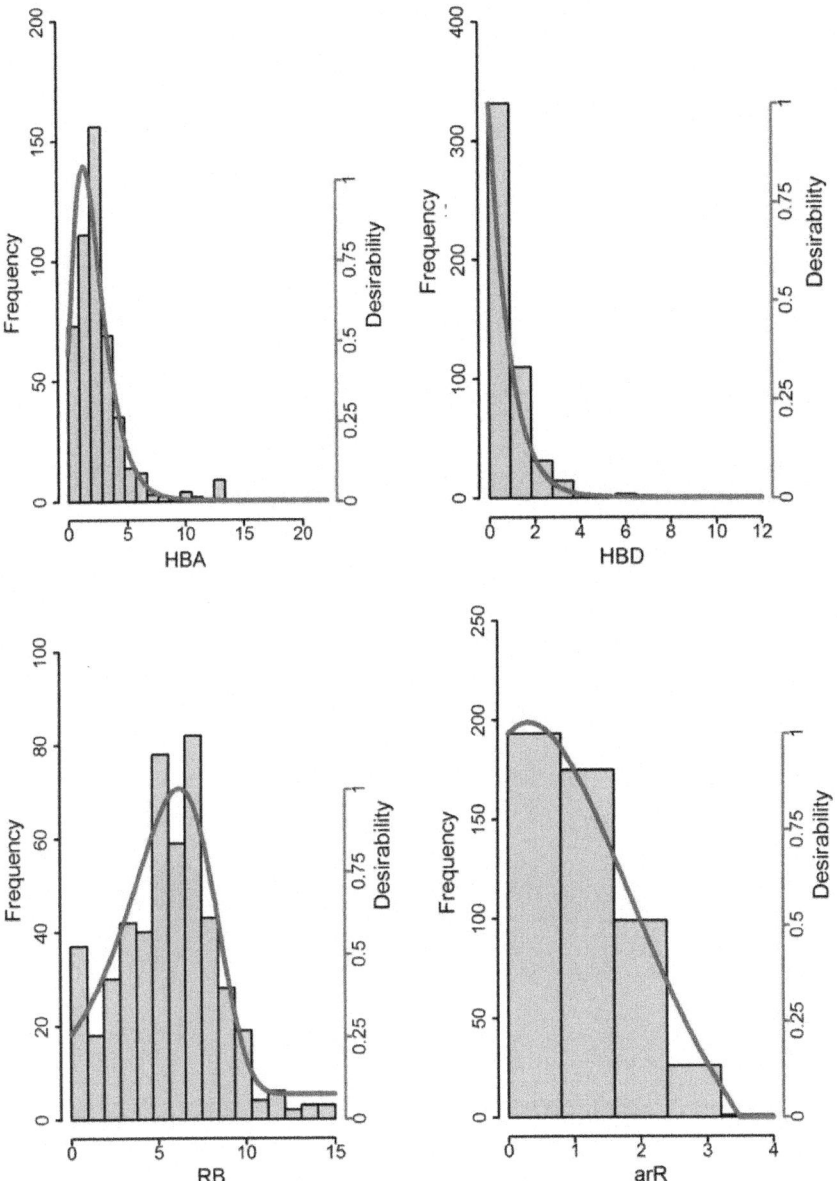

Figure 2. Frequency counts and desirability function plots of insecticides. Histograms and desirability functions (red curve, see right scale) of six molecular descriptors, i.e., MW (molecular weight), LogP (log of the octanol–water partition coefficient), HBA (number hydrogen bond acceptors), HBD (number hydrogen bond donors), RB (number of rotatable bonds), arR (number of aromatic rings), computed for the insecticides subset.

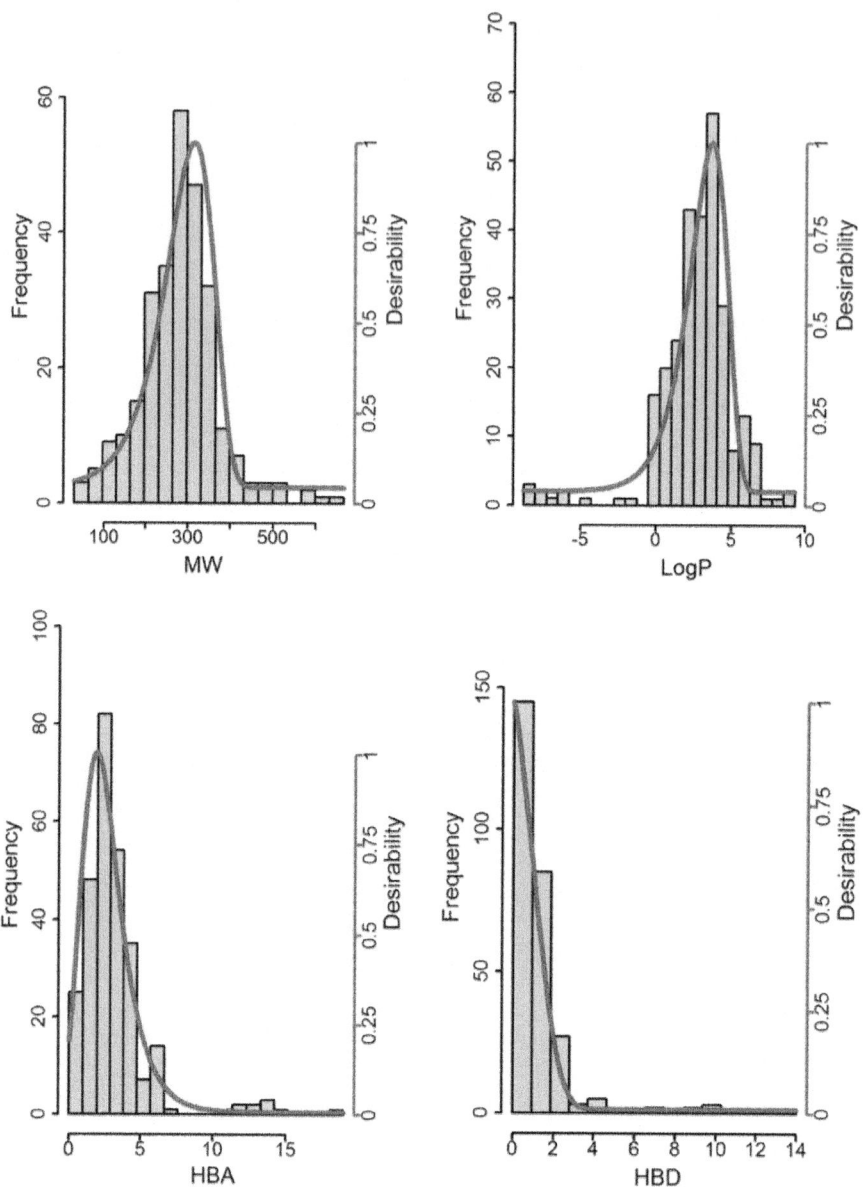

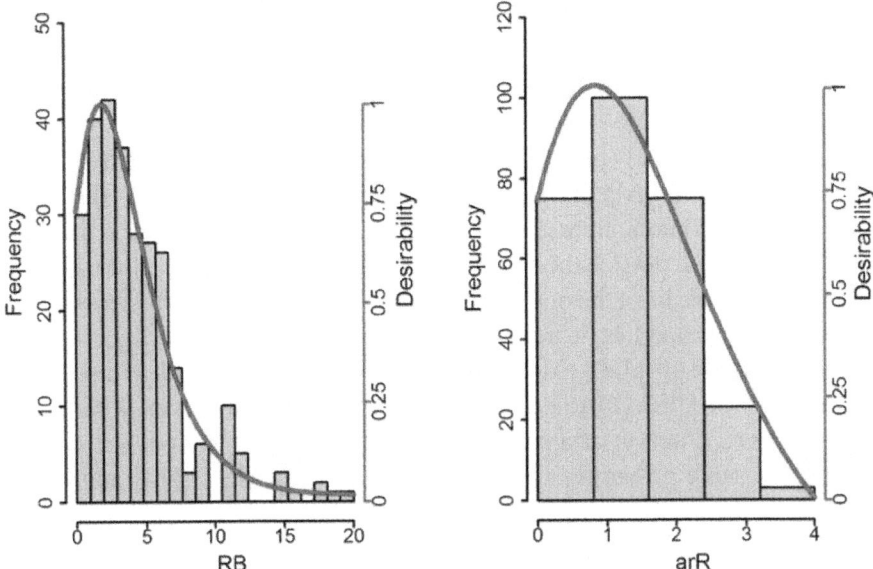

Figure 3. Frequency counts and desirability function plots of fungicides. Histograms and desirability functions (red curve, see right scale) of six molecular descriptors, i.e., MW (molecular weight), LogP (log of the octanol–water partition coefficient), HBA (number hydrogen bond acceptors), HBD (number hydrogen bond donors), RB (number of rotatable bonds), arR (number of aromatic rings), computed for the fungicides subset.

The individual df_i (i molecular descriptor) were joined accordingly for each pesticide-class by computing geometric means. This can be expressed by logarithmic identities, as the exponent of the arithmetic mean of the logarithm transformed dfs (see eq. 2). As argued by Derringer and Suich [18] the geometric mean exhibits several advantages in this case: (i) zero to one range, (ii) output values will increase as the balance of the properties becomes more favorable, (iii) if any $df_i = 0$ (is unacceptable) the geometric mean is null, i.e., if a property is unacceptable the compounds becomes unacceptable.

$$QEX = e^{\frac{1}{n}\sum_{i=1}^{n} \ln df_i}, \text{ for } df_i > 0; \text{ if } df_i \leq 0,$$

$$QEX = 0, \text{ where } X = \{\text{``}H\text{''}, \text{``}I\text{''}, \text{``}F\text{''}\} \quad (2)$$

We denominate the resulted scoring functions as quantitative estimates of herbicide-likeness (QEH), insecticide-likeness (QEI) and fungicide-likeness

(QEF), according to the pesticide class. These functions reflect the probability of a molecule to exhibit desirable characteristics as a pesticide. Thereby, we obtained an intuitive quantitative indicator of the likeness of a molecule to match the physicochemical profile of pesticides.

In order to model specific properties of large data sets, predictive models often use many descriptors limiting the applicability domains of the model. The more descriptors are used, the greater is the likelihood that a candidate molecule will fall outside the limits of one or more of these descriptors [19]. In our approach, we limit the number of descriptors to six basic physicochemical, independent, properties, correlated with pesticide bioavailability, solubility and stability [3],[9],[20],[21]. These descriptors are included also in the formulation of QED [8] to define drug-likeness, and moreover, with a slight variation, i.e., count of aromatic rings – arR – replaced by count of aromatic bonds, the same properties were are encountered in Hao's [14] approach to identity pesticides (see Table 1).

Pesticide Class Scorings

The three main classes of pesticides are: herbicides (against weeds), insecticides (against harmful insect pests), and fungicides (against harmful diseases) [12],[14],[22]. In this section, we will describe the way the above established pesticide class-specific desirability functions relate to each other.

In Figure 4 we plotted herbicide, insecticide and fungicide desirability functions against each variable separately. Differences between the three classes can be observed for all descriptors. In the case of MW ranging between 400 and 500, herbicides and insecticides can receive considerable higher scores compared to fungicides. One can observe that insecticides span over a broader range of LogP values. A considerable drop in scoring herbicides and fungicides can be noted at LogP > 5.5, whilst insecticides reach maximal desirability around this LogP value. The more hydrophilic nature of herbicides (and fungicide), in comparison to insecticides, is further consistently underlined in the HBA and HBD plots. More noticeable differences are present in the number of rotatable bounds plot: the peaks of the functions are reached at 2 RB for fungicides, 5 RB for herbicides and 6 RB for insecticides, but considerable area overlap can be observed. Finally, non-aromatic molecules provide major scoring variations between pesticide-classes: herbicides are poorly scored and, in contrast, insecticides gain maximum desirability scores.

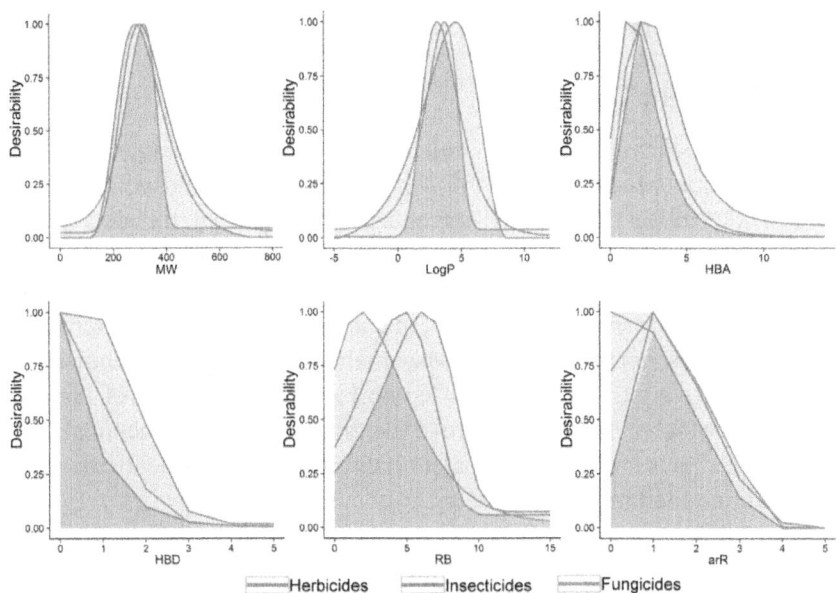

Figure 4. Comparative representation of desirability functions. Desirability function curves describing the three classes of pesticides: herbicides, insecticides and fungicides, in terms of MW (molecular weight), LogP (log of the octanol–water partition coefficient), HBA (number hydrogen bond acceptors), HBD (number hydrogen bond donors), RB (rotatable bonds), and arR (number of aromatic rings); dark grey –overlapping area described by the three curves; light grey – maximum area described by the three curves.

The recent analysis, conducted by Hao et al. [14], concerning the distributions of herbicides, insecticides and fungicides as described by six molecular descriptors, i.e., MW, ClogP, HBA, HBD, RB, number of aromatic bonds, indicated CLogP, HBD, and the number of aromatic bonds to be important constitutive properties to distinguish between the three classes of pesticides. Furthermore, the same study, describes RB distributions of herbicides and fungicides to be similar, with lower values compared to insecticides [14]. We note that, for the most part, our *df*s agree with previous findings, and slight variations in the distributions might be reasoned by the various datasets employed.

EXPERIMENTAL

AgroSAR Patent Database

GVKBio agrochemical patents collection (AgroSAR) comprises ~ 59 k (58915) unique structures and ~ 413 k (413103) SAR end-points measured in ~110 k (109733) assays. A percentage of 38.7% of the data has been published in the seventies, 29.6% in the eighties and 28.67% in the nineties up to 2005. AgroSAR gathers herbicides, insecticides, fungicides, acaricides, nematocides, bactericides, algaecide, plant growth, biocides, microbiocides and rodenticides in a relational database, manually curated and annotated, easy to query and subset. This database comprises large amounts of unexplored patent data, which can help to improve the discovery of agrochemicals. To our knowledge, this is the only SAR patent database built specifically from patent specifications filed in the agro sector.

We selected a subset of potent herbicides, insecticides and fungicides available in AgroSAR, as defined by more than 50% activity obtained at concentrations of 4.5 lb/acre (0.826 kg/ha) for herbicides, 125 ppm for insecticides and 100 mg/L for fungicides (cutoffs established by the medians of the activity data available per class). Hence, after removing marketed pesticides, we retrieved 1105 herbicides, 8983 insecticides and 9371 fungicides (Table 2). In this study, we will employ only these sets to assess the pesticide-likeness by various methods.

Table 2. Pesticide sets extracted from AgroSAR

Class	Herbicides	Insecticides	Fungicides	Pesticides
Num. of compounds	1105	8983	9371	19459
RoS (%)	97.29%	73.56%	91.55%	83.65%

The class of Pesticides comprises compounds merged from the Herbicide, Insecticide and Fungicide sets; RoS (%) - percentages of compounds passing Lipinski's RoS with no violation.

Basic statistics to describe the AgroSAR database are reported in Table 3 (and individual statistics of pesticide-class sets are reported in Additional file 1: Table S4). Additionally, a graphical description of the pesticide class-distributions in AgroSAR is shown in Figure 5. One can observe a slight shift towards higher molecular weight and LogP values in the case of insecticides compared to fungicides and herbicides. The latter two seem to exhibit more similarities, however, in term of arR, most herbicides display a smaller number of aromatic rings compared to insecticides and fungicides.

Table 3. Statistics of the pesticides extracted from AgroSAR

Properties	5% quantile	95% quantile	Median	Mean	SD
MW	228.3	553.3	354.8	370.1	108
LogP	1.2	7.2	4.1	4.2	1.8
HBA	1	7	3	3.3	2
HBD	0	2	0	0.5	0.8
RB	2	11	6	6.1	3.1
arR	0	3	2	1.8	1

SD - standard deviation; MW - molecular weight; LogP - hydrophobicity; HBA - number of hydrogen bond acceptors; HBD - number of hydrogen bond donors; RB - number of rotatable bonds; arR - number of aromatic rings.

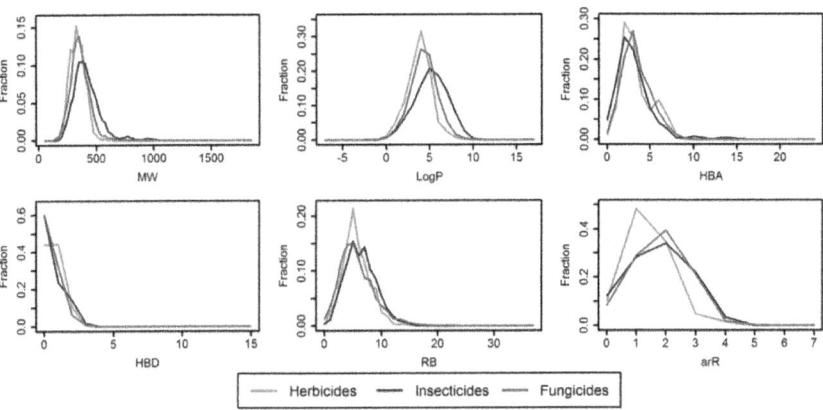

Figure 5. Basic molecular properties of herbicides, insecticides and fungicides selected from AgroSAR. Comparative distribution plots of AgroSAR selected herbicides (green), insecticides (blue) and fungicides (red), in terms of MW (molecular weight), LogP (log of the octanol–water partition coefficient), HBA (number hydrogen bond acceptors), HBD (number hydrogen bond donors), RB (rotatable bonds), and arR (number of aromatic rings).

Rule-based methods are widely used in the field of agrochemicals to identify chemicals with desirable properties. Based on a minimum set of easy-to-compute and interpretable molecular descriptors, we recall the efforts of Tice [11] and, more recently, Hao [14] to define herbicide- and insecticide-likeness and pesticide-likeness, respectively, as shown in Table 1. We evaluated the AgroSAR database, correspondingly, by means of these rules. We found that a percentage of 69.68% of the AgroSAR herbicides pass Tice's filter for herbicides (with zero violations) and 67.96% of AgroSAR insecticides pass Tice's filter for insecticides (with zero violations). We merged the AgroSAR pesticide-classes and applied Hao's rules for pesticide-likeness. The results indicate that 59.61% of the molecules are recognized (passed with no violation) as pesticides (Figure 6a).

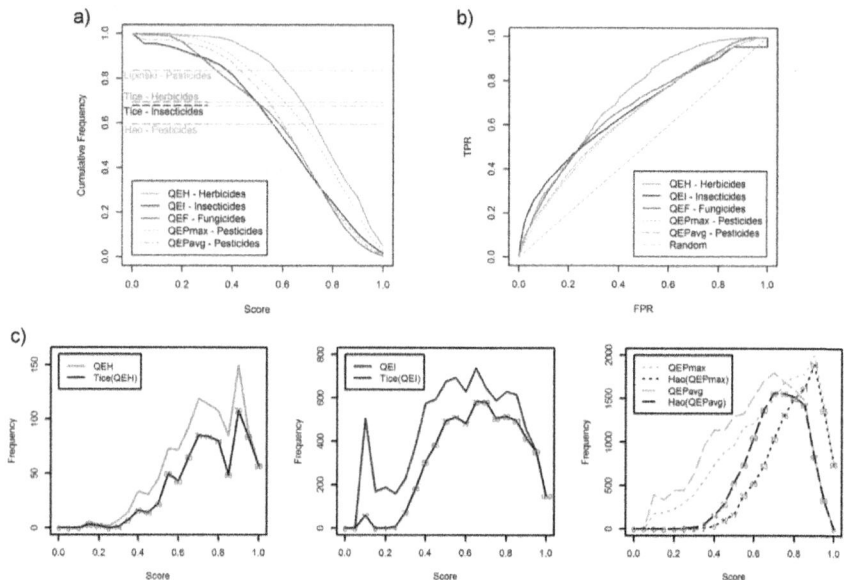

Figure 6. Evaluation of AgroSAR pesticides. (a) Cumulative frequencies of AgroSAR pesticide sets (herbicides – green, insecticides – blue, fungicides – red, pesticides – orange) plotted against quantitative estimates scores and performance of Tice's, Hao's and Lipinski's rule-based approaches as describes in Table 1 (rule-type performances are represented independent from the x-axis score values)(b); ROC curves showing the discriminative power of the scoring functions (c); frequency distributions of herbicides (left), insecticides (middle) and pesticides (right) in terms of quantitative estimates scores and frequencies corresponding to compounds passing rule-based models (in red percentages of compounds passing rule-based filters *per* cutoff). In the panels: QEH, Quantitative estimate of herbicide-likeness; QEI, Quantitative estimate of insecticide-likeness; QEF, Quantitative estimate of fungicide-likeness; QEP, Quantitative estimate of pesticide-likeness; QEPmax and QEPavg, - the maximum and the average of QEH, QEI and QEF values, respectively.

The field of drug discovery is closely related to that of agrochemical-discovery. The development of new medicine offered by agrochemicals and *vice-versa* may benefit upon the similarities between agrochemical and pharmaceutical research [22]. Similar to drugs, modern-day pesticides are optimized for low mammalian toxicity and act *via* a single target at *nano*-molar concentrations. Herbicides and fungicides were reported to generally meet the Lipinski's Ro5 criteria for drug-like compounds [12]. This observation is strongly confirmed also by AgroSAR pesticide database: 97.29% of the herbicides and 91.55% of the fungicides pass Ro5 (with zero violation). In the case of insecticides, 73.56% of the molecules were recognized as drug-like (Table 2). We encountered similar results also for the marketed pesticide set (see Additional file1: Table S5). As described above, insecticides exhibit a slightly different profile, compared to herbicides and fungicides, mainly consistent with increased hydrophobicity. Future explorations of these datasets can significantly contribute to improve the pesticide discovery and development programs.

Scoring AgroSAR Pesticide Database

In this section, we will report and discuss the capabilities of the hereby-proposed scoring functions to quantitatively define pesticide-likeness. In addition to the quantitative estimates of class-specific pesticide-likeness, we explored two data fusion rules to provide quantitative estimates of pesticide-likeness. Hence, we define QEP_{max} and QEP_{avg}, as the maximum and the average, respectively, of QEH, QEI and QEF values. The two fusion rules use QEH, QEI and QEF outcomes in different manners, i.e., the 'max-value'- rule reflects only the highest pesticide-class score whilst the 'average-value'-rule takes into account the contribution of all pesticide classes averaging the scores. Thus, in this section we will evaluate AgroSAR pesticides by means of QEH, QEI, QEF, QEP_{max} and QEP_{avg}.

In Figure 6a, we show the cumulative frequency counts of herbicides, insecticide, fungicides and pesticides plotted against the scores assigned by the corresponding quantitative estimate function, i.e., QEH - herbicides, QEI - insecticides, QEF - fungicides, QEP_{max} - and QEP_{avg} - pesticides. The highest scores can be observed in the case of QEH scoring herbicides. According to the pesticide-class, half of the molecules received QEH scores ≥ 0.72 (herbicides), QEI scores ≥ 0.57 (insecticides), QEF score ≥ 0.6 (fungicides), $QEP_{max} \geq 0.7$ and $QEP_{avg} \geq 0.6$ (pesticides). These results, further supported by the cutoff values corresponding to 25% and 75% of the datasets (see Additional file 1: Table S6), confirm the ability of the scoring functions to assign high scores to the equivalent pesticide-class.

In Figure 6c, we show the distribution of herbicides, insecticides and pesticides against the corresponding scoring functions values, i.e., QEH, QEI, QEP_{max} and QEP_{avg}. In order to see how these scores relate to well known rule-based models we plotted, correspondingly, the frequency counts of molecules passing Tice's filters for herbicides and insecticides, and Hao's filter for pesticides. One can observe a consistent trend between higher scores and increased percentages of compounds passing rule-based filters (Figure 6c).

To be marketed as pesticides, candidates need to meet a series of criteria, which cannot be fully addressed by the six molecular descriptors employed in QEPest-SFs. A number of 406 insecticides, 31 fungicides and 37 pesticides received null scores by the corresponding QEPest-SFs. On the other side, Figure 7, shows the chemical representation of the six best scored herbicides, insecticides and fungicides in AgroSAR database. One can observe the more hydrophobic insecticides and also the abundance of halogens (more noticeable for the exemplified fungicides) underlines the observation of Jeschke P [23] according to which modern agrochemicals tend to be more halogenated. The equivalently poorest scored molecules (ignoring zero scored representatives) fall clearly outside the acceptable limits of most scoring functions (see Additional file 1: Figure S2) and were scored consequently.

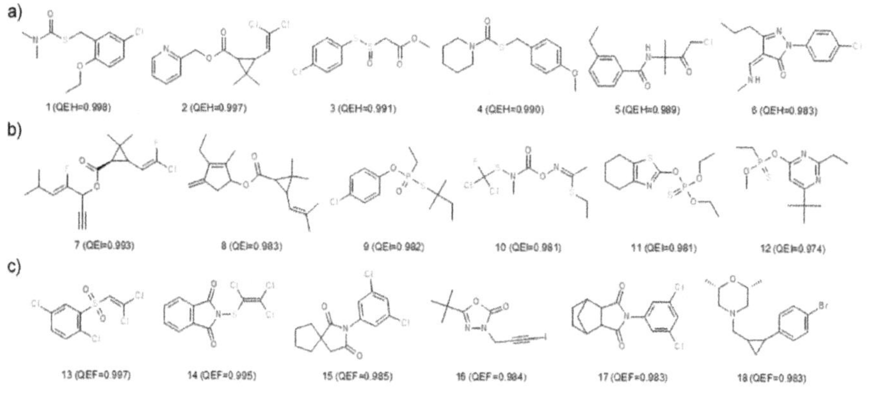

Figure 7. Examples of highly scored AgroSAR pesticides. Chemical representation of AgroSAR herbicides (a), insecticides (b) and fungicides (c) and quantitative estimation scores in parenthesis.

Simple rule-based methods that define pesticide-likeness are applied in the early stages of pesticide-discovery programs. Due to their simplicity, these methods serve to trim large chemical libraries to smaller sets, which are supplied to more computational-expensive approaches. In this sense, a challenging exercise for QEPest-SFs would be to recognize pesticides from a larger set of decoys. In consequence, ten times larger sets of randomly chosen representatives from PubChem Compounds (http://pubchem.ncbi.nlm.nih.gov/; 46.75 million molecules downloaded on December 10, 2013) were assembled for each pesticide class. Using the same six molecular properties, we computed QEH, QEI, QEF, QEP_{max} and QEP_{avg} also for the decoys sets (the decoys assembled for the pesticide-classes were merged for the evaluation of QEP_{max} and QEP_{avg}).

In Figure 6b, we show the ROC (receiver operating curve [24] – see *Performance measure* section in Methods) plots describing the capacity of QEH, QEI, QEF, QEP_{max} and QEP_{avg} to recognize the corresponding pesticide sets. A barely increased early enrichment can be seen in the case of QEI retrieving insecticides and, in contrast, QEH retrieved more lately herbicides. The discriminative performance was numerically assessed by AUC (area under the ROC [25] – see *Performance measure* section in Methods) values as reported in Additional file1: Table S7. With the exception of QEH (AUC > 0.7), we encountered relative poor separation capabilities. However, these functions are not meant to be as accurate as virtual screening tools but rather estimative indicators of compounds showing desirable pesticide-like physicochemical properties. Moreover, the decoys employed here were not experimentally demonstrated to not qualify as pesticides. Thus, these results must be seen in the light of the purpose and utility of the scoring functions as described above.

QEPest-SFs have the ability to rank compounds whether they fail pesticide-likeness rules or not. In consequence, different cutoffs for the scoring functions provide various levels of sensitivity and specificity. One might be tempted to find optimal cutoffs values for these scoring functions. The results of such an approach are reported in Additional file 1: Table S8 and Figure S3. However, as underlined by Bikerton et al. [8] in the case of QED, the usage of any threshold is discouraged as this results in qualitative outcomes, similar to rule-based approaches. A practical application of the hereby-proposed scoring functions would be to rank compounds by their scores and select the number of top ranking compounds required.

CONCLUSIONS

In this study, we have demonstrated that QEPest-SFs are able to rank compounds according to their herbicide-, insecticides-, fungicide- or pesticide-likeness. These scoring functions are based upon six simple molecular descriptors and a single type of function, parameterized accordingly to provide desirability scores. These quantitative assessments provide increased flexibility compared to traditional rule-based methods. For example, large chemical libraries can be reduced to desirable sizes, profiling pesticide-like molecules at various levels. In the usual pipeline of a drug and agrochemical discovery programs the resulted sets are supplied to more accurate virtual screening methods to increase cost-effectiveness in further experimental steps. For this purpose, we provide a simple Java-based program ("QEPest.jar") to compute QEH, QEI and QEF (see Additional file 2).

METHODS

Marketed Pesticide set

A set of 1685 pesticides (585 herbicides, 495 insecticides and 278 fungicides) was assembled from The Pesticide Manual [26] and Compendium of Pesticide Common Names [27]. For standardization (structure canonicalization and transformation – see Additional file 1: Table S9) the molecules were supplied to ChemAxon's Standardizer module (JChem 6.0.0, 2013, ChemAxon, http://www.chemaxon.com). The marketed pesticide set was used to derive quantitative estimate scoring functions for herbicide-likeness (QEH), insecticide-likeness (QEI), fungicide-likeness (QEF) and overall pesticide-likeness (QEP).

Molecular Descriptors

Molecular descriptors were computed with ChemAxon's structure database management software Instant JChem (JChem 6.0.0, 2013, ChemAxon, http://www.chemaxon.com). Six descriptors, i.e., molecular weight (MW), molecular hydrophobicity (log of the octanol–water partition coefficient; LogP), number of hydrogen bond acceptors (HBA), number of hydrogen bond donors (HBD), rotatable bonds (RB), aromatic rings (arR) were used to derive desirability functions for QEPest-SFs. Other hydrophobicity estimation metrics such as MLogP [15] and ClogP [16] were computed with Dragon (for Windows, Software for Molecular Descriptor Calculations, version 5.5, 2007 Talete srl, http://www.talete.mi.it) and BioByte (ClogP for Windows, version 1.0.0, 1995, BioByte Corp., http://www.biobyte.com/), respectively, and were used accordingly, as required by rule-based methods (Table 1).

Distribution of Data

For the assessment of the desirability functions we computed the frequency counts for each class of pesticides, according to the descriptor type-values, i.e., for continuous values (MW and LogP) the optimum bin size was computed with *Web Application for Bin-width Optimization* - Ver. 2.0 (http://176.32.89.45/~hideaki/res/histogram.html, accessed on Sep 21 2013) [28], and for discreet values (HBA, HBD, RB, arR) we used a bin-size of one (R 2.14.2) [29].

Curve Fitting

The frequency counts and bins computed for each molecular descriptor served as input for curve fitting processed by means of ZunZun.com*Online Curve Fitting and Surface Fitting Web Site* (http://zunzun.com/, accessed on Aug 6, 2013). Depending on the data to be modeled, up to 573 non-linearly, and 23 linearly equations, were fitted.

Performance Measure

The discriminative power of QEPest-SFs was assessed graphically and numerically by means of receiver operating curve (ROC) [24] and the area under the ROC (AUC) [25]. The ROC plot describes the true positive rate (TPR = sensitivity) *versus* the false positive rate (FPR = 1- specificity) according to the ranked list. AUC values indicate the ability of a scoring method (or prediction models, in general) to discriminate between two classes of elements, e.g., actives and inactives, and is defined by the area under the ROC. Values range from 0 to 1 (perfect separation), 0.5 suggesting a random spread of the representatives of the two classes.

ADDITIONAL FILES

Table S1. Molecular descriptors (generated with Instant JChem) evaluated for desirability functions assessment

Number	Abbreviation	Molecular descriptor
1	AlphA	Aliphatic atom count
2	ArA	Aromatic atom count
3	ArB	Aromatic bond count
4	ArR	Aromatic ring count
5	chB	Side-chain bond count
6	HBA	Number of H-bond acceptors
7	HBD	Number of H-bond acceptors
8	LogP	Octanol/water particion coefficient
9	MW	Molecular mass
10	nC	Number of carbon atoms
11	nHl	Number of halogens
12	nHA	Number of heteroatoms
13	RB	Number of rotatable bonds
14	RC	Ring count
15	TPSA	Topological surface area

JChem 6.0.0, 2013, ChemAxon, http://www.chemaxon.com)

Table S2. Coefficients and maximum values used to compute property-specific desirability functions (dfs) for herbicides, insecticides and fungicides

Pesticide class	Property	a	b	c	o	Max
Herbicides	MW	7.077E+001	2.830E+002	8.497E+001	-1.185E+000	69.5849922
	LogP	9.381E+001	3.077E+000	1.434E+000	6.164E-001	94.4228257
	HBA	1.176E+002	2.409E+000	1.567E+000	7.155E+000	120.4572352
	HBD	2.334E+002	4.535E-001	-1.480E+000	4.470E+000	228.1589796
	RB	8.470E+001	4.758E+000	-2.423E+000	5.437E+000	89.7012502
	arR	3.018E+002	1.101E+000	8.869E-001	-2.281E+001	276.9634213
Insecticides	MW	7.638E+001	2.983E+002	8.364E+001	1.912E+000	78.2919965
	LogP	7.427E+001	4.555E+000	-2.193E+000	-2.987E+000	71.2829691
	HBA	1.394E+002	1.363E+000	1.283E+000	5.341E-001	133.9224801
	HBD	6.706E+002	-1.163E+000	7.856E-001	7.951E-001	331.170104
	RB	6.549E+001	6.219E+000	-2.448E+000	5.318E+000	70.5540709
	arR	2.875E+002	3.050E-001	1.554E+000	-8.864E+001	193.0023343
Fungicides	MW	5.103E+001	3.142E+002	-5.631E+001	2.342E+000	53.3719946
	LogP	5.073E+001	3.674E+000	-1.238E+000	2.067E+000	52.773116
	HBA	7.379E+001	1.841E+000	1.326E+000	5.158E-001	73.7976536
	HBD	1.647E+002	-9.762E-001	-2.027E+000	1.384E+000	144.9887053
	RB	4.091E+001	1.822E+000	2.582E+000	6.235E-001	41.4385926
	arR	1.344E+002	8.383E-001	1.347E+000	-3.117E+001	102.3024319

Table S3. R^2 values of the fitted desirability functions for the three classes of pesticides

Class	MW	LogP	HBA	HBD	RB	arR
Herbicides	0.93	0.93	0.95	1.00	0.95	0.98
Insecticides	0.97	0.97	0.95	1.00	0.88	1.00
Fungicides	0.97	0.93	0.96	1.00	0.96	0.99

Table S4. Statistics of pesticides extracted from AgroSAR sets

Sets	Properties	5% Quantile	95% Quantile	Median	Mean	SD[a]
Pesticide	MW	228.3	553.3	354.8	370.1	108.2
	LogP	1.2	7.2	4.1	4.2	1.8
	HBA	1	7	3	3.3	2
	HBD	0	2	0	0.5	0.8
	ROT	2	11	6	6.1	3.1
	arR	0	3	2	1.8	1
Herbicides	MW	225.2	430.7	320.8	324.3	67.7
	LogP	0.8	5.4	3.4	3.3	1.4
	HBA	1	7	3	3.4	1.8
	HBD	0	2	1	0.7	0.7
	ROT	2	10	5	5.6	2.4
	arR	0	3	1	1.4	0.8
Insecticides	MW	245.3	608.6	384	403.5	123.4
	LogP	1.4	7.7	4.7	4.6	1.9
	HBA	1	7	3	3.1	2.2
	HBD	0	2	0	0.6	0.8
	ROT	2	11	6	6.4	2.8
	arR	0	3	2	1.8	1
Fungicides	MW	216.3	480.8	336.3	343.6	84.7
	LogP	1.2	6.4	3.8	3.8	1.6
	HBA	1	7	3	3.4	1.7
	HBD	0	2	0	0.5	0.7
	ROT	2	12	5	5.8	3.3
	arR	0	3	2	1.8	1

[a] SD standard deviation

Table S5. Percentages of marketed pesticides used to develop QEPest-SFs passing rule-based filters

Class	Number of pesticides	Lipinski	Tice [1]	Hao [2]
Herbicides	585	97.61 %	64.44 %	-
Insecticides	495	88.46 %	65.79 %	-
Fungicides	278	92.75 %	-	-
Pesticides	1319	93.71 %	-	77.35 %

accurate ClogP values were computed for 78.7% of the pesticides

Table S6. QEPest scores assign to AgroSAR sets at cumulative frequency (CF) values of 0.25, 0.5, 0.75

Scoring function	CF 0.25	CF 0.5	CF 0.75
QEH	0.859	0.72	0.595
QEI	0.746	0.574	0.404
QEF	0.733	0.601	0.39
QEP_{max}	0.835	0.696	0.515
QEP_{avg}	0.734	0.600	0.419

Table S7. Area under the receiver operating curve (AUC)[3] and standard deviation (SD) values achieved by the QEPest scoring functions

Scoring function	QEH	QEI	QEF	QEP_{max}	QEP_{avg}
AUC (±SD)	0.721 (±0.007)	0.668 (±0.003)	0.677 (±0.003)	0.643 (±0.002)	0.651 (±0.002)

Table S8. Optimum cutoffs for QEPest-SFs as suggested by Matthews correlation coefficients and sensitivity-specificity versus cutoff (Se-Sp Cutoff) plots

Metric	QEH	QEI	QEF	QEP_{max}	QEP_{avg}
Matthews correlation coefficients	0.608	0.784	0.607	0.702	0.639
Se-Sp Cutoff	0.650	0.500	0.510	0.630	0.530

Table S9. Standardization of pesticides before molecular properties generation;

Number	Standardization step
1	Remove fragment
2	Strip salts
3	Remove water
4	Transform nitro
5	Transform nitroso
6	Transform Azide
7	Transform Isocyanate
8	Transform Aromatic N-Oxide
9	Transform Tertiary N-Oxide
10	Transform Phosphoric
11	Transform Phosphonium Ylide
12	Transform Sulfoxide
13	Transform Sulfon
14	Disconnect Metal Atoms
15	Neutralize
16	Clean Isotones
17	Tautomerize

ChemAxon's Standardizer module - JChem 6.0.0, 2013, ChemAxon, http://www.chemaxon.com

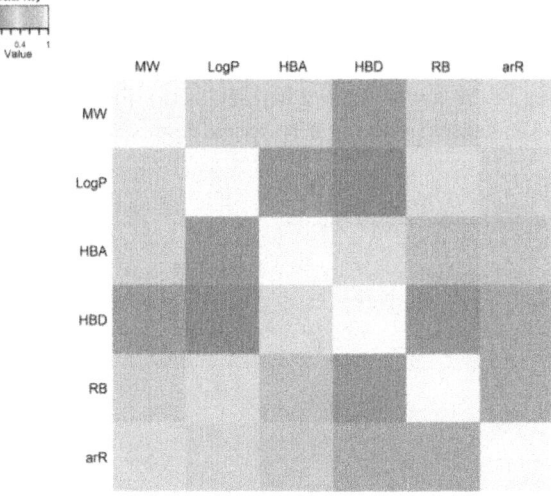

Figure S1. Kendall tau [4] values describing the pair wise correlation between MW, LogP, HBA, HBD, RB and arR, in the marketed pesticide set (see paper).

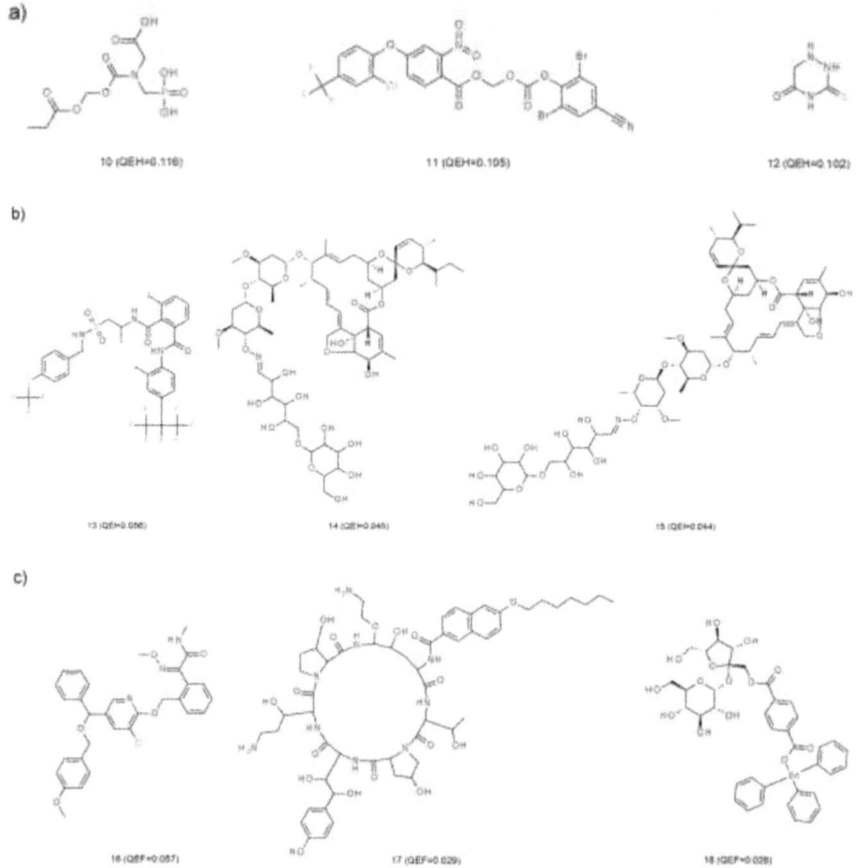

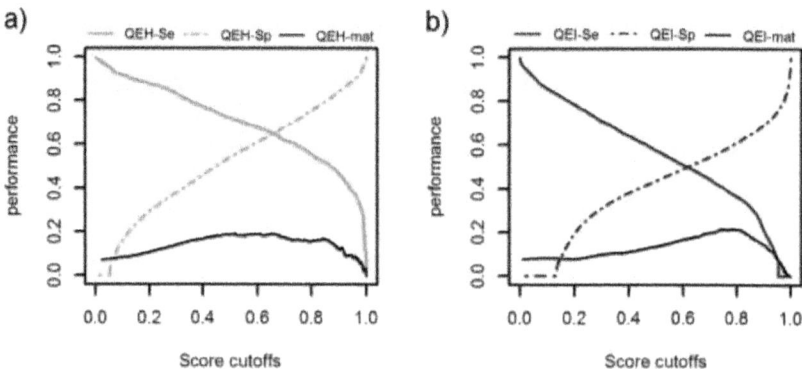

Figure S2. Three of the lowest scored herbicides (a), insecticides (b) and fungicides (c) from AgroSAR dataset (ignoring zero scored representatives).

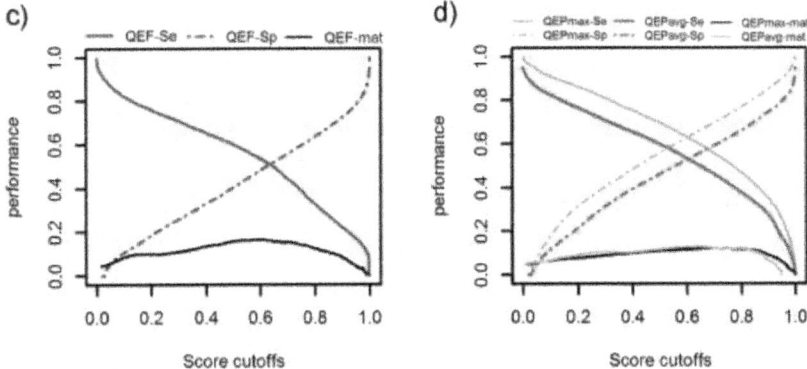

Figure S3. Sensitivity(Se)-specificity(Sp) and Matthews correlation coefficient versus score cutoffs plots of AgroSAR (a) herbicides, (b) insecticides, (c) fungicides, (d) pesticides.

AUTHORS' CONTRIBUTIONS

SM initiated and supervised the project. SA carried out the calculations, implemented and tested the scoring functions, developed the Java-based program and prepared the manuscript. SFT and AB contributed to data preparation for model development and validation and drafted the manuscript. SRC and AKM provided the AgroSAR patent database and corresponding annotations. All authors read and approved the final manuscript.

ACKNOWLEDGEMENTS

This project was financially supported by Project 1.1 of the Institute of Chemistry Timisoara of the Romanian Academy. The authors are indebted to ChemAxon Ltd for access to JChem software and to Alan Wood (http://www.alanwood.net/pesticides/index.html) for maintaining the Compendium of Pesticide Common Names.

REFERENCES

1. Oprea TI, Davis AM, Teague SJ, Leeson PD: Is there a difference between leads and drugs? A historical perspective. J Chem Inf Comput Sci. 2001, 41: 1308-1315. 10.1021/ci010366a.
2. Hann MM, Oprea TI: Pursuing the leadlikeness concept in pharmaceutical research. Curr Opin Chem Biol. 2004, 8: 255-263. 10.1016/j.cbpa.2004.04.003.

3. Lipinski CA, Lombardo F, Dominy BW, Feeney PJ: Experimental and computational approaches to estimate solubility and permeability in drug discovery and development settings. Adv Drug Deliv Rev. 1997, 46: 3-25. 10.1016/S0169-409X(00)00129-0.

4. Ursu O, Oprea TI: Model-free drug-likeness from fragments. J Chem Inf Model. 2010, 50: 1387-1394. 10.1021/ci100202p.

5. Oprea TI: Property distribution of drug-related chemical databases. J Comput Aided Mol Des. 2000, 14: 251-264. 10.1023/A:1008130001697.

6. Ertl P, Rohde B, Selzer P: Fast calculation of molecular polar surface area as a sum of fragment-based contributions and its application to the prediction of drug transport properties. J Med Chem. 2000, 43: 3714-3717. 10.1021/jm000942e.

7. Cumming JG, Davis AM, Muresan S, Haeberlein M, Chen H: Chemical predictive modelling to improve compound quality. Nat Rev Drug Discov. 2013, 12: 948-962. 10.1038/nrd4128.

8. Bickerton GR, Paolini GV, Besnard J, Muresan S, Hopkins AL: Quantifying the chemical beauty of drugs. Nat Chem. 2012, 4: 90-98. 10.1038/nchem.1243.

9. Veber DF, Johnson SR, Cheng H-Y, Smith BR, Ward KW, Kopple KD: Molecular properties that influence the oral bioavailability of drug candidates. J Med Chem. 2002, 45: 2615-2623. 10.1021/jm020017n.

10. Ghose AK, Viswanadhan VN, Wendoloski JJ: A knowledge-based approach in designing combinatorial or medicinal chemistry libraries for drug discovery. 1: a qualitative and quantitative characterization of known drug databases. J Comb Chem. 1999, 1: 55-68. 10.1021/cc9800071.

11. Tice CM: Selecting the right compounds for screening: does Lipinski's Rule of 5 for pharmaceuticals apply to agrochemicals?. Pest Manag Sci. 2001, 57: 3-16. 10.1002/1526-4998(200101)57:1<3::AID-PS269>3.0.CO;2-6.

12. Clarke ED, Delaney JS: Physical and molecular properties of agrochemicals: an analysis of screen inputs, hits, leads, and products. Chim Int J Chem. 2003, 57: 731-734. 10.2533/000942903777678641.

13. Clarke ED: Beyond physical properties-application of Abraham descriptors and LFER analysis in agrochemical research. Bioorg Med Chem. 2009, 17: 4153-4159. 10.1016/j.bmc.2009.02.061.

14. Hao G, Dong Q, Yang G: A comparative study on the constitutive properties of marketed pesticides. Mol Inform. 2011, 30: 614-622. 10.1002/minf.201100020.

15. Moriguchi I, Hirono S, Liu Q, Nakagome I, Matsushita Y: Simple method of calculating octanol/water partition coefficient. Chem Pharm Bull. 1992, 40: 127-130. 10.1248/cpb.40.127.
16. Leo AJ: Calculating log Poct from structures. Chem Rev. 1993, 93: 1281-1306. 10.1021/cr00020a001.
17. Harrington ECJ: The desirability function. Ind Qual Control. 1965, 21: 494-498.
18. Derringer G, Suich R: Simultaneous optimization of several response variables. J Qual Technol. 1980, 12: 214-219.
19. Clark RD, Waldman M: Lions and tigers and bears, oh my! Three barriers to progress in computer-aided molecular design. J Comput Aided Mol Des. 2012, 26: 29-34. 10.1007/s10822-011-9504-3.
20. Ritchie TJ, Macdonald SJF: The impact of aromatic ring count on compound developability-are too many aromatic rings a liability in drug design?. Drug Discov Today. 2009, 14: 1011-1020. 10.1016/j.drudis.2009.07.014.
21. Akamatsu M: Importance of physicochemical properties for the design of new pesticides. J Agric Food Chem. 2011, 59: 2909-2917. 10.1021/jf102525e.
22. Delaney J, Clarke E, Hughes D, Rice M: Modern agrochemical research: a missed opportunity for drug discovery?. Drug Discov Today. 2006, 11: 839-845. 10.1016/j.drudis.2006.07.002.
23. Jeschke P: The unique role of halogen substituents in the design of modern agrochemicals. Pest Manag Sci. 2010, 66: 10-27. 10.1002/ps.1829.
24. Fawcett T: An introduction to ROC analysis. Pattern Recognit Lett. 2006, 27: 861-874. 10.1016/j.patrec.2005.10.010.
25. Hanley A, Mcneil J: The meaning and use of the area under a Receiver Characteristic (ROC) curve. Radiology. 1982, 143: 29-36. 10.1148/radiology.143.1.7063747.
26. Tomlin CDS: The Pesticide Manual. 2000, The British Crop Protection Council, Farnham, UK
27. Wood A: Compendium of pesticide common names. 1995–2014, [http://www.alanwood.net/pesticides/index.html]
28. Shimazaki H, Shinomoto S: A method for selecting the bin size of a time histogram. Neural Comput. 2007, 19: 1503-1527. 10.1162/neco.2007.19.6.1503.
29. R: A Language and Environment for Statistical Computing. 2012, R Foundation for Statistical Computing, Vienna, Austria

30. Tice CM (2001) Selecting the right compounds for screening: does Lipinski's Rule of 5 for pharmaceuticals apply to agrochemicals? Pest Manag Sci 57: 3–16.
31. Hao G, Dong Q, Yang G (2011) A Comparative Study on the Constitutive Properties of Marketed Pesticides. Mol Inform 30: 614–622.
32. Hanley A, Mcneil J (1982) The meaning and use of the area under a Receiver Characteristic (ROC) curve. Radiology 143: 29–36.
33. Kendall MG (1976) Rank Correlation Methods. 4th Ed. Griffin, London.

Chapter 6

EVALUATION OF A DRY EXTRACT SYSTEM INVOLVING NIR SPECTROSCOPY (DESIR) FOR RAPID ASSESSMENT OF PESTICIDE CONTAMINATION OF FRUIT SURFACES

Umesh Kumar Acharya, Phul Prasad Subedi, and Kerry Brian Walsh

Central Queensland University, Institute for Resource Industries and Sustainability, Plant Science Group, Rockhampton, Australia

ABSTRACT

The dry-extract system for (near) infrared (DESIR) technique was implemented using reflectance near-infrared spectroscopy in context of detection of contact pesticide residues on fruit. Based on chemical structure, spectra features and regression statistics for PLSR models, a product containing metiram and pyraclostrobin was chosen from six pesticides for further consideration. Regression models based on spectra of dry extracts of aqueous solutions and either acetone or water washes of contaminated fruit were encouraging (RMSECV of approximately 0.03 - 0.06 mg a.i.). This level of analytical performance would support the use of the technique as a rapid screening tool, with suspect samples then subject to the reference GC-MS analysis method. However, the PLSR model performance was poor across populations of fruit, suggesting that matrix changes in the solvent wash between sets of fruit is problematic. Further work is required to establish whether sufficient variation can be built into a calibration set to overcome this issue, without degrading model performance to the point where it loses practical application.

INTRODUCTION

The use of chemicals in production horticulture is increasingly regulated. Only certain designated chemicals are approved for use with a given crop, and strict conditions are placed on the concentration, frequency and timing of application, particularly with respect to the last application before harvest. Maximum residue levels in product that is destined for human consumption have been set. The front line of defence in terms of quality control on such standards is through audit of spray diaries. Analytical assessment of pesticide levels occurs very occasionally, given the cost and timing of analysis. Japan probably has the highest enforcement level, and the best example of analytical capacity to monitor compliance in imports from other countries.

As Sanchez and co-workers [1] point out, there is increasing public interest in food safety issues, and thus there is a driver for increased compliance testing. Thus there is a need for analytical techniques that enable swift and low cost screening for pesticide residues in food in general and fruit in particular. Techniques in current use include gas chromatography, gas and liquid chromatography combined with mass spectrometry, capillary electrophoresis and immunoassay [2]. However such analysis is relatively expensive, time consuming and destructive of the sample, and thus enable sampling of only a few samples per batch.

Analytical methods based on near infrared spectroscopy (NIRS; 750 - 2500 nm) are generally very rapid and require little sample preparation. For example, comercial fruit grading systems are available with the capacity of grading up to 10 pieces of fruit per second on a pack-line [3]. These systems utilise short-wave near infrared spectroscopy to non-invasively assess the soluble sugar or dry matter content of the fruit. Also, the analysis of active ingredient content of pesticide formulations using FT-NIR of a solvent extract of sample was noted to be "10 times faster", with less solvent use than the chromatographic procedure [4].

Near infrared spectroscopy is capable of the determination of organic pesticides, given the presence of dipolar bonds in these chemicals. A number of studies, notably by de la Guardia and co-workers, have considered the use of NIRS in quality control of pesticide formulations.

This work has included consideration of number of scans and resolution for the assessment of Iprodione (a postharvest fungicide) content of agrochemical product, based on peak areas [5], and the assessment of Buprofezin (an insecticide), Diuron (a herbicide) and Daminozide (a plant growth regulator), based on PLS regression of NIR spectra (with RMSEP of 1.1%, 1.7% and 0.7% w/w for the latter three chemicals, respectively) [6].

Most recently, two studies on the use of NIRS to determine pesticide levels were published in 2011. One study describes the development of a PLSR model based on DESIR, using pure (aqueous) pesticide solutions (1.25 to 400 mg·kg^{-1}), with spectra (10,000 to 4000 cm^{-1}) acquired using an FT-NIR unit [7]. The use of multiplicative scatter correction with first derivative of absorbance data was recommended, with a R of 0.899 and a RMSECV of 42.3 mg·kg^{-1} reported. A support vector machine (SVM) was used to establish a classification model, however because of the low sample number used the model could easily overfit the data. Further, issues of matrix and model robustness were not considered in this study. The second study also considered determination of pesticide in pure methanol/water solutions, using a 1 mm transmittance cell and a Foss NIRSystem 6500 spectrophotometer [8]. A limit of detection of 12.6 and 46.4 mg·kg^{-1} for Alachlor and Atrazine respectively, was reported. Further verification of this result is recommended.

NIRS technology is generally associated with the assessment of concentrations at the %, not ppm, level. In those applications where ppm level determination has been reported, the assessment may represent measurement of an attribute that was correlated to the constituent of interest. This indirect correlation may be to either another chemical present in the sample at % levels, or to physical properties of the sample. For example, ergosterol in cereal grains may be detected at ppm levels because of changes in the physical (light scattering properties), or through indirect correlation with another chemical constituent [9]. The concentration of pesticides on fruit is typically at the ppm (mg a.i. kg^{-1}) level (**Table 1**).

As for any analytical method, detection improvement can be achieved by simplifying the background matrix and/or by concentrating the constituent of interest. For example, the concentration of chlorinated pesticides in water was determined using a polymer with affinity for the analyte, coated on an ATR crystal and detected using FTIR spectroscopy [10]. In another report, Diuron was determined to a limit of detection 13 mg·kg^{-1} in pesticide formulations based on pesticide extraction with acetonetrile, followed by transmittance measurement (peak area measurement between 2021 and 2047 nm, corrected by baseline established at 2071 nm) [11]. More recently, a method was proposed for analysis of dithiocarbamate residues at ppm levels in foodstuff, based on the generation of carbon disulphide, and subsequent trapping of this gas into tetrachloroethylene and assessment by transmission infrared spectroscopy [12].

In 1987, Meurens et al. [13] introduced the "dry-extract system for infrared" (DESIR). In this procedure, a liquid containing the constituent of interest is dried onto a solid substrate with low absorptivity (e.g. sugar in the range 0.0% to 5.0% w/v, onto a glass fibre filter [13]). The drying step allows for concentration of the analyte, and removal of the solvent (which typically has strong absorption features in the NIR and IR). Other sample preparation procedures can be added to further simplify the sample matrix, or concentrate the constituent of interest in the liquid sample, prior to its addition to the glass filter paper. The technique can be used in combination with either reflectance infrared or near infrared spectroscopy. For example, good prediction of carbon in lake water at 0.2 to 1 ppm was reported based on a DESIR procedure involving filtering 150 mL of lakewater through a glass fibre and analysis using SWNIRS [14].

Table 1. Descriptions of six chemicals used in this exercise.

Commercial name/Manufacturer	Function	Active chemical	WHP[a] (week)	Mode of action	% a.i.[b]	Recommended dose (% w/v)	MRL[c] (mg·kg^{-1})	Target pest
Aero/Nufarm	fungicide	Metiram and Pyraclostrobin	2	non systemic locally systemic	55.0 5.0	0.30	5 (mango) 0.3, 0.05 (tomato, mango)	Anthracnose
Octave/Bayer Crop Science	fungicide	Prochloraz	4	non-systemic	46.2	0.20	5 (mango)	Anthracnose and blossom blight
Penncozeb/Titan Crop Protection	fungicide	Mancozeb	3	non-systemic and contact	75.0	0.20	5 (mango) 3 (tomato)	Anthracnose
Amistar/Syngenta	fungicide	Azoxystrobin	2	systemic	25.0	0.08	0.5 (tomato and mango)	Anthracnose and stem end rot
Applaud/Dow AgroSciences	insecticide	Buprofezin	4	contact	44.0	0.06	0.2 (mango) 2 (tomato)	scale and mealy bug
Lorsban/Dow AgroSciences	insecticide	Chlorpyrifos	3	contact	50.0	0.10	0.05 (mango) 0.5 (tomato)	mango scale other insects

[a]WHP refers to the required withholding period between last application and consumption; [b]% a.i. refers to the concentration of active ingredient (%w/w); [c]MRL (Maximum Residue Limit) from Codex Alimentarus (http://www.apvma.gov.au/residues/standard.php).

The first use of DESIR for the detection of pesticides on fruit was made by Saranwong and Kawano in 2005 [15]. In their procedure, the fruit was rinsed with acetone, and a sample of the acetone rinse introduced to the glass fibre filter. This work involved set of 95 tomatoes that was sprayed with different concentrations of Euparin (containing 50% dichlofluanid, N-dichlorofluoromethy-thioN',N'-dimethyl-N-phenyl-sulfamide, CAS No. 1085-98- 9), allowed to dry, then washed with 25 mL acetone followed by 15 mL acetone (with acetone used as the solvent to suit the reference HPLC method). An aliquot of the wash solution was dried onto 37 mm glass fibre (GF/A Whatman, Maidstone, UK) at 45°C for 1 h, then scanned with reflectance optics using a FOSS NIRSystems 6500 (1100 - 2500 nm) spinning cup accessory, with a ceramic plate as a reflector and a reference. A RMSECV of 6.6 ppm was achieved with pure solutions, and 7.9 ppm with tomato wash (n = 45, 2 to 90 ppm at 2 ppm intervals, mean = 40.2 ppm, SD = 27.6 ppm). Given the use of 40 mL wash solution and 200 g tomato fruit, the authors calculated that this was equivalent to a detection limit of 1.6 ppm (mg dichlofluanid per kg fresh weight of tomato) by the standard procedures, and quite acceptable relative to the acceptable limit of 15 ppm set by the Japanese government. The authors were encouraged by this result, and recommended further work, including the use of water, rather than acetone, as the wash solution. Note however that only one set of tomato fruit was used in this exercise, giving a similar background to all samples, and thus a rather ideal application. Prediction across populations differing in origin (and thus in surface extractables) can be expected to suffer increased bias and RMSEP.

Further work was undertaken by the same researchers (published in 2007 [16]), with consideration of pure acetone based solutions of three pesticides (O,S-dimethylyN-acethylphosphoramidothioate [CAS 30560-19-1], Ndichlorofluoromethythio-N',N'-dimethyl-N-phenyl-sulfamide [CAS] and tetrach-loro-isophthalonitrile [CAS 1085- 98-9], known as acephate, dichlofluanid and TPN, respectively, and marketed as Ortran, Euparen and Daconil/Chlorothalonil, respectively). All three chemicals were used over the range 0 to 48 ppm (w/v). A RMSECV of 2.1, 5.3 and 9.3 ppm was reported for solutions of the three pesticides, respectively, and the detection limited ascribed to the number of groups with strong dipole moment groups in the chemical (3 CH_3 and 1 NH in tetrachloro-isophthalonitrile; 2 CH_3 in dichlofluanid; none in tetrachloro-isophthalonitrile). The SEP was noted to increase in mixtures of two pesticides. The result for TPN was deemed unacceptable for use in fruit quality assessment. The result for dichlofluanid was noted to be similar to the previous study, while the RMSECV for acephate was calculated to represent an accuracy of 1 ppm on a fruit weight basis, and deemed acceptable relative to the acceptable limit of 5.0 ppm set for fruit and vegetables in Japan. A collaborative study using acephate was implemented over three instruments; with similar RMSECV values obtained using the different instruments and technicians. A larger collaborative study involving multiple laboratories and presumably using fruit based samples was foreshadowed, however this did not occur (S. Saranwong, pers.comm.), and the group has since disbanded, with retirement of Sumio Kawano.

The above procedures are relevant to contact pesticides, which remain on the sample surface. The detection of systemic pesticides, in which the chemical is dispersed through the tissue, was indicated to be possible using near infrared spectroscopy by a study involving pesticide (chlorpyrifos) addition to vegetable juice extracts (R^2 0.981, RMSECV 0.15 mg·kg^{-1}) [17]. However, this study contained quite low sample numbers and a single juice matrix, so the model robustness in prediction is questionable. Indeed the authors concluded that the work was exploratory and that validation was required. Another report indicated a RMSEP of 0.1% w/v was achievable for detection of Chlorpyrifos (an insecticide) in minced white radish, using FTNIR (sample presentation method not given) [18]. However this exercise involved agrochemical addition to a single set of radish fruit, and used a random allocation of samples to calibration and validation sets. Thus, again, model robustness is questionable. In a 2010 publication, Sanchez et al. [1] report the detection of systemic pesticides in intact raw capsicum fruit using three separate methods—intact fruit scanned with a diode array spectrometer (1100 - 1700 nm), fruit homogenised and scanned with a tilting grating spectrophotometer (FOSS NIRSystems 6500, 1100 - 2200 nm) and homogenate used in a DESIR exercise (glass fibre discs soaked in juice, then dried at 40°C 24 h) using the tilting grating unit. Fruit were sourced from commercial farms, with 659 intact pepper and

717 crushed pepper samples scanned, and then analysed for a broad range of chemicals (organophosphates, organochlorides, carbamates, pyrethoids, pyrimidine compounds, dicerboximides, thiazoles and naturalytes by GS/MS working in tandem MS/MS mode. A validation set of samples, not included in the calibration set, was used, however the validation set was selected on the basis of a Mahalanobois distance related measure to represent the calibration set. Thus the validation set does not represent a truly independent test set. A PLS discriminant analysis was undertaken using samples classified as either above or below the maximum residue limit (as set by EC 396/2005) for any of the assessed pesticides. It was claimed that differences in the first derivative spectra were visible for the two classes (pesticide levels above/below MRL) in all three sample presentations, and that the best classifications (at around 75% correct) were achieved with the intact fruit/scanning diode array procedure. Obviously such a capability would have wide ramifications in the fruit and vegetable industry. However this study does have limitations, and the authors cautioned that the results "must be considered as a feasibility study". For example, it is not clear if all high residue samples were associated with fruit of a different variety to that associated with low residue samples, or from a different growing location (differing in growing conditions), and as the validation set was not independent of the calibration set, the validation result will be optimistic in terms of practical application. Certainly the ability to develop a model that can detect the presence of these agrochemicals on intact fruit seems optimistic.

Given the tremendous application potential for a rapid, cost effective assessment of pesticides in fruit, further assessment of the DESIR sample presentation method is certainly warranted. In the current study we attempt to extend the previous considerations [15,16] of detection of contact pesticides using a DESIR procedure. From our consideration of the published work to date, we considered that an exercise involving multiple populations of fruit was warranted, to address the issue of matrix variation and model robustness.

MATERIAL AND METHODS

DESIR Sample Preparation and Spectra Acquistion

A 47 mm diameter glass microfiber filter (GF/A) (Whatman International Ltd., UK, cat. No. 1820 047) was placed into a 50 mm diameter glass Petri dish for each sample. An aliquot (0.5 mL) of solution was gently delivered onto the filter paper using a pipette. This volume was found to just saturate the filter. The filter was held for 12 h in a fan forced oven at 31°C and then stored in a desiccator to avoid interference of water.

The filters were inserted into spinning cup modules, with their upper surface facing the quartz window of the cup. Reflectance spectra (400 - 2500 nm) were acquired using a spinning module on a NIRSystems 6500 spectrophotometer, scanning the internal ceramic plate as a white reference prior to each sample. Duplicate spectra were acquired of each sample.

Pesticide Solutions

Six pesticide products in common use in agricultural systems of local significance, in either pre or post-harvest stages of mango and tomato, were considered (**Table 1**). Pure solutions of pesticides of various concentrations were made using deionised water. The concentrations were 0, 0.1, 0.5, 1, 2 and 10 times the recommended rate of application, which were 0.3, 0.20, 0.20, 0.08, 0.06 and 0.10 (% a.i.) for the products Aero, Octave, Penncozeb, Amistar, Applaud and Lorsban, respectively (**Table 1**). Maximum residue limits, drawn from the Codex Alimentarus (http://www.codexali-mentarius.org), are also presented in **Table 1**. The chemical structure of the active ingredient of these products, indicating functional groups that should display overtone and combination bands in the near infrared, is provided as **Figure 1**. The product Aero was chosen for further work, as described below.

Subsequently, a greater range of concentrations of the product Aero was considered (16 concentrations, from 0.225% to 0.60% w/v in 0.025% steps). This exercise was repeated at a later date, making fresh stocks of all concentration (from the same product source). DESIR preparations of each product and concentration were prepared and scanned as described above.

Fruit Treatment and Analysis

Populations of fruit (1 set of mango fruit, Mangifera indica, var. Kensington Pride, and 3 sets of apple fruit, Malus pyrus, var. Granny Smith) were sourced from a fruit retail outlet. One set of tomato (Lycopersicum esculentum, var. Gomerg) fruit were harvested from a commercial farm. The tomato fruit were visibly externally contaminated with white spray residue. Each population represented a different consignment of fruit. An aliquot (10 mL) of the pesticide Aero at various concentrations (0, 0.1%, 0.2%, 0.3%, 0.4% and 0.5% active ingredient) was sprayed onto fruit contained in separate open ziplock polyethylene (18 cm × 18 cm) bags, and allowed to dry for 4 h at room temperature. Five replicate fruit were treated per concentration.

In one exercise, mango fruit were rinsed first with 30 mL, then with 10 mL, of 95% acetone. The two extracts were combined in a beaker, and allowed to evaporate at room temperature. The solution volume was adjusted to 0.5 mL and transferred to the glass fibre filter paper. NIR spectra acquisition followed using the procedure indicated above.

The utility of water as an extractant, rather than acetone, was tested through the following exercises. In the first exercise, 10 mL of Aero solution of various concentrations were sprayed into empty bags (i.e. without fruit), then dried and extracted with two water washes (30 followed by 10 mL), instead of 95% acetone. Otherwise the experimental design and procedure was as above. This process was repeated, with inclusion of apple fruit (given that mango were no longer commercially available), using three sets of apple fruit (sourced at different times). Again, water was used as the rinse solvent rather than acetone. The DESIR sample preparation and NIR scan methods was then followed as per first experiment. In another exercise, tomato fruit were used, again with water used as the wash solvent.

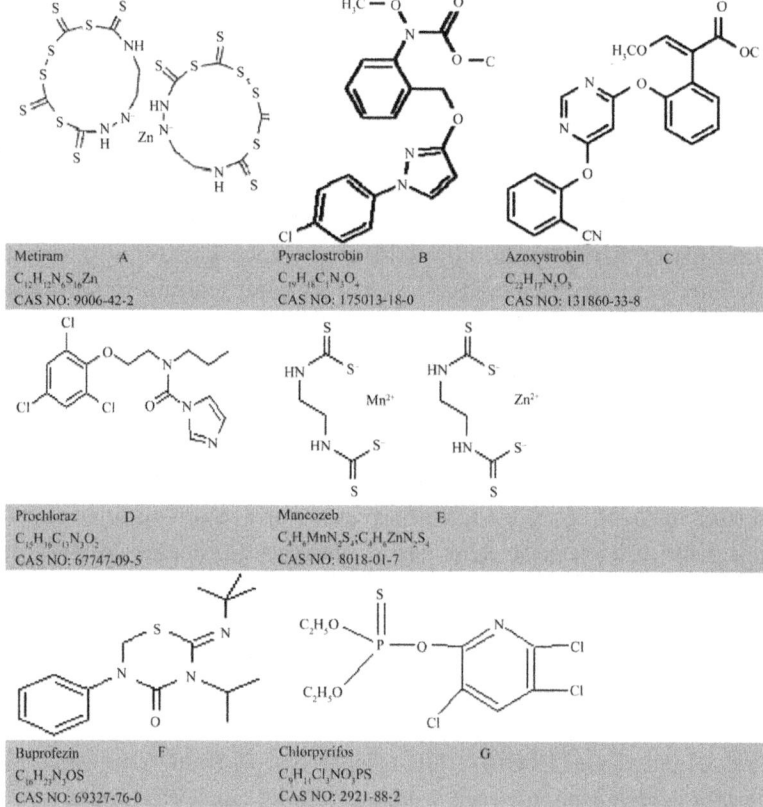

Figure 1. Chemical structure of six pesticides: (A) Metiram; (B) Pyraclostrobin; (C) Prochloraz; Panel; (D) Mancozeb; (E) Azoxystrobin; (F) Buprofezin and (G) Chlorpyrifos.

Chemometrics (Data Processing and Analysis)

Spectra acquired from DESIR samples were processed using The UnScrambler chemometrics software, V9.1. Difference spectra were calculated in an Excel (Microsoft) spread sheet and the spectral window was optimised using a MatLab PLS toolbox (Eigenvector) and an inhouse developed script [19]. Processing of spectra with a Savitsky Golay second derivative involved a second order polynomial fit, with a left and right interval of nine points each.

RESULTS AND DISCUSSIONS

Methodology

Sample processing to assess pesticide contamination of fruit surfaces involved washing the fruit, reducing the volume of the wash solvent, loading to a glass filter disc, drying the disc, loading the disc into a spinning disc accessory and scanning in a NIRSystems 6500 spectrophotometer. The total effort involved in sample analysis was considerably less than GC-MS methodology, with less skill required of the operator. Effective processing time per sample was approximately 30 min.

Experiment 1. Comparison of Pesticide Spectra

Relative to the featureless spectra of the blank glass fibre, all pesticide treated fibre displayed unique spectra (Figure 2). The spectral features can be ascribed to overtone and combination bands of various C-H and N-H bonds within these molecules. Pyraclostrobin has 10 CH, 1 CH_2, and 1 CH_3 bonds, azoxystrobin has 11 CH, 1 CH_3 and 1 C=O bond, prochloraz has 4 C-H, 4 CH_2, 1 CH_3, and 1 C=O, buprofezin has 1 C=O and 5 CH_3, mancozeb has 2 NH and 2 CH_2, chlorpyrifos has 2 CH_2 and 1 CH_3, and metiram has two NH bonds (**Figure 1**), and presumably the extinction coefficient of these chemicals would decrease in the order presented (i.e. absorption per unit concentration in order: pyraclostrobin = azoxystrobin > prochloraz > mancozeb = buprofezin > chlorpyrifos > metiram). If the NH bond is in a position that is unhindered within the chemical structure, there is a high likelihood it will produce a sharp, strong absorption band, which should improve detection of metiram and mancozeb. The NH absorption features are expected at about 2100 - 2200 nm (NH combinations), 1500 nm (first overtone) and 1000 - 1100 nm (second overtone). Absorption features around 1400, 1900 and 2300 are consistent with a strong O-H feature. These observations suggest that the product Aero might be a good candidate for detection using near infrared spectroscopy.

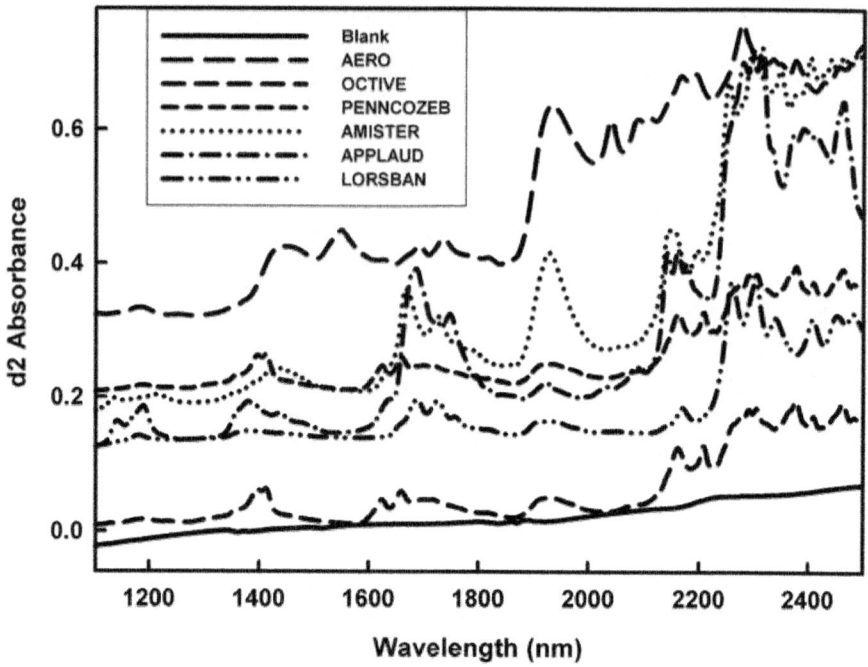

Figure 2. Reflectance (log1/R) spectra of DESIR preparations of 0.5 mL of neat solutions of the pesticide products Octave, Penncozeb, Amistar, Applaud and Lorsban (at 500, 500, 800, 60, 100 mg/kg of a.i., respectively), and Aero powder on glass fibre (at 300 mg/kg of a.i.). To assist visualisation spectra have been offset (0.3, 0.25, 0.2, 0.15, 0.1 and 0.05 units added to spectra of Aero, Penncozeb, Amistar, Applaud, Lorsban and Octave, respectively).

In practice, the highest PLS regression model R^2 was obtained for the model of Aero concentration, although higher SDR values were obtained for models of Penncozeb and Amistar (**Table 2**). Note that R^2 is related to RMSEC and SD, as $R^2 = 1 - (RMSEC/SD)^2$. However, in this exercise, SDR was calculated as SD/RMSECV. The difference between RMSECV and RMSEC was lower in the Penncozeb and Amistar models than was the case for the Aero model. Nonetheless, Aero was chosen for further work, based on this calibration result, on the interpretation that its structure should allow for NIR based assessment, and as this chemical is of particular importance to local industry.

Experiment 2. Further Consideration of Aero as Pure Solutions

With the product Aero selected as a good candidate for detection using near infrared spectroscopy, DESIR spectra were acquired of a range of concentrations (SD = 0.14, **Table 3**). Spectral features were related to concentration at several wavelengths (e.g. from around 2000 to 2100 nm, as seen in the d2log1/R spectra, **Figure 3**(a), and in the difference spectra, **Figure 3**(b)). The part of the d2log1/R spectrum carrying information about pesticide concentration was revealed in a plot of R^2 against wavelength for regressions on pesticide concentration based on reflectance values at each wavelength (**Figure 3**(c)). A moving window PLS regression method was also adopted (following the procedure of Guthrie et al. [19]), trialling all combinations of start and stop wavelengths between 400 and 2500 nm (data not shown). This analysis, together with the observed spectral differences (Figures 2, 3), guided the selection of the wavelength range 1850 - 2048 nm for model development.

For a PLSR model developed using this wavelength region, a R^2 of over 0.95, a RMSECV of around 0.03% a.i. and an SDR of around 4, was obtained in each of two replicate exercises (**Table 2**; also see scatter plot, Figures 4, 5). This result is encouraging for the use of near infrared spectroscopy for determination of Aero product, at least in pure solution or constant matrix conditions.

Table 2. PLSR calibration results for six pesticides, based on second derivate of log1/R spectra of DESIR samples of aqueous solutions of each pesticide, using the wavelength range 1850 - 2048 nm in all cases.

Product	Mean (% a.i.)	SD (% a.i.)	# factors	R^2_{cv}	RMSECV (% a.i.)	SDR
Aero	9.180	22.40	3	0.998	4.08	5.5
Octave	6.120	16.20	2	0.996	6.72	2.4
Penncozeb	0.390	0.723	2	0.954	0.09	8.0
Amistar	0.136	0.274	3	0.990	0.029	9.4
Applaud	0.117	0.217	3	0.956	0.078	2.8
Lorsban	6.420	17.60	3	0.774	14.90	1.2

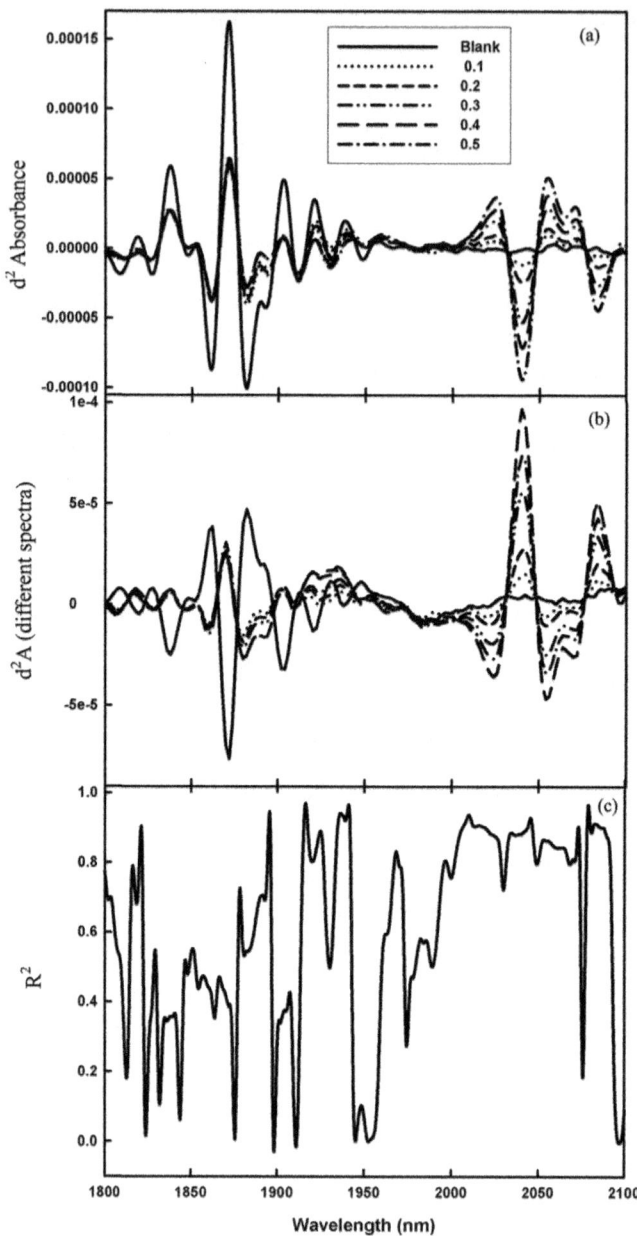

Figure 3. Second derivative of log1/R spectra (d^2A) of DESIR preparations of 0.5 mL aliquots of Aero solutions varying between 0 and 0.5% w/v (a), difference spectra (subtracting the spectra of the blank glass fibre) (b), and the correlation coefficient of determination (R^2) between the second derivative of log1/R and the concentration of pesticide, at each wavelength (c).

A detection limit of 0.1% a.i. in a 0.5 mL sample is equivalent to 0.5 mg a.i. If this amount of product was washed from a fruit we would have 0.5 mg a.i. derived from, say, a 200 g fruit (following previous logic [15]), and thus the analytical equivalence of detection of 2.5 ppm (w/fw). However, the MLR for pyraclostrobin is 0.3 and 0.05 mg·kg^{-1} (ppm) for tomato and mango respectively, while that for metiram is 5 mg/kg (**Table 1**). Thus the near infrared spectroscopy technique would not replace GC-MS as the analytical method of choice, however as a relatively rapid and low cost technique, it could be of value for high throughput screening exercises, with suspect samples subject to GC-MS analysis.

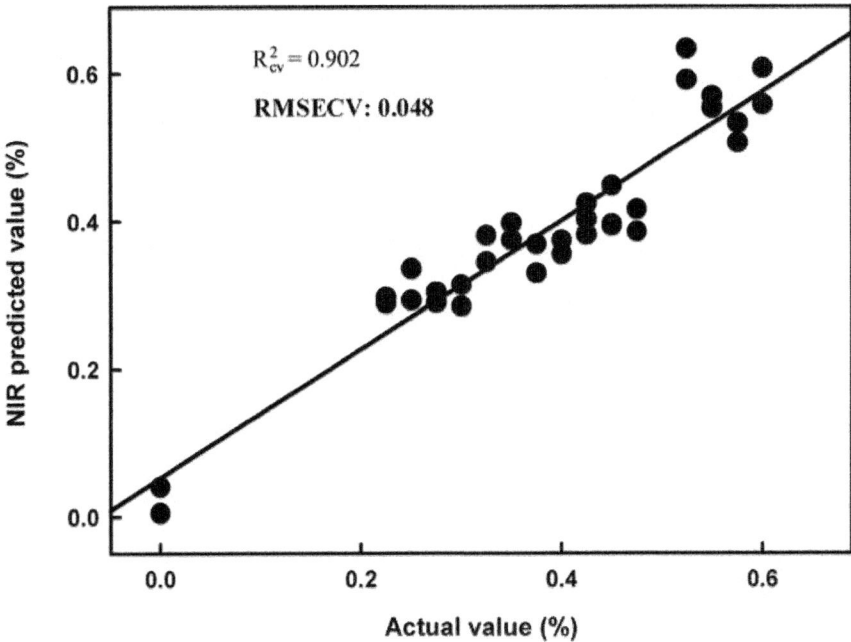

Figure 4. Scatter plot of PLS calibration regression model cross validation results for second derivative log1/R spectra of dry extract on glass fibre of Aero chemical at various concentrations (data of Population 1).

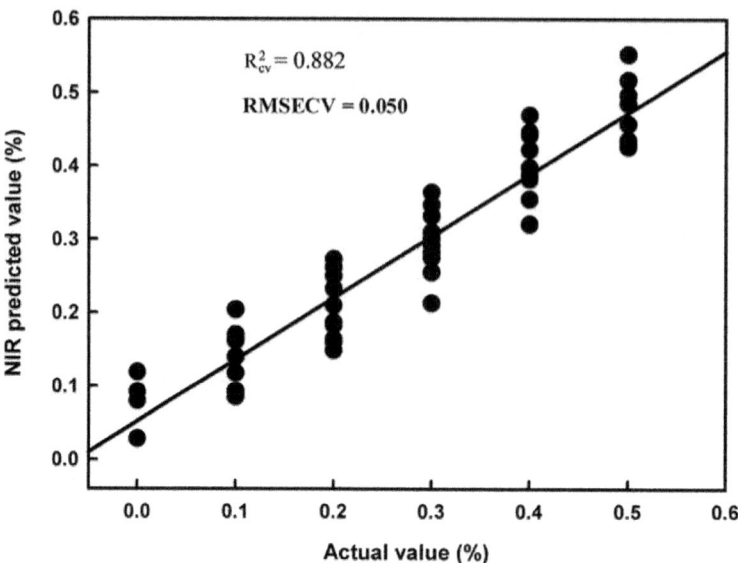

Figure 5. Scatter plot of PLS calibration regression model cross validation results for second derivative log1/R spectra of dry extract on glass fibre discs of Aero chemical at various concentrations sprayed onto mango and collected using an acetone wash (data of Population 3).

Experiment 3. Consideration of Aero Contamination of Fruit

Fruit were sprayed with a known volume of a known concentration of Aero fungicide while contained in a polyethylene bag. Thus all active ingredient contained into the bag. As a first exercise, Aero solution were sprayed onto mango fruit, dried, and recovered using an acetone wash, to follow earlier procedure [15]. Calibration statistics of R^2_{cv} 0.88, RMSECV 0.0487% a.i. and SDR 2.94 were achieved (**Table 3**(a)). A decrease in model performance was expected, compared to the model based on DESIR using straight Aero solutions, given the acetone wash will carry a range of other chemicals (e.g. dissolved cuticular wax) that could vary in quantity and composition from fruit to fruit.

Table 3. Calibration (a) and prediction (b) results for PLS models based on reflectance spectra of DESIR samples of aqueous solutions of Aero chemical, and of wash from contaminated fruit. Wash solvent is indicated in brackets (95% acetone or water). R^2 value greater than 0.75 displayed in bold.

(a) Calibration Statistics

Commodity	POP #	Sample	Mean	SD	R_{cv}^2	RMSECV	Slope
Aero solutions-1	1	32	0.409	0.115	**0.931**	0.029	0.92
Aero solutions-2	2	30	0.200	0.144	**0.953**	0.031	0.96
Aero solutions-1,2	1 + 2	62	0.356	0.140	**0.852**	0.052	0.87
Mango (acetone)	3	50	0.300	0.144	**0.882**	0.049	0.87
Bag (H_2O)	4	30	0.280	0.175	**0.841**	0.069	0.86
Apple-1 (H_2O)	5	38	0.300	0.144	**0.823**	0.062	0.87
Apple-2 (H_2O)	6	60	0.260	0.160	**0.843**	0.063	0.86
Apple-3 (H_2O)	7	50	0.300	0.143	**0.929**	0.039	0.95
Apple-1,2 (H_2O)	6 + 7	88	0.300	0.144	0.624	0.091	0.73
Apple-1,2,3 (H_2O)	5 + 6 + 7	198	0.300	0.149	0.545	0.101	0.59
Aero-Tomato (H_2O)	8	56	0.232	0.164	**0.949**	0.037	0.95

(b) Prediction Statistics

Cal set	Predict set	SD	R_p^2	RMSEP	Bias	Slope
Pop 1-2	Pop 3	0.159	0.160	0.530	0.374	1.05
Pop 1-3	Pop 4	0.175	0.260	0.195	0.079	0.55
Pop 1-4	Pop 5	0.144	0.058	0.376	0.229	0.38
Pop 4	Pop 5	0.160	0.416	0.438	0.417	0.67
Pop 1-5	Pop 6	0.160	0.260	0.175	0.109	0.22
Pop 4+5	Pop 6	0.143	**0.760**	**0.399**	0.309	2.26
Pop 5	Pop 6	0.143	**0.752**	2.725	2.640	4.82
Pop 6	Pop 7	0.164	0.274	0.199	0.143	0.23
Pop 1-6	Pop 7	0.143	0.593	0.893	0.809	2.60
Pop 1-7	Pop 8	0.164	0.717	0.119	0.070	0.55

Indeed the wash was quite discoloured, suggestion extractions of some pigments from the mango skin. However, the results were comparable to that of the pure aqueous solutions, and on this basis further trials were undertaken, using water rather than acetone as the wash solvent.

As a control, Aero solutions (10 mL) were sprayed into empty bags, dried, and washed with water. The wash was delivered onto the glass fibre discs. Again, PLS regression models were encouraging (R_{cv}^2 0.84; **Table 3**(a)), indicating that consistent recovery of material could be achieved using water rather than acetone as the wash solvent.

Aero solutions were sprayed onto three sets of apples, dried, and the water rinse dried onto glass fibre discs. Again, PLS regression models were encouraging ($R_{cv}^2 > 0.8$; **Table 3**(a)), indicating that consistent recovery of material was achieved using water rather than acetone as the wash solvent.

Aero solutions were also sprayed onto field collected tomato fruit which were heavily contaminated with other sprays, dried, then water rinsed for DESIR. Again, PLS regression models were encouraging (R_{cv}^2 0.95; **Table 3**(a)), despite the known contamination of the tomato fruit surface by other chemicals.

However, PLS regression calibration statistics for combinations of these populations were less encouraging (**Table 3**(a)). Further, plots of principal component 1 against 2 revealed that differences existed between the populations (data not shown). These observations are consistent with a change in matrix from the solvent wash of the fruit. The loss of model performance in combining two data sets of Aero chemical in solution cannot be ascribed to a matrix change, but rather may represent change in instrument performance or sample presentation. Thus, not surprisingly, a model based on a single population tended to perform poorly in predictions of independent sets (**Table 3**(b)). Models based on combinations of populations did improve in predictive ability (**Table 3**(b)), although the results were still below that required for practical application (an R^2 of 0.75 is equivalent to an RMSEP = SD/2, i.e. at best suited to grading to 2 groups).

CONCLUSION

The calibration results achieved are consistent with previous reports (e.g. [15,16]) on the sensitivity of NIRS based on dry extract sample preparation, and support the use of water as the solvent for fruit washing, rather than acetone, for the detection of metiram and pyraclostrobin on fruit. However, this level of analytical performance would support the use of the technique only as a rapid screening tool, with suspect lots then subject to the reference GC-MS analysis method. Further, matrix variation in the solvent wash between sets of fruit was demonstrated to be problematic for model predictive performance. Thus previous reports of the utility of the DESIR technique for assessment of pesticides on fruit have been over optimistic. Further work is required to establish whether sufficient variation can be built into a calibration set to overcome this issue, as well as any sample presentation or instrument variation, without degrading model performance to the point where it loses practical application.

ACKNOWLEDGEMENTS

We acknowledge the funding support of Hortical P/L and Horticulture Australia Ltd. We thank Roy Collis, Giru and SP Exports, Bundaberg for provision of mangoes and tomatoes, respectively.

REFERENCES

1. M. T. Sanchez, K. Flores-Rojas, E. G. Jose, A. G. Varo and D. Perez-Marin, "Measurement of Pesticides Residues in Peppers by Near-Infrared Reflectance Spectroscopy," Pest Management Science, Vol. 66, No. 6, 2010, pp. 580-586. doi:10.1002/ps.1910

2. L. Alder, K. Greulich, G. Kempe and B. Veith, "Residue Analysis of 500 High Priority Pesticides: Better by GCMS or LC-MS/MS?" Mass Spectrometry Reviews, Vol. 25, No. 6, 2006, pp. 838-865. doi:10.1002/mas.20091

3. M. Golic, K. B. Walsh and P. Lawson, "Short-Wavelength Near Infrared Spectra of Sucrose, Glucose, and Fructose with Respect to Sugar Concentration and Temperature," Applied Spectroscopy, Vol. 57, No. 2, 2003, pp. 64A-85A. doi:10.1366/000370203321535033

4. J. Moros, S. Armenta, S. Garrigues and M. de la Guardia, "Univariate near Infrared Methods for Determination of Pesticides in Agrochemicals," Analytica Chimica Acta, Vol. 579, No. 1, 2006, pp. 17-24. doi:10.1016/j.aca.2006.07.009

5. S. Armenta, S. Garrigues and M. de la Guardia, "Optimization of Transmission near Infrared Spectrometry Procedures for Quality Control of Pesticide Formulations," Analytica Chemica Acta, Vol. 571, No. 2, 2006, pp. 288- 297. doi:10.1016/j.aca.2006.05.003

6. S. Armenta, S. Garrigues and M. de la Guardia, "Partial Least Squares-Near Infrared Determination of Pesticides in Commercial Formulations," Vibrational Spectroscopy, Vol. 44, No. 2, 2007, pp. 273-278. doi:10.1016/j.vibspec.2006.12.005

7. J. Chen, Y. Peng, Y. Li, W. Wang and J. Wu, " A Method for Determining Organophosphate Pesticide Concentration Based on Near Infra-Red Spectroscopy," Transactions of the ASABE, Vol. 54, 2011, pp. 1025-1030.

8. A. Gowen, Y. Tsuchisaka, C. O'Donnell and R. Tsenkova, "Investigation of the Potential of Near Infrared Spectroscopy for the Detection and Quantification of Pesticides in Aqueous Solution," American Journal of Analytical Chemistry, Vol. 2, 2011, pp. 53-62. doi:10.4236/ajac.2011.228124

9. F. E. Dowell, M. S. Ram and L. M. Seitz, "Predicting Scab, Vomitoxin, and Ergosterol in Single Wheat Kernels Using Near-Infrared Spectroscopy," Cereal Chemistry, Vol. 76, No. 4, 1999, pp. 573-576. doi:10.1094/CCHEM.1999.76.4.573

10. F. Regan, M. Meaney, J. G. Vos, B. D. MacCraith and J. E. Walsh, "Determination of Pesticides in Water Using ATR-FTIR Spectroscopy on PVC/Chloroparaffin Coatings," Analytica Chimica Acta, Vol. 334, No. 1-2, 1996, pp. 85-92. doi:10.1016/S0003-2670(96)00259-0

11. J. Moros, S. Armenta, S. Garrigues and M. de la Guardia, "Near Infrared Determination of Diuron in Pesticide Formulations," Analytica Chimica Acta, Vol. 543, No. 1-2, 2005, pp. 124-129. doi:10.1016/j.aca.2005.04.045

12. A. Gonzalvez, S. Garrigues, S. Armenta and M. de la Guardia, "Determination at Low ppm Levels of Dithiocarbamate Residues in Foodstuff by Vapour Phase-Liquid Phase Microextraction-Infrared Spectroscopy," Analytica Chemica Acta, Vol. 688, No. 2, 2011, pp. 191-196. doi:10.1016/j.aca.2010.12.037

13. M. Meurens, O. V. D. Eydne and M. Vanbelle, "Fine Analysis of Liquids by NIR Reflectance Spectroscopy of Dry Extract on Solid Support (DESIR) in Near Infrared Diffuse Reflectance/Transmittance Spectroscopy," In: J. Hollow, K. J. Kaffka and J. L. Gonczy, Eds., Akademiai Kiado, Bundapest, Hungary, 1987, pp. 297-302.

14. D. F. Malley, P. C. Williams, M. P. Stainton and B. W. Hauser, "Application of Near-Infra-Red Reflectance Spectroscopy in the Measurement of Carbon, Nitrogen and Phosphorus in Seston from Oligotrophic Lakes," Canadian Journal of Fisheries and Aquatic Sciences, Vol. 50, No. 8, 1993, pp. 1779-1785. doi:10.1139/f93-199

15. S. Saranwong and S. Kawano, "Rapid Determination of Fungicide Contaminated on Tomato Surfaces Using the DESIR-NIR: A System for ppm—Order Concentration," Journal of Near Infrared Spectroscopy, Vol. 13, No. 3, 2005, pp. 169-175. doi:10.1255/jnirs.470

16. S. Saranwong and S. Kawano, "The Reliability of Pesticide Determinations Using Near Infrared Spectroscopy and the Dry-Extract System for Infrared (DESIR) Technique," Journal of Near Infrared Spectroscopy, Vol. 15, No. 4, 2007, pp. 227-236.doi:10.1255/jnirs.740

17. J. Wu, C. Liu, Y. Chen, Y. Chen and Y. Xu , " Study on Detection Technology of Pesticide Residues in Vegetables Based on NIR," Computer and Computing Technologies in Agriculture II, Vol. 295, 2009, pp. 2217-2222. doi:10.1007/978-1-4419-0213-9_73

18. Y. Zhou B. Xiang, Z. Wang and C. Chen, "Determination of Chloropyrifos Residue by Near-Infrared Spectroscopy in White Radish Based on Interval Partial Least Square (iPLS) Model," Analytical Letters, Vol. 42, No. 10, 2009, pp. 1518-1526.doi:10.1080/00032710902961032

19. J. A. Guthrie, K. B. Walsh, D. J. Reid and C. J. Liebenberg, "Assessment of Internal Quality Attributes of Mandarin Fruit. 1. NIR Calibration Model Development," Australian Journal of Agricultural Research, Vol. 56, No. 4, 2005, pp. 405-416.doi:10.1071/AR04257

Chapter 7

INTRODUCTION AND TOXICOLOGY OF FUNGICIDES

Dr. Rachid Rouabhi

Doctor of Toxicology and Ecotoxicology Larbi Tebessi University, Biology department, Tebessa Algeria

INTRODUCTION

Fungicides are either chemicals or biological agents that inhibit the growth of fungi or fungal spores. Modern fungicides do not kill fungi, they simply inhibit growth for a period of days or weeks. Fungi can cause serious damage in agriculture, resulting in critical losses of yield, quality and profit. Fungicides are used both in agriculture and to fight fungal infections in animals. Chemicals used to control oomycetes, which are not fungi, are also referred to as fungicides as oomycetes use the same mechanisms as fungi to infect plants (Latijnhouwers et al., 2000).

Fungicides can either be contact, translaminar or systemic. Contact fungicides are not taken up into the plant tissue, & only protect the plant where the spray is deposited; translaminar fungicides redistribute the fungicide from the upper, sprayed leaf surface to the lower, unsprayed surface; systemic fungicides are taken up & redistributed through the xylem vessels to the upper parts of the plant. New leaf growth is protected for a short period.

Most fungicides that can be bought retail are sold in a liquid form. The most common active ingredient is sulfur, present at 0.08% in weaker concentrates, and as high as 0.5% for more potent fungicides. Fungicides in powdered form are usually around 90% sulfur and are very toxic. Other active ingredients in fungicides include neem oil, rosemary oil, jojoba oil, and the bacterium Bacillus subtilis.

Fungicide residues have been found on food for human consumption, mostly from postharvest treatments (Brooks & Roberts, 1999).

Some fungicides are dangerous to human health, such as vinclozolin, which has now been removed from use (Hrelia, 1996), FCX and DFB that are used as pesticides to control pests and they have many side effects on natural non-target organisms (Rouabhi et al., 2009). In this chapter, we will develop the fungicides and their toxicity on biological and ecological systems.

CLASSIFICATION OF FUNGICIDES

Different authors have differing classification systems according to chemical structure, which somewhat complicates and confuses both the presentation and the discussion of fungicides. Several classification systems based on structure appear more of a web organization than a rationalized listing. In addition to classification by chemical structural grouping, fungicides can be categorized agriculturally and horticulturally according to the mode of application (use). According to the origin of fungicides, we can classify them in two major groups of fungicides

- Biologically based fungicides (biofungicides): Contain living microorganisms (bacteria, fungi) that are antagonistic to the pathogens that cause turf disease. Examples: Ecoguard contains Bacillus licheniformis; Bio-Trek 22G contains Trichoderma harzianum. In the case of a biofungicide, the Latin name of the microbe that it contains is the generic name of the fungicide.
- Chemically based fungicides: Synthesized from organic and inorganic chemicals, most of the fungicides that are sold throughout the world are chemically-based. They can be recognized according to similarities in three groups:

Chemical Structure

There are 29 generic names (active ingredients) associated with turf grass fungicides, shown in the table1.

Table 1. Fungicides generic names according to their chemical structure (Burpee, 2006).

1. propiconazole	2. triadimefon
3. myclobutanil	4. fenarimol
5. triticonazole	6. tetraconazole
7. fluoxastrobin	8. trifloxystrobin
9. azoxystrobin	10. pyraclostrobin
11. flutolanil	12. boscalid
13. polyoxin D	14. thiophanate-methyl
15. iprodione	16. vinclozolin
17. mefenoxam	18. propamocarb
19. fosetyl aluminum	20. phosphonate
21. quintozene	22. chloroneb
23. ethazole	24. mancozeb
25. thiram	26. hydrogen dioxide
27. chlorothalonil	28. fludioxonil
29. cyazofamid	30. Biofungicides

These 29 names represent 16 groups that have similar chemical structures (table 2). It is important to know which fungicides are chemically related to one another. For example, you should know that azoxystrobin, trifloxistrobin, and pyraclostrobin are chemically related to each other. However, they differ chemically from a fungicide, such as propiconazole, which is in a different chemical group. Because (i) all fungicides in a chemical group generally, control the same diseases. For example, the strobilurin fungicides provide good to excellent control of anthracnose, brown patch, gray leaf spot and summer patch. If you have purchased one strobilurin fungicide for control of these diseases, it is probably not necessary to purchase another. (ii) Since all fungicides in a chemical group control the same diseases, it does not make sense to tank-mix fungicides that represent a common chemical group in order to expand the scope of control. For example, to control anthracnose and dollar spot, tank-mixing two strobilurin fungicides will not work well because the strobilurins provide poor to weak control of dollar spot. (iii) If a pathogen develops resistance to one fungicide in a chemical group, the pathogen is usually resistant to all fungicides in that particular group. In Georgia, the fungi that cause dollar spot, Pythium blight and anthracnose have developed resistance to fungicides in one or more chemical groups.

Table 2. Chemical groups of fungicides according the Generic names (Burpee, 2006).

Generic Names	Chemical Group	
propiconazole triadimefon myclobutanil triticonazole tetraconazole	triazoles	DMI (demethylation Inhibitors fungicides)
fenarimol	pyrimidines	
fluoxastrobin trifloxystrobin azoxystrobin pyraclostrobin	strobilurins	
polyoxin D	polyoxins	
thiophanate-methyl	benzimidazoles	
iprodione vinclozolin	dicarboxamides	
mefenoxam	phenylamides	
propamocarb	carbamates	
fosetyl aluminum phosphonate	phosphonates	
mancozeb thiram	dithiocarbamates	

quintozene chloroneb ethazole	aromatic hydrocarbons
hydrogen dioxide	peroxides
chlorothalonil	nitriles
fludioxonil	phenylpyrolles
cyanofamid	cyanoimidazole
flutolanil boscalid	carboxamides
Ecoguard Sonata Soilguard	Biofungicides

Topical Activity

Fungicides can be placed into one of four groups based topical activity:

Contact Fungicides

Contact fungicides act only on plant surfaces. They are not absorbed by leaves, stems or roots and cannot inhibit fungal development inside plants. Example: dithiocarbamates, nitriles, aromatic hydrocarbons, peroxides, phenylpyrolles, cyanoimidazoles.

Localized Penetrants

Localized penetrant fungicides are absorbed by leaves and move short distances within a treated leaf, they do not move from one leaf to another and they are not absorbed by roots. These fungicides inhibit fungi on treated plant surfaces and inside treated leaves. Example: dicarboximides, strobilurins (except azoxystrobin and fluoxastrobin)

Acropetal Penetrants

Acropetal penetrants can penetrate plants through roots, shoots and leaves. These fungicides are absorbed by the xylem and move upward (acropetally) in plants. Acropetal penetrants inhibit fungi on and in treated plant surfaces and inside plant parts that lie above the treated surface. Example: benzimidazoles, triazoles, pyrimidines, carboximides, acylalanines, plus the strobilurins azoxystrobin and fluoxastrobin.

Systemic Fungicides

Systemic fungicides are the only fungicides that are absorbed into xylem and phloem and moves up and down in plants. These fungicides inhibit fungi on and in treated plant surfaces and inside plant parts that lie above or below the treated surfaces. Example: phosphonates.

Mode of Action

The body or thallus of most fungi exists as microscopic tubes called hyphae (Fig. 1)

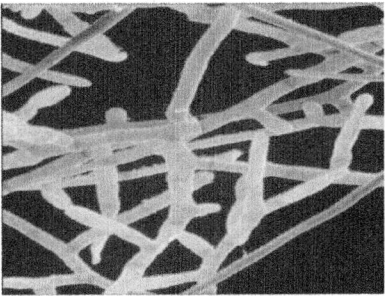
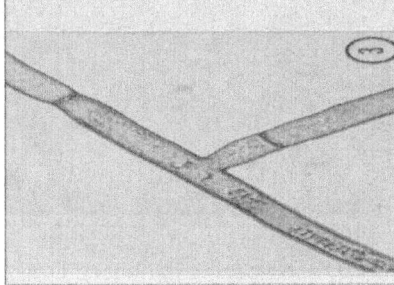

Figure 1. Hyphae of a fungi (Burpee, 2006).

A fungal cell contains many of the same organelles as other eukaryotes (Fig.2).

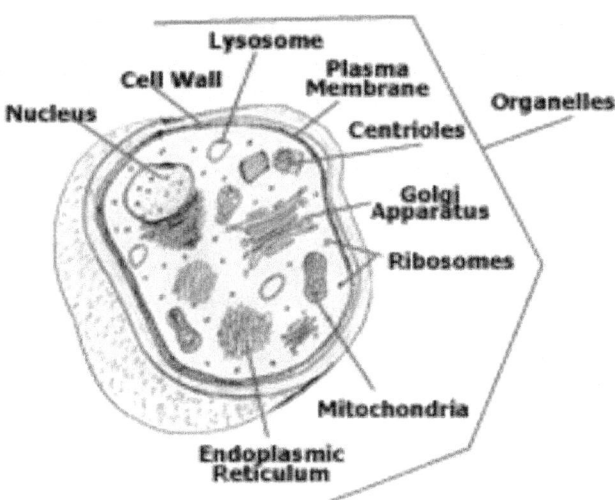

Figure 2. Fungal cell with organelles (Foster and Smith, 2010).

Fungicides can be divided into 2 groups based on mode of action in fungal cells:

Site-specific Inhibitors

Site-specific inhibitors target individual sites within the fungal cell (Fig. 3).

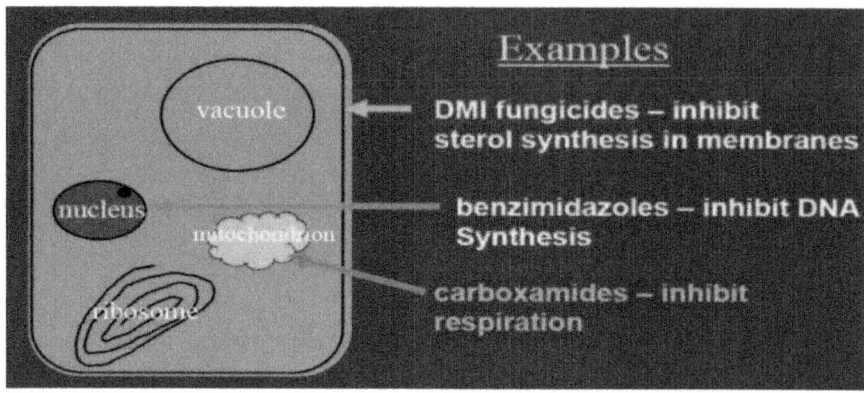

Figure 3. Site-Specific Inhibitors. DMI: demethylation inhibitors fungicides (Burpee, 2006).

Multi-site Inhibitors

Multisite inhibitors target many different sites in each fungal cell (fig. 4)

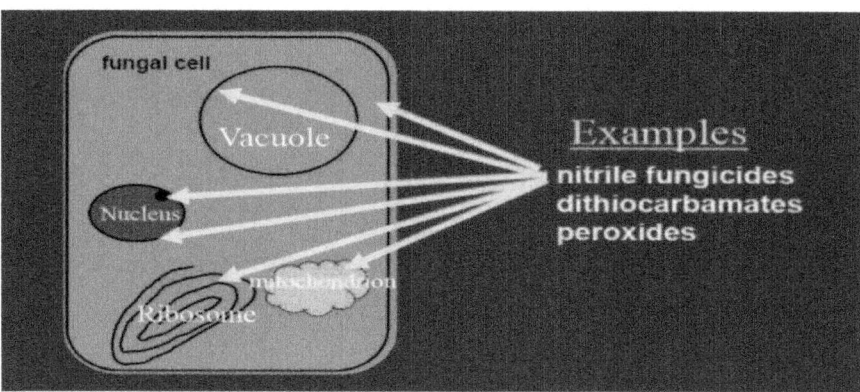

Figure 4. Multi-site Inhibitors (Burpee, 2006).

TOXICOLOGY OF FUNGICIDES

In general, fungicides are of low to moderate mammalian toxicology, although they are believed to have a higher overall incidence than other pesticides to cause developmental toxicology and oncogenesis (Costa, 1997). It has, for example, been estimated that more than 80 per cent of all oncogenic risk from the use of pesticides comes from a few fungicides (NAS, 1987). However, fungicides usually are responsible for only a small proportion of pesticide-related deaths, and account for only about 5 per cent or less of human pesticide exposures reported to Poison Control Centers (Blondell, 1997; Hayes and Vaughn, 1977; Litovitz et al., 1994). It has been noted that since fungi differ significantly in morphology and physiology from other forms of life, they may be successfully combated by compounds of low toxicity to other organisms, notably mammals (Edwards et al., 1991). However, since the mechanism of injury to pathogenic fungi may be different to that for injury to mammalian systems, it is possible that the two properties may co-exist in a given fungicide molecule (Marrs and Ballantyne, 2004).

It has been noted (Phillips, 2001) that the ideal fungicide should have the following characteristics: (a) low mammalian toxicity, (b) low ecotoxicity, (c) low phytotoxicity, (d) high penetration rates for spores and mycelia, and (e) limited biodegradation on the plant surface. Many fungicides combine several of these characteristics but few approach optimum for all of them.

Triazoles

This chemical family of fungicides, introduced in the 1980s, consists of numerous members: difenoconazole, fenbuconazole, myclobutanil, propiconazole, tebuconazole, tetraconazole, triadimefon, and triticonazole. They are important tools against diseases of turfgrasses, vegetables, citrus, field crops and ornamental plants. Homeowner products are available for use as well, and may be readily obtained at garden and nursery retail centers. They are applied as foliar sprays and seed treatments, but are diverse in use, as they may be applied as protectant or curative treatments. If applied as a curative treatment, triazole applications must be made early in the fungal infection process. Once the fungus begins to produce spores on an infected plant, the triazoles are not effective. Although the triazoles do not have the degree of systemic movement of many herbicides, they are xylem-mobile. They are readily taken up by leaves and move within the leaf. The triazoles are very specific in their mode of action – they inhibit the biosynthesis of sterol, a critical component for the integrity of fungal cell membranes. Because their site of action is very specific, there are resistance concerns (Fishel, 2005).

Toxicology

By the oral route of exposure, these triazoles would be considered as having low toxicity. Inhalation of dusts can cause irritation of the nose, lungs, and throat. For mycobutanil, in animals, effects were reported on the following organs: testes, adrenal gland, kidney, and thyroid. Mycobutanil did not cause cancer or birth defects; only doses that caused significant toxicity to parent animals caused reproductive effects on laboratory animals. Increased incidence of liver tumors at extremely high doses was reported in laboratory studies involving male mice who had been exposed to propiconazole or tebuconazole. There were no reproductive, developmental or chronic effects reported with either propiconazole or tebuconazole. Additionally, tebuconazole is considered to not cause any mutagenic or genotoxic effects; however, EPA has classified it as a "possible human carcinogen" because of the liver effects seen with mice. The main concern with triadimefon is its potential to cause birth defects, although data suggest that in humans such effects would occur only at moderate to high doses of exposure. Ecologically, the main concern with the triazoles is with fish and other aquatic organisms. Their labels will carry statements expressing this concern in the Environmental Hazards section. Of this pesticide family, only difenoconazole is considered to be highly toxic to fish. Most of the triazoles are considered to be practically nontoxic to birds and bees. Mammalian toxicities for the triazole fungicides are shown in Table 3. Table 4 lists the toxicities to wildlife by the common name of the pesticide (Fishel, 2005).

Table 3. Triazole fungicide mammalian toxicities (mg/kg of body weight).

Common name	Rat oral LD 50	Rabbit dermal LD50
Difenoconazole	1,453	2,010
Fenbuconazole	>2,000	>5,000 (rat)
Mycobutanil	1,600	>5,000 (rat)
Propiconazole	1,517	>4,000
Tebuconazole	1,700	>2,000
Triadimefon	569	2,000
Triticonazole	>2,000	>2,000

Table 4. Triazole fungicide wildlife toxicity ranges.

Common name	Bird acute oral LD 50 (mg/kg)*	Fish (ppm)**	Bee***
Difenoconazole	PNT	HT	PNT
Fenbuconazole	ST	PNT	PNT
Mycobutanil	PNT	MT	PNT
Propiconazole	PNT	MT	PNT
Tebuconazole	PNT	MT	PNT
Triadimefon	PNT - ST	ST	PNT
Triticonazole	PNT	ST	PNT

*Bird LD 50: Practically nontoxic (PNT) = >2,000; slightly toxic (ST) = 501 – 2,000; moderately toxic (MT) = 51 – 500; highly toxic (HT) = 10 – 50; very highly toxic (VHT) = <10.

**Fish LC 50: PNT = >100; ST = 10 – 100; MT = 1 – 10; HT = 0.1–1 ; VHT = <0.1.

***Bee: HT = highly toxic (kills upon contact as well as residues); MT = moderately toxic (kills if applied over bees); PNT = relatively nontoxic (relatively few precautions necessary).

Toxicology of an Example of the Family "Cyproconazole"

The primary dissipation routes of Cyproconazole (Fig. 5) in surface soil are microbial degradation and plant uptake. Soil photolysis (breakdown by sunlight) and volatilization are not significant routes of degradation. The major breakdown product (degradate, metabolite) of cyproconazole is further broken down to intermediate metabolites, carbon dioxide and bound material. There has been no evidence that Cyproconazole accumulates in the soil.

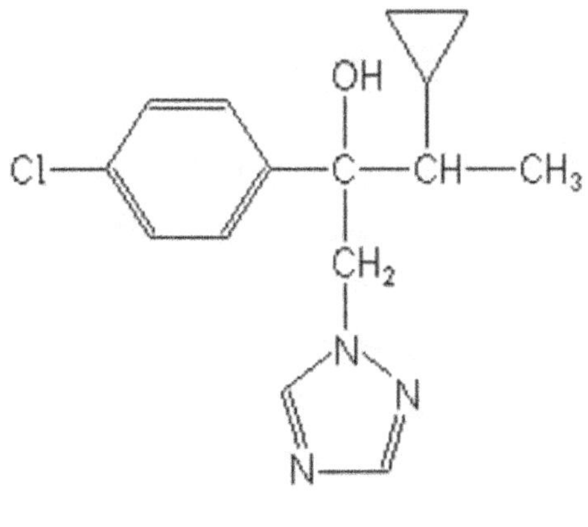

Figure 5. Cyproconazole.

Toxicology

Cyproconazole is of low risk to birds, mammals, bees and other non-target terrestrial organisms. Cyproconazole is moderately to slightly toxic to most aquatic organisms, and because of the low use rates the exposure potential is low and hence poses minimal risk of adverse effects (Envirofacts, 2005).

Oral LD50s are 1020mg/kg for male rats, 1333 mg/kg for female rats, 200 mg for male mice, and 218 mg/kg for female mice. The percutaneous rabbit LD50 is >2000 mg/kg. The rat 4-h inhalation is LC50 > 5.65 mg/L. The major plant residue is Cyproconazole. There is moderately rapid soil degradation; DT50 is about 3 months. Avian acute oral LD50 for Japanese quail is 150 mg/kg. Eight-day dietary LC50s are 816mg/kg (diet) for Japanese quail and 1197 mg/kg (diet) for mallard duck. Aquatic organism 96-h LC50 toxicity values include 18.9 mg=L for carp, 19 mg/L for trout, and 21 mg/L for bluegill sunfish. In Daphnia the 48-h LC50 is 26 mg/L. For bees, the contact LD50 is >0.1 mg/bee and the peroral LD50 is >1mg/bee (Marrs and Ballantyne, 2004). Plants absorb cyproconazole and rapidly degrade it to multiple metabolites. Studies have shown that cyproconazole residues may exist at harvest, but the levels are insignificant and well under the safety margins for human and environmental risks as established by regulatory authorities in many countries, including the US EPA (Envirofacts, 2005).

Aromatics Hydrocarbons Fungicides

Major fungicides in this group include chlorothalonil, tecnazine, chloroneb, dichloran, hexachlorobenzene, quintozene, pentachlorophenol, and sodium pentachlorophenate. Many of these are, or metabolized to, uncouplers of oxidative phosphorylation. This can lead to excessive heat production, hyperpyrexia, liver damage, and corneal opacities.

Toxicology of an Example of the Family "Chloroneb"

Chloroneb (Fig. 6) is a broad spectrum systemic fungicide taken up by the roots, and used on various fruit and vegetable crops as a wet or dry application powder or dust. Mechanism of fungal toxicity may be related to inhibition of DNA polymerization (Phillips, 2001). Principal use in Soil systemic and supplemental seed treatment forseedling diseases of beans, sugar beets, turf and soybeans. Excellent against damping-off (Harding, 1979-80). Used for the treatment of turfgrass to control snow mold (Typhula) and Pythium blight (Meister et al., 1994).

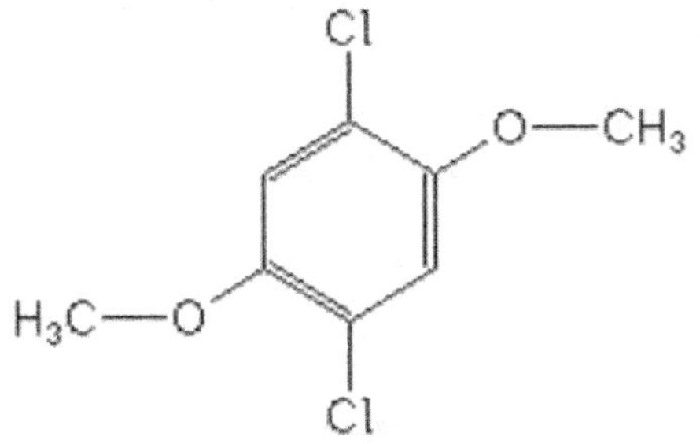

[Chemical structure: 1,4-dichloro-2,5-dimethoxybenzene]

Figure 6. Chloroneb.

Toxicology

a. *Acute Toxicity*

Dermal: LD50 = >5000 mg/kg (rabbit). A 50% aqueous suspension of the 65% caused no irritation to guinea-pigs and repeated applications did not result in skin sensitization (Worthing, 1979).

Oral: LD50 = >11,000 mg/kg (rat) (Wothing, 1979)

b. *Environmental Considerations*

Hazardous to fish and wildlife. Nonphytotoxic when used as directed. Does not leach from the soil (Harding, 1979-80). The material was not toxic to bluegill Sunfish at 4,200 ppm in the 48-hour exposure.

Dithiocarbamate Fungicides

The dithiocarbamate fungicides: ferbam, mancozeb, maneb, nabam, thiram, zineb and ziram were evaluated at the Joint FAO/WHO Meeting in 1967. Although the biochemical data were limited, temporary acceptable daily intakes (ADI's) were established for all of these compounds, but it was pointed out that these ADI's are to be applicable to the parent compounds only (FAO/WHO, 1968).

Toxicology of an Example of the Family "Thiram"

Thiram (fig. 7) is a dimethyl dithiocarbamate compound used as a fungicide to prevent crop damage in the field and to protect harvested crops from deterioration in storage or transport. Thiram is also used as a seed protectant and to protect fruit, vegetable, ornamental, and turf crops from a variety of fungal diseases. In addition, it is used as an animal repellent to protect fruit trees and ornamentals from damage by rabbits, rodents, and deer. Thiram is available as dust, flowable, wettable powder, water dispersible granules, and water suspension formulations, and in mixtures with other fungicides. Thiram has been used in the treatment of human scabies, as a sunscreen, and as a bactericide applied directly to the skin or incorporated into soap.

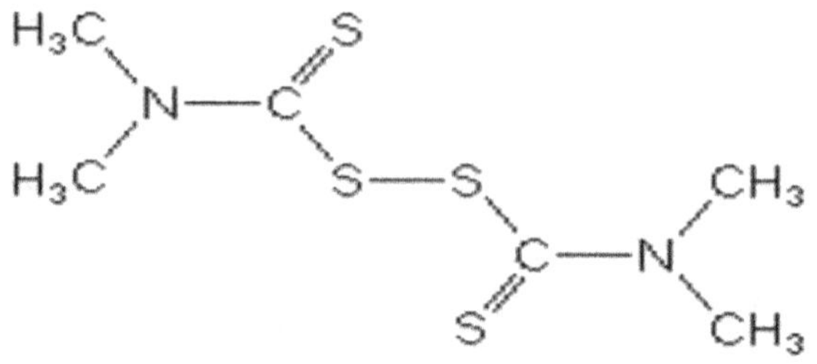

Figure 7. Thiram.

Toxicology

a. Acute Toxicity

Thiram is slightly toxic by ingestion and inhalation, but it is moderately toxic by dermal absorption. Acute exposure in humans may cause headaches, dizziness, fatigue, nausea, diarrhea, and other gastrointestinal complaints. In rats and mice, large doses of thiram produced muscle incoordination, hyperactivity followed by inactivity, loss of muscular tone, labored breathing, and convulsions. Most animals died within 2 to 7 days. Thiram is irritating to the eyes, skin, and respiratory tract. It is a skin sensitizer. Symptoms of acute inhalation exposure to thiram include itching, scratchy throat, hoarseness, sneezing, coughing, inflammation of the nose or throat, bronchitis, dizzines, headache, fatigue, nausea, diarrhea, and other gastrointestinal complaints. Persons with chronic respiratory or skin disease are at increased risk from exposure to thiram (U.S. National Library of Medicine, 1995). Ingestion of thiram and alcohol together may cause stomach pains, nausea, vomiting, headache, slight fever, and possible dermatitis. Workers exposed to thiram during application or mixing operations within 24 hours of moderate alcohol consumption have been hospitalized with symptoms. The 4-hour inhalation LC50 for thiram is greater than 500 mg/L in rats. Reported oral LD50 values for thiram are 620 to over 1900 mg/kg in rats; 1500 to 2000 mg/kg in mice; and 210 mg/kg in rabbits (Edwards et al., 1991; Kidd and James, 1991). The dermal LD50 is greater than 1000 mg/kg in rabbits (U.S. National Library of Medicine, 1995) and in rats (Edwards et al., 1991; Kidd and James, 1991).

b. Chronic Toxicity

Symptoms of chronic exposure to thiram in humans include drowsiness, confusion, loss of sex drive, incoordination, slurred speech, and weakness, in addition to those due to acute exposure. Repeated or prolonged exposure to thiram can also cause allergic reactions such as dermatitis, watery eyes, sensitivity to light, and conjunctivitis (Edwards et al., 1991). Except for the occurrence of allergic reactions, harmful chronic effects from thiram have been observed in test animals only at very high doses. In one study, a dietary dose of 125 mg/kg/day thiram was fatal to all rats within 17 weeks. Oral doses of about 49 mg/kg/day to rats for 2 years produced weakness, muscle incoordination, and paralysis of the hind legs. Rats fed 52 to 67 mg/kg/day for 80 weeks exhibited hair loss, and paralysis and atrophy of the hind legs. Symptoms of muscle incoordination and paralysis from thiram poisoning have been shown to be associated with degeneration of nerves in the lower lumbar and pelvic regions. Day-old white leghorn chicks fed 30 and 60 ppm for 6 weeks exhibited bone malformations. At doses of about 10% of the LD50 for 15 days, thiram reduced blood platelet and white blood cell counts, suppressed blood formation, and slowed blood coagulation in rabbits (Edwards et al., 1991).

c. Organ Toxicity

Studies have shown evidence of damage to the liver by thiram in the form of decreased liver enzyme activity and increased liver weight (Edwards et al., 1991). Thiram may also cause damage to the nervous system, blood, and kidneys (U.S. National Library of Medicine, 1995).

d. Ecological Effects

Effects on birds: Thiram is practically nontoxic to birds. The reported dietary LC50 of thiram in Japanese quail is greater than 5000 ppm (Hill and Camardese, 1986). Reported dietary LC50 values in pheasants and mallard ducks are 2800 ppm and 673 ppm, respectively (Hudson et al., 1984). The LD50 for the compound in red-winged blackbirds is greater than 100 mg/kg (Kidd and James, 1991).

Effects on aquatic organisms: Thiram is highly toxic to fish (U.S. National Library of Medicine, 1995). The LC50 for the compound is 0.23 mg/L in bluegill sunfish, 0.13 mg/L in trout, and 4 mg/L in carp (Mayer and Ellersieck, 1986). Thiram is not expected to bioconcentrate in aquatic organisms (Howard, 1989).

Benzimidazoles Fungicides

This class is confused since the individual classes are closely related. Sometimes, however, the benzimidazoles are classified separately but in an overlapping manner, creating confusion. They are nitrogen heterocyclic compounds, with parent structures of thiabendazole and/or benzimidazole. Included in this overall group are benomyl, thiabendazole, thiophanate, thiophanate–methyl, mebendazole, carbedazim, imazalil, and fuberidazole. Benomyl, carbendazim, thiophanate, and thiophanate–methyl are sometimes referred to (and classified) as benzimidazoles carbamates. Many of these fungicides inhibit mitochondrial fumarate reductase, reduce glucose transport, and uncouple oxidative phosphorylation. Inhibition of microtubule polymerization by binding to -tubulin is a primary action, and specific high affinity binding to host α-tubulin occurs at significantly lower concentrations than mammalian protein binding (Phillips, 2001).

Toxicology of an Example of the Family "Benomyl"

Benomyl was first reported as a fungicide in 1968 and introduced onto the UK market in 1971 by the US Company Du Pont (Tomlin, 1994). It is a systemic benzimidazole fungicide that is selectively toxic to microorganisms and to invertebrates, especially earthworms (Extoxnet, 1994).

Benomyl and its main metabolite carbendazim bind to microtubules (an essential structure of all cells) and therefore interfere with cell functions such as cell division and intracellular transportation. The selective toxicity of benomyl as a fungicide is possibly due to its heightened effect on fungal rather than mammalian microtubules (WHO/PCS, 1994).

Benomyl is used as a pre-harvest systemic fungicide, and as a post-harvest dip or dust. It combats a wide range of fungal diseases of arable and vegetable crops, apples, soft fruit, nuts, ornamentals, mushrooms, lettuce, tomatoes and turf. It is also available widely for amenity and amateur garden use (Whitehead, 1996).

Toxicology

a. Acute Toxicity

Benomyl is of such a low acute toxicity to mammals that it has been impossible or impractical to administer doses large enough to establish an LD50. It therefore has an arbitrary LD50 that is 'greater than 10,000 mg/kg/day for rats'. However, skin irritation may occur with workers exposed to benomyl (Extoxnet, 1994). It is a mild to moderate eye irritant and is a skin sensitizer. Florists, mushroom pickers and flower growers have reported allergic reactions to benomyl (MAFF, 1992).

In 1992, benomyl exposure caused adverse occupational health effects (headaches, diarrhoea and sexual dysfunction) in agricultural workers in Florida (Agrow, 1992).

b. Chronic Toxicity

In a laboratory study, dogs fed benomyl in their diets for three months developed no major toxic effects but did show evidence of altered liver function at the highest dose (150 mg/kg). With longer exposure, more severe liver damage occurred including cirrhosis after two years (Extoxnet, 1994).

c. Carcinogenic Effects

The US Environmental Protection Agency classified benomyl as a possible human carcinogen (Office of Pesticide Programs, 1996). There is an element of doubt in this classification because carcinogenic studies have produced conflicting results. A two year experimental mouse study has shown it probably caused an increase in liver tumors. The Ministry of Agriculture Fisheries and Food (MAFF) takes the view that this was bought about by the hepatoxic effect of benomyl (MAFF, 1992).

d. Reproductive Effects

Tests on laboratory animals have shown benomyl can have an effect on reproduction. In one rat study, where the mothers were fed 1,000 mg/kg/day for four months, the offspring showed a decrease in viability and fertility (WHO, 1993). In studies to investigate the effects of benomyl on male reproductive performance, fertility was reduced at all dose levels tested. In another study, a no-effect level of 15mg/kg/day was established based on testicular abnormalities (MAFF, 1992).

Permanent reductions occurred in the size of testes and male accessory glands in 100 dayold offspring from female laboratory rats receiving 31.2 mg benomyl/kg body weight per day. Rats developed a reduced sperm activity following acute inhalation exposure, acute and sub-chronic oral exposure. The same effect occurred in dogs following a single fourhour inhalation exposure (MAFF, 1992).

e. Environment

Benomyl binds strongly to soil and does not dissolve in water largely. When applied to turf, it has a half-life of three to six months, and when applied to bare soil the half-life is six to 12 months (Extoxnet, 1994).

Piperazines Fungicides

Triforine Toxicology

Triforine (fig. 8) is a piperazine derivative used as a systemic fungicide with protectant, eradicant and curative characteristics. It is used for control of powdery mildew, rusts, black rot and scab on cereals, fruit, ornamentals, and vegetables (Royal Society of Chemistry, 1983; Worthing, 1983). Triforine is also active against storage diseases of fruit and suppresses red spider mite activity (Worthing, 1983). Because of its low hazard to beneficial insects, triforine may be used in Integrated Pest Management (IPM) programs. Triforine comes in emulsifiable concentrates, liquid seed treatments, and wettable powder formulations. Triforine is miscible with common insecticides and herbicides in the recommended manner of use (Royal Society of Chemistry, 1983).

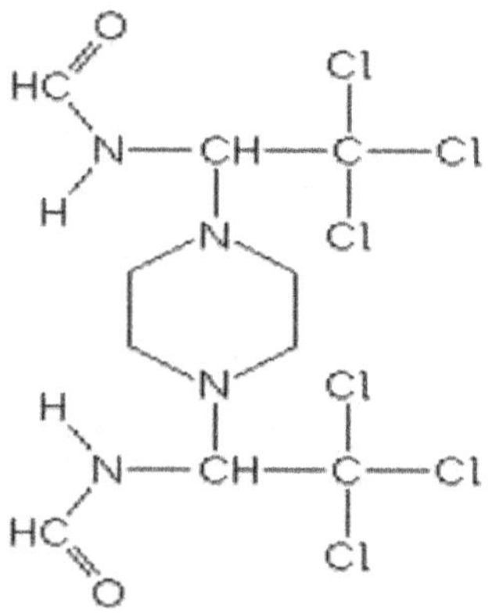

Figure 8. Triforine.

Toxicological Effects

   a.   Acute Toxicity

Triforine and the formulated product Saprol have a low acute and dermal toxicity and have a moderate acute inhalation toxicity. The acute oral LD50 for triforine in rats is greater than 16,000 mg/kg body weight. The acute percutaneous LD50 for rats is greater than 10,000 mg/kg. Acute dermal LD50 for rats is greater than 10,000 mg/kg body weight. The onehour acute inhalation LC50 for triforine in rats is greater than 4.5 mg/l air (Worthing, 1983). This compound is rapidly absorbed and metabolized by the rat (OHS Database, 1994). The acute oral LD50 for the formulated product Saprol in rats is 5,273 mg/kg body weight. The acute dermal LD50 for Saprol in rats is 4,186 mg/kg body weight. The acute inhalation LC50 for Saprol in rats is greater than 5,288 mg/m3. Saprol is considered an irritant to the skin. The acute oral LD50 for triforine in mice is greater than 6,000 mg/kg; and greater than 2,000 mg/kg in dogs. The acute dermal LD50 for rabbits is greater than 10,000 mg/kg body weight (Thomson, 1990).

    b.   Chronic Toxicity

In two-year feeding studies, the No-effect-level (NEL) for triforine in dogs was 100 mg/kg diet and 625 mg/kg diet for rats (Worthing, 1983).

Reproductive Effects

A decreased number of fetuses and an increased number of resorptions were observed in a study of pregnant rats fed triforine at a dietary level of 1,600 mg/kg (OHS Database, 1994). The formulated product Saprol does not affect reproduction and development. In another developmental study, rabbits were fed doses of 0, 5, 25 and 125 mg/kg/day of triforine. The maternal No-observable-effect-level (NOEL) was 5 mg/kg/day; the maternal Lowest-effectlevel (LEL) was 25 mg/kg/day, rabbits exhibited reduced food intake and loss of body weight. The fetotoxic NOEL was 5 mg/kg/day; the fetotoxic LEL was 25 mg/kg/day, decreased average relative weight was observed (U.S Environmental Protection Agency, 1993).

Teratogenic Effects

In a developmental study, rabbits were fed doses of 0, 5, 25 and 125 mg/kg/day of triforine. The teratogenic NOEL was greater than 125 mg/kg/day. The formulated product Saprol is not considered a teratogen (U.S Environmental Protection Agency, 1993).

Mutagenic Effects

The formulated product Saprol is not considered a mutagen.

Carcinogenic Effects

In short and long-term studies of the formulated product Saprol, no irreversible or carcinogenic effects were observed.

    c.   Ecological Effects

- *Effects on Birds*: The acute oral LD50 for triforine in bobwhite quail is greater than 5,000 mg/kg (Worthing, 1983). The formulated product Saprol is practically non-toxic to birds by acute oral exposure and only slightly toxic by dietary exposure. The acute oral LD50 for Saprol in bobwhite quail is greater than 5,000 mg/kg. The dietary LC50 for bobwhite quail is 1,850 ppm in the diet. Mallard ducks had a dietary LC50 of greater than 4,640 ppm in the diet.

- *Effects on Aquatic Organisms*: At 50 mg/l in water, there are no signs of poisoning in Lebistes reticulatus. Rainbow trout and bluegill sunfish tolerate 1,000 mg/l in water for 96 hours without symptoms (Royal Society of Chemistry, 1983). The 96- hour LC50 for rainbow trout and bluegill sunfish is greater than 1,000 mg/l (Worthing, 1983). The formulated product Saprol is of low hazard to fish and aquatic invertebrates. Both rainbow trout and bluegill sunfish had a 96-hour LC50 of greater than 500 mg/l. The aquatic invertebrate Daphnia (water flea) had a 48-hour EC50 of greater than 25 mg/l. Saprol was also noted to be of low hazard to Scenedesmus subspicatus (aquatic algae). The 96-hour EC50 was greater than 380 mg/l.

- *Effects on Other Animals (Nontarget species)*: No toxic effect was observed in honeybees at less than or equal to 1,000 mg/kg diet (Worthing, 1983). Triforine and the formulated product Saprol are considered of low hazard to honeybees and to the predatory mite Typhlodromus pyrii. It is also of low hazard to earthworms at recommended dose rates (Meister et al., 1994).

Aliphatic Aldehydes Fungicides

Several aliphatic aldehydes are used as fungicides, amongst them formaldehyde and formaldehyde releasers, which have been considered in detail elsewhere (Feinman, 1988).

Toxicology of Acrolein

Acrolein reacts with SH groups. It is formulated as a liquid (fig. 9).

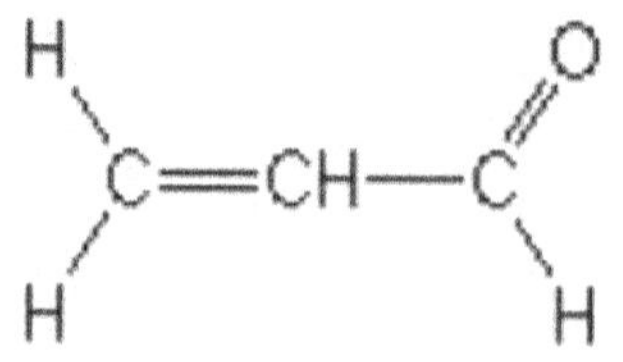

$$\begin{array}{c} H \\ \\ C=CH-C \\ \\ H \end{array} \begin{array}{c} O \\ \\ \\ H \end{array}$$

Figure 9. Acrolein.

Toxicology

a. Acute Toxicity

The rat oral LD50 is 29mg=kg and the mouse oral LD50 is 13.9mg=kg (males) and 17.7mg=kg (females). The percutaneous rabbit LD50 is 231mg=kg. By inhalation the rat 1-h LC50 is 65mg=m3 (males) and 60mg=m3 (females): the rat 4-h LC50 is 18.5mg=m3 (males) and 22mg=m3 (females). The rat 30-min LC50 is 131 ppm and the 10-min LC50 is 355 ppm (Ballantyne et al., 1989).

b. Short-term and Subchronic Toxicology

Hamsters, rats, and rabbits were exposed to acrolein vapor at 0, 0.4, 1.4, and 4.9 ppm for 6 h/day for 5 days/week for 13 weeks. At 4.9 ppm there was mortality, ocular and nasal irritation, depression of growth, and inflammation, necrosis, hyperplasia, and metaplasia of the respiratory tract epithelium. A no-effect concentration was not established for the rat (Feron et al., 1978).

c. Chronic Toxicology

Sprague-Dawley rats were given acrolein by gavage at daily dosages of 0, 0.05, 0.5, and 2.5mg/kg up to 102 weeks. The only effects noted were decreased serum creatinine kinase and increased early cumulative mortality. There were no significant increases in neoplastic or non-neoplastic histopathology (Parent et al., 1992a).

Carcinogenicity

Given by intraperitoneal injection to male Fischer 344 rats, Acrolein had an initiating activity for urinary bladder carcinogenesis (Cohen et al., 1992).

Reproductive Toxicology

Male and female rats were incubated and given 70 daily doses of Acrolein at 0, 1, 3, or 6mk=kg. The F0 generation was assigned to a 21-day period of co-habitation and dosing of females continued through co-habitation, gestation, and lactation. F1 generation pups were similarly treated. In general, reproductive indices were unaffected, with the exception of reduced pup weights in the F1 generation at the high dose. Gastric lesions were consistently found in the high dose and some mid-dose animals; erosions of the glandular mucosa and hyperplasia=hyperkeratosis of the fore stomach were the most frequent lesions. Relative to the controls, mortality and body weight gain decreases were noted for high dosage animals (Parent et al., 1992b).

Biological Fungicides (Biofungicides)

Biofungicides are microorganisms (microbial pesticides) and naturally occurring substances that control diseases (biochemical pesticides) that are approved for organic production. Biofungicides are widely used by organic vegetable growers to control selected foliar and soilborne diseases of vegetable crops (see Table 5). Biofungicides can be applied as a standalone treatment to control a target disease, provided the application is made before the disease starts (Francis and Keinath, 2010). In the case of a biofungicide, the Latin name of the microbe that it contains is the generic name of the fungicide.

Toxicology of EcoGuard

EcoGuard is a concentrated suspension of spores of Bacillus licheniformis SB3086 that has been found effective as a natural inhibitor of a variety of agronomically important fungal diseases - particularly dollar spot and anthracnose. EcoGuard allows you to control the disease and significantly improve overall turf quality at the same time. The activity of EcoGuard is due to the synthesis of powerful anti-fungal compounds that inhibit fungal growth. As a primary benefit, EcoGuard can be used just like other fungicides. As a secondary benefit, when EcoGuard is integrated with conventional turf management practices, you will see a noticeable and often dramatic improvement in the health and vigor of your turf. In addition, the turf will also recover more quickly from diseases with improved color and increased density in the damaged areas (Novozymes Biologicals Inc., 2007).

B. licheniformis SB3086 is a naturally occurring, ubiquitous bacterium originally isolated from United States farm soil. Consequently, the United States Environmental Protection Agency (USEPA) required limited data for federal registration of EcoGuardTM Biofungicide. The data from acute toxicity/pathogenicity studies on the active ingredient indicate that Bacillus licheniformis SB3086 is not toxic, infective or pathogenic via the oral or inhalation routes of exposure (tested at 1×10^8 Colony Forming Units (CFU) per animal), or via intravenous injection (tested at 1×10^7 CFU/animal). The end product was not very acutely toxic via oral, dermal, or inhalation routes of exposure. It was also not irritating to the eyes (tested on rabbits) or a dermal sensitizer (tested on guinea pigs), but was a slight dermal irritant (tested on rabbits).

Table 5. List of biofungicides (biological) used to control selected vegetable crop diseases (Francis and Keinath, 2010)

Product	Active Ingredient	Disease	Treatment Site
Ballad	*Bacillus pumilus*	Several (Foliar)	Foliar
Bio-Save	*Pseudomonas syringe*	Post-harvest	Irish and sweet potatoes in storage
Contans	*Coniothyrium minitans*	White Mold	Soil applied
Kodiak	*Bacillus subtilis*	*Pythium, Rhizoctonia, Fusarium*	Seed treatment, beans only
Mycostop	*Streptomyces griseoviridis*	Several	Greenhouse; Soil applied
Regalia	Plant extract	Powdery mildew	Foliar
RootShield Granules, RootShield WP	*Trichoderma harzianum*	*Pythium, Rhizoctonia, Fusarium*	Soil applied
Serenade	*Bacillus subtilis*	Powdery mildew, other foliar diseases	Foliar
T22-HC	*Trichoderma harzianum*	*Pythium, Rhizoctonia, Fusarium*	Soil applied
Surround	Kaolin	Powdery mildew	Foliar
Trilogy	Neem Oil	Powdery mildew	Foliar
Actinovate AG	*Streptomyces lydicus*	*Fusarium, Rhizoctonia, Pythium, Phytophthora*	Soil applied Foliar
SoilGard	*Gliocladium virens*	Damping off	Greenhouse-transplants, soil applied

The USEPA waived the requirement for subchronic, chronic, developmental, reproductive toxicity, genotoxicity and oncogenicity studies for federal registration of EcoGuardTM Biofungicide. Instead, the USEPA used reports from the scientific literature to evaluate this product. The data from these reports suggest that B. licheniformis is occasionally associated with infections in individuals who have significant preexisting health problems such as severe immune system depression, cancer or trauma. In addition, it has been associated with reproductive failures (spontaneous abortions and inflammation of the placenta) in cattle, sheep and swine, usually in association with the ingestion of moldy hay. A search of the toxicological literature did not find any additional significant information on B. licheniformis. A quantitative worker risk assessment was not provided in the registration package. However, B. licheniformis has been used in various industrial fermentation processes for a number of years and, according to the registrant, no pathogenicity, toxicity or hypersensitivity has been reported among these workers. Given the use pattern of the EcoGuardTM product, exposure of applicators would likely be less than that of fermentation workers. Also, the product label requires applicators and other handlers to wear a long-sleeved shirt, long pants, and shoes plus socks. In addition, handlers must also use a non-powered air purifying NIOSH approved respirator.

The limited toxicity data required to support the federal registration of EcoGuardTM Biofungicide indicate that this product is not very toxic following acute exposures. The active ingredient B. licheniformis also appears to have a low degree of pathogenicity and infectivity to animals and humans. Although there are no animal study data on longer-term exposure, significant risks to workers or the general public from EcoGuardTM use are not expected given the use pattern and the required personal protective equipment. The required personal protective equipment also should protect against the slight dermal irritation that EcoGuardTM may cause.

Ecological Risk

B. licheniformis SB3086 is not toxic, infective, or pathogenic to mammals when administered orally, by inhalation, direct tracheal injection, dermally, and by intravenous injection. A significant decrease in weight gain in young mallard ducks was observed when administered EcoGuardTM formulation at a rate of approximately 1.0 ml/kg body weight. The effect was attributed to formulation ingredients other than the active ingredients. B. licheniformis SB3086 appears not to be toxic, infective, or pathogenic to mallards; there were no mortalities.

The 30-day EcoGuardTM rainbow trout $LC50$ is greater than 1.1×10^6 CFU/ml which is roughly I I 7X (times) the expected environmental concentration (EEC) when the maximum application rate is applied directly to the surface of six-inch deep water body. The daphnia No Observable Adverse Effect

Concentration (NOAEC) is 120X the EEC. There was no sign of infection or pathogenicity in either study.

The EcoGuardTM formulation had no effect on honeybee larva when exposed to B. licheniformis SB3086 at 1.6 x106 CFU/ml, roughly 2/3 full strength formulation, in their diet. No adverse behavioral or developmental abnormalities were observed in emerged adult honeybees that had been exposed as larva. All marine/estuarine organism and nontarget plant testing was waived by the USEPA.

Bacillus licheniformis is a ubiquitous soil organism. While numbers of B. licheniformis in soil are unknown and likely vary from soil to soil, the total number of Bacillus organisms is estimated to be 10' CFU/g soil. The added soil density of B. licheniformis from the proposed use rates would be 0.42% to 1.5%. This is a very small proportion of the naturally occurring bacilli in soil and is not expected to add substantially to the effects of the normally occurring Bacillus populations. No adverse effects to fish or wildlife resources are expected from use of EcoGuardTM Biofungicide when used as labeled. There are reports in the literature of B. licheniformis being a sporadic mammal pathogen. B. licheniformis diseases appear to be limited to cows, sheep, and swine as very unusual events associated with the ingestion of moldy hay. Wild ruminants exposure to the combination of conditions seemingly implicated in livestock disease should be minimal. The IBA contained in EcoGuardTM poses no risk to fish or wildlife resources: The IBA concentration in EcoGuardTM is very low, application at the maximum label rate results in an IBA application rate of roughly 30 milligrams/acre (Serafini, 2003).

CONCLUSION

It is to note that this study showed a less or more toxicity of all category of chemical fungicides, contrarily, to biofungicides that showed a little or note side effects on human and ecosystems.

REFERENCES

1. Agrow, (1992). More problems for Benlate? 13 March 1992, 13 p.
2. Ballantyne, B.; Dodd, D.E.; Pritts, I.M.; Nachreiner, D.J. & Fowler, E.H. (1989). Acute vapour inhalation toxicity of acrolein and its influence as a trace contaminant in 2- methoxy-3,4-dihydro-2H-pyran. Hum. Toxicol., Vol., 8: 229–235.
3. Blondell, J. (1997). Epidemiology of pesticide poisoning in the United States, with special reference to occupational cases. Occup. Med., State of the Art Rev., Vol., 12:209–220.

4. Brooks, G.T. & Roberts, T.R. (1999). Pesticide Chemistry and Bioscience. Published by the Royal Society of Chemistry.
5. Burpee, L. (2006). Integrated disease management, an introduction to Fungicides. Courses support.
6. Cohen, S.M.; Garland, E.M.; St John, M.; Okamura, T. & Smith, R.A. (1992). Acrolein initiates rat urinary bladder carcinogenesis. Cancer Res.,Vol., 52: 3577–3581.
7. Costa, L.G. (1997). Basic toxicology of pesticides. Occup. Med. State of the Art Rev. Vol., 12:251–268.
8. Edwards, I.R., Ferry, D.G. & Temple, W.A. (1991). Fungicides & related compounds, In: Handbook of Pesticide Toxicology. Hayes, W.J. & Laws, E.R., Eds. Academic Press, New York, NY,. Vol., 3, pp: 1409–1470
9. Envirofacts, (2005). CYPROCONAZOLE The Active Ingredient in Alto® One of the Active Ingredients in Quadris Xtra®. Syngenta Crop Protection Inc., Greensboro, NC 27419-8300.
10. Extoxnet, (1994). Pesticide Management Education Program, Cornell University, NY.
11. FAO/WHO. (1968). Evaluations of some pesticide residues in food. FAO/PL: 1967/M/11/1; WHO/Food Add./68.30.
12. Feinman, S.E. (1988). Formaldehyde Sensitivity and Toxicity. CRC Press, Boca Raton, FL.
13. Feron, V.J.; Kruysse, A.; Til, H.P. & Immel, H.R. (1978). Repeated exposure to Acrolein vapour: sub-acute studies in hamsters, rats and rabbits. Toxicology, Vol., 9: 47–57.
14. Fishel, F.M. (2005). Pesticide Toxicity Profile: Triazole Pesticides. University of Florida, IFAS extension. PI68.
15. Foster, and Smith, (2010). Germs: Viruses, Bacteria, and Fungi. Veterinary & Aquatic Services Department Foster & Smith, Inc. Wisconsin. USA.
16. Francis, R. and Keinath, A. (2010). Biofungicides and chemicals for managing diseases in organic vegetable production. CLEMSON Cooperative extension. Information leaflet 88.
17. Harding, W.C. (1979-80). Pesticide profiles, part two: fungicides and nematicides. Univ. Maryland, Coop. Ext. Service Bull. 283, 22 pp.
18. Hayes, W.J and Vaughn, W.K. (1977). Mortality from pesticides in the United States from 1973 to 1976. Toxicol. Appl. Pharmacol. Vol., 42: 235–252.
19. Hill, E.F. & Camardese, M.B. (1986). Lethal Dietary Toxicities of Environmental Contaminants to Coturnix, Technical Report Number 2.

U.S. Department of Interior, Fish and Wildlife Service, Washington, DC., 4-37

20. Howard, P. H., (1989). Handbook of Environmental Fate and Exposure Data for Organic Chemicals: Pesticides. Lewis Publishers, Chelsea, MI., 4-20

21. Hrelia, R. (1996). The genetic and non-genetic toxicity of the fungicide. Vinclozolin Mutagenesis. Vol., 11: 445-453.

22. Hudson, R.H.; Tucker, R.K. & Haegele, M.A. (1984). Handbook of Acute Toxicity of Pesticides to Wildlife, Resource Publication 153. U.S. Department of Interior, Fish and Wildlife Service, Washington, DC., 4-15

23. Kidd, H. and James, D.R., (1991). The Agrochemicals Handbook, Third Edition. Royal Society of Chemistry Information Services, Cambridge, UK.

24. Latijnhouwers, M.; de Wit, P.J. and Govers, F. (2000). Oomycetes and fungi: similar weaponry to attack plants. Trends in Microbiology. Vol., 11: 462-469.

25. Litovitz, T.A.; Felberg, L. & Soloway, R.A. (1994). Annual reports of the American Association of Poison Control Centers. Toxic exposure surveillance system. Am. J. Emerg. Med. Vol., 13: 551–597.

26. MAFF, (1992). Benomyl evaluation No. 57, July 1992, pp: 91-111.

27. Marrs, C.T. & Ballantyne, B. (2004). Pesticides toxicology and international regulation. John Wiley and Sons, Ltd. 554 pages.

28. Mayer, F. L. and Ellersieck, M. R. (1986). Manual of Acute Toxicity: Interpretation and Data Base for 410 Chemicals and 66 Species of Freshwater Animals. Resource Publication 160. U.S. Department of Interior, Fish and Wildlife Service, Washington, DC,. 4-18

29. Meister, R.T.; Berg, G.L.; Sine, C.; Meister, S. & Poplyk, J. (1994). Farm Chemicals Handbook, 70th Eds. Eds. Meister Publishing Co., Willoughby, OH.

30. NAS, (1987). Regulating Pesticides in Food. The Delaney Paradox. Report of Committee on Scientific and Regulatory Issues. Unlikely Pesticide Use Patterns. National Academy of Sciences, National Academy Press, Washington, DC

31. Novozymes Biologicals, Inc. (2007). Roots EcoGuard biofungicide. www.novozymes.com/roots.

32. Office of Pesticide Programs, (1996). List of Chemicals Evaluated for Carcinogenic Potential, US EPA, Washington, US.

33. OHS Database. 1994. Occupational Health Services, Inc. 1994. MSDS for Vernolate. OHS Inc., Secaucus, NJ.
34. Parent, R.A.; Caravello, H.E. and Hiberman, A.H. (1992b). Reproduction study of Acrolein on two generations of rats. Fund. Appl. Toxicol., Vol., 29: 228–237.
35. Parent, R.A.; Caravello, H.E. and Long, J.E. (1992a). Two-year toxicology and carcinogenicity study of acrolein in rats. J. Appl. Toxicol., Vol., 12: 131–139.
36. Phillips, S.D. (2001). Fungicides and biocides. In: Clinical Environmental Health and Toxic Exposures, Sullivan, J.B. & Krieger, G.R., Eds. Lippincott Williams and Wilkins, Philadelphia, 2nd Eds. pp:1109–1125.
37. Rouabhi, R.; Djebar, H. & Djebar, M.R. (2009). Toxic Effects of Combined Molecule from Novaluron and Diflubenzuron on Paramecium caudatum. Am-Euras. J. Toxicol. Sci. Vol., 1(2). September 2009. 74-80. ISSN: 2079-2050; EISSN: 2079-2069.
38. Royal Society of Chemistry (1983). The Agrochemicals Handbook. The University, Nottingham, England.
39. Serafini, P.M. (2003). Bacillus licheniformis SB3086 NYS DEC Letter - New Active Ingredient Registration 8/03. New York State Department of Environmental Conservation.
40. Thomson, W.T. (1990) Agricultural Chemicals, Book IV: Fungicides. Thomson Publications, Fresno, CA.
41. Tomlin, C., (1994). The Pesticide Manual, 10th Edition, British Crop Protection Council/Royal Society of Medicine.
42. U.S. Environmental Protection Agency, (1993). Office of Pesticides. TOX Oneliners - Triforine. April, 1993.
43. U.S. National Library of Medicine, (1995). Hazardous Substances Data Bank. Bethesda, MD., 4-5
44. Whitehead, R. (1996). The UK Pesticide Guide, British Crop Protection Council/CAB International.
45. World Health Organization, (1993). Benomyl, Environmental Health Criteria No 148, Geneva, Switzerland.
46. World Health Organization, (1994). Data sheet on benomyl, WHO/PCS/94.87, Geneva.
47. Worthing, C.R. (1979). The Pesticide Manual: A World Compendium, 6th Eds. The British Crop Protection Council, Croydon, England. 655 pp.
48. Worthing, C.R. (1983). The Pesticide Manual: A World Compendium. Seventh edition. Published by The British Crop Protection Council.

Chapter 8

GENETIC TOXICOLOGICAL PROFILE OF CARBOFURAN AND PIRIMICARB CARBAMIC INSECTICIDES

Sonia Soloneski and Marcelo L. Larramendy

Faculty of Natural Sciences and Museum, National University of La Plata, Argentina

INTRODUCTION

It's well known that the pesticide usages in agriculture have led increase in food production worldwide. Although the benefits of conventional agricultural practices have been immense, they utilize levels of pesticides and fertilizers that can result in a negative impact on the environment (WHO, 1988). Only for the 2006-2007, the total world pesticide amount employed was approximately 5.2 billion pounds (www.epa.gov). Their application is still the most effective and accepted method for the plant and animal protection from a large number of pests, being the environment consequently and inevitably exposed to these chemicals. Herbicides accounted for the largest portion of total use, followed by other pesticides, like insecticides and fungicides (www.epa.gov). The goal in pesticide investigation and development is identifying the specificity of action of a pesticide toward the organisms it is supposed to kill (Cantelli-Forti et al., 1993). Only the target organisms should be affected by the application of the product. However, because pesticides are designed and selected for their biological activity, toxicity on non-target organisms frequently remains a significant potential risk (Cantelli-Forti et al., 1993). The benefits in using pesticides must be weighed against their deleterious effects on human health, biological interactions with non-target organisms, pesticide resistance and/or accumulation of these chemicals in the environment (WHO, 1988). Pesticides are high volume, widely used environmental chemicals and there is continuous debate concerning their probable role in both acute and chronic human health effects (Cantelli-Forti et al., 1993; Hodgson & Levi, 1996). Among the potential risk effects of agricultural chemicals, carcinogenesis is of special concern. The genetic toxicities of pesticides have been determined by numerous factors like their biological accumulation or degradation in the environment, their metabolism in humans, and their action in cellular components such as DNA, RNA and proteins (Shirasu, 1975). It seems essential the determination of the genotoxic risks of these pesticides before they are used in agriculture. Therefore, the carcinogenic and mutagenic potential of a large amount of pesticides has been the object of an extensive and wide investigation (WHO, 1990). These results have great predictive value for the carcinogenicity of several pesticides (IARC, 1987). The International Agency for Research on Cancer (IARC) has reviewed the potential carcinogenicity of a wide range of insecticides, fungicides, herbicides and other similar compounds. Fifty-six pesticides have been classified with carcinogenic potential in different laboratory animals (IARC, 2003). Among them, and as a brief example, chemicals compounds as phenoxy acid herbicides, 2,4,5-trichlorophenoxyacetic acid (2,4,5-T), lindane, methoxychlor, toxaphene, and some organophosphates have been reported with a carcinogenic potential in human studies (IARC, 2003).

Numerous well known pesticides have been tested in a wide variety of mutagenicity as well as DNA, chromosomal, and cellular damage endpoints (IARC, 2003). Several investigations have been reported positive associations between exposure and pesticide risk (Shirasu, 1975; Bolognesi et al., 1993, 2009, 2011; Pavanello & Clonfero, 2000; Bolognesi, 2003; Clark & Snedeker, 2005; Castillo-Cadena et al., 2006).

CARBAMIC INSECTICIDES

The carbamates are chemicals mainly used in agriculture as insecticides, fungicides, herbicides, nematocides, and/or sprout inhibitors (IARC, 1976). These chemicals are part of the large group of synthetic pesticides that have been developed, produced, and used on a large scale within the last 50 years. Additionally, they are used as biocides for industrial or other applications as household products including gardens and homes (IARC, 1976).

During the last decades, considerable amounts of pesticides belonging to the class of carbamates have been released into the environment. Humans may be exposed to carbamates through food and drinking water around residences, schools, and commercial buildings, among others (IARC, 1976). Consequently, carbamates are potentially harmful to the health of different kinds of organisms (EPA, 2004). Among all classes of pesticides, carbamates are most commonly used compounds because organophosphates and organochlorines are extremely toxic and possess delayed neurotoxic effects (Hour et al., 1998). They share with organophosphates the ability to inhibit cholinesterase enzymes and therefore share similar symptomatology throughout acute and chronic exposures. Likewise, exposure can occur by several routes in the same individual due to multiple uses, and there is likely to be additive toxicity with simultaneous exposure to organophosphates (IARC, 1976).

The N-methyl carbamates are a group of closely related pesticides employed in homes, gardens and agriculture that may affect the functioning of the nervous system (EPA, 2007). Toxicological characteristics of the N-methyl carbamates involve maximal cholinesterase enzyme inhibition followed by a rapid recovery, typically from minutes to hours (EPA, 2007). Several compounds namely aldicarb, carbaryl, carbofuran, formetanate HCl, methiocarb, methomyl, oxamyl, pirimicarb, propoxur, and thiodicarb are included as members of the N-methyl carbamate class (EPA, 2007).

CARBOFURAN. GENOTOXICITY AND CYTOTOXICITY PROFILES

Carbofuran (2,3-dihydro-2,2-dimethylbenzofuran-7-yl methylcarbamate; CASRN: 1563-66-2) is one of the most widely granular employed N-methyl carbamate esters with both contact and systemic activity. Carbofuran is a derivative of carbamic acid being its chemical structure formula shown in Fig. 1.

Carbofuran is a relatively unstable compound that breaks down in the environment within weeks or months (www.inchem.org). It is registered on a variety of agricultural uses to control soil-dwelling and foliar-feeding insects, mites and nematodes on a variety of field, fruit, forage, grain, seed, and fiber crops (EPA, 2006). Carbofuran is a systemic, broad spectrum insecticide and nematocide registered N-methyl carbamate for control of soil and foliar pests. It has been reported for 2006 that nearly one million pounds of carbofuran was applied worldwide (EPA, 2006). The most sensitive and appropriate effect associated with the use of carbofuran is its toxicity following acute exposure (HSDB, 2011). On the basis of its acute toxicity, it has been classified as a highly hazardous member (class Ib) by WHO (2009) and highly toxic compound (category I) by EPA (2006) based on its potency by the oral and inhalation exposure routes. In spite of the recommendation and regulation proposed by the United States Environmental Protection Agency (EPA) concerning the use of this carbamate within the United States of America, its application has been recently cancelled all over the Northern country by the same organization since 2009 (www.epa.gov). However, the contamination of environment with this compound can by far occur, particularly taking into consideration those countries where it is still in use and the probability of long-term low dose exposure becomes increased. Due to its extensive employment in agriculture and household, contamination of food, water and air has become serious and undesirable health problem for humans, animals and wildlife. Large quantities of this carbamate are particularly applied to different environments worldwide.

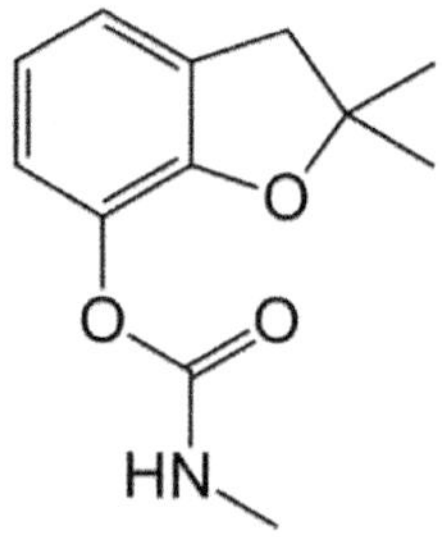

Figure 1. Chemical structure of carborufan. Source: INCHEM (www.inchem.org).

Metabolism of carbofuran has been extensively studied in plants and animals (Dorough & Casida, 1964; Metcalf et al., 1968). In mammals, it reversibly inhibits acetylcholinesterase by carbamylation as well as others non-specific serine-containing enzymes, such as carboxylesterases and butyrylcholinesterases (Gupta, 1994). This results in accumulation of acetylcholine at nerve synapses and myoneural junctions leading to cholinergic signs and causing toxic effects (Karczmar, 1998). Epidemiological studies suggested that exposure to carbofuran may be associated with increased risk of gastrointestinal, neurological, cardiac dysfunction, and retinal degeneration (Cole et al., 1998; Kamel et al., 2000; Peter & Cherian, 2000). Carbofuran represents an acute poison when absorbed into the gastrointestinal tract by inhalation of dust and spray mist and minimally poison thought the intact skin contact (Gupta, 1994). In summary, carbofuran is reported to be teratogenic, embryotoxic and highly toxic to mammals (Gupta, 1994; WHO-FAO, 2004, 2009; WHO, 2009).

Genotoxicity and cytotoxicity studies have been conducted with this N-methyl carbamate member using several end-points on different cellular systems. A summary of the results reported so far is presented in Table 1.

The compound produced both conflicting and inconclusive results in mutagenicity tests varying according to either the end-point assessed (WHO, 1988, 2000-2002, 2009; WHOFAO, 2004, 2009). When mutagenic activity was assessed in bacterial systems either positive or negative results have been reported. Carbofuran has been found to be non-mutagenic in Salmonella typhimurium since negative or weak positive response were observed in the number of mitotic recombinants regardless of the presence or absence of a rat liver metabolic activation system (Blevins et al., 1977a; Gentile et al., 1982; Waters et al., 1982; Haworth & Lawlor, 1983; Hour et al., 1998; Yoon et al.,

2001). These results indicate that carbofuran cannot be considered mutagenic in bacterial systems. However, it was active in Salmonella typhimurium TA1538 and TA98 strains in the presence or absence of S9 metabolic system (Gentile et al., 1982; Moriya et al., 1983; Hour et al., 1998). Whereas the insecticide did not induce reverse mutations in Escherichia coli (Simmon, 1979), it has been claimed as a relatively weaker mutagen with the repair defective Ames Escherichia coli K-12 test (Saxena et al., 1997). Similarly, positive results have been found after exposure in Vibrio fischeri regardless the absence or presence of S9 metabolic system (Canna-Michaelidou & Nicolaou, 1996). When DNA damage and repair assays were performed, carbofuran was also negative in both Escherichia coli and Bacillus subtilis bacterial systems (SRI, 1979). Similar negative results were also found after carbofuran exposure in Saccharomyces cerevisiae mitotic recombination assay (Simmon, 1979).

The mammalian in vitro gene mutation assay systems generated results consistent with the microbial gene mutation assays, although they were generally more responsive. When a mammalian cell system was employed for mutagenic screening, carbofuran was found to be positive in V79 cells (Wojciechowski et al., 1982). Similar results were reported for the cell mutation assay in mouse lymphoma L5178 cells (Kirby, 1983a, b). Unscheduled DNA synthesis was monitored in human fibroblasts and primary rat hepatocytes following treatment with the insecticide with and without S9 fraction. Both negative and positive results were obtained for the same endpoint in human primary fibroblasts regardless of the presence or absence of a rat liver metabolic activation system (Simmon, 1979; Gentile et al., 1982) but negative results were obtained in primary rat hepatocyte cultures (SRI, 1979). Single-strand breaks detected by alkaline comet assay were induced in in vitro human peripheral lymphocytes (Das et al., 2003; Naravaneni & Jamil, 2005). The induction of DNA fragmentation on human skin fibroblasts have been found to be enhanced after in vitro carbofuran treatment (Blevins et al., 1977b).

As opposed to mutation assays that detect specific gene defects, the chromosomal assays evaluate the structure of the whole chromosome. Five studies of carbofuran have evaluated the induction of sister chromatid exchanges in mammalian cell cultures. In one of the first studies, carbofuran was negative in Chinese hamster ovary cells regardless of the presence or absence of S9 fraction (Thilagar, 1983b). However, other authors reported positive results for the same cell system (Gentile et al., 1982; Thilagar, 1983c; Lin et al., 2007; Soloneski et al., 2008) as well as human lymphocytes (Georgian et al., 1985). Similarly, the effects on chromosomal structure following exposure to carbofuran were investigated in Chinese hamster ovary and primary human lymphocytes cells. While carbofuran did not induce in vitro chromosome damage in Chinese hamster ovary cells with or without metabolic system activation (Thilagar, 1983a), positive results were reported to occur not only in the same cellular system (Lin et al., 2007) but also in human lymphocytes in vitro (Pilinskaia & Stepanova, 1984; Das et al., 2003). However, inconclusive response for this endpoint has been also reported to occur in the latter system after carbofuran exposure (Naravaneni & Jamil, 2005). Positive results have been also reported for the ability of carbofuran to induce micronuclei in both Chinese hamster ovary cells and human lymphocytes in vitro with and without S9 metabolic fraction (Soloneski et al., 2008; Mladinic et al., 2009).

Several assays have been developed to assess the ability of carbofuran to cause cytotoxic effects on different cellular systems. Negative response was observed in both Escherichia coli and Bacillus subtilis bacterial systems (Simmon, 1979). When the analysis of cell-cycle progression on mammalian cells was studied, carbofuran gave negative results in Chinese hamster ovary cells and lung fibroblasts (Yoon et al., 2001; Lin et al., 2007).

Table 1. Evaluation of carbofuran-induced genotoxicity and cytotoxicity on different target systems. a, expressed as reported by authors; b, exposed to pesticide mixture containing carbofuran; * , from agricultural workers occupationally exposed to carbofuran.

End-point/Test System	Concentration[a]	Results	References
In vitro assays			
Ames test			
Salmonella typhimurium, S9 +/-	100 – 10 000 μg/plate	+/-	Blevins et al., 1977a; Waters et al., 1982; Haworth & Lawlor, 1983; Yoon et al., 2001
Salmonella lactam assay, S9 +/-	1 – 10 000 μg/plate	-	Hour et al., 1998
	0.1 – 100 μg/plate	+	Gentile et al., 1982; Moriya et al., 1983; Hour et al., 1998
Pol A reverse mutation			
Escherichia coli (WP$_2$), S9 +/-	1 – 5 000 μg/plate	-	Simmon, 1979
Escherichia coli (K-12)	1 – 5 000 μg/plate	+	Saxena et al., 1997
Mutatox test			
Vibrio fischeri (M169), S9 +/-	175 μg/plate	+	Canna-Michaelidou & Nicolaou, 1996
DNA damage and repair			
Escherichia coli (W3110-p3478)	0 – 5 mg/6-mm disk	-	SRI, 1979
Bacillus subtilis (H17-M45)	0 – 5 mg/6-mm disk	-	SRI, 1979
Mitotic recombination			
Saccharomyces cerevisiae (D3), S9 +/-	1 – 50 mg/ml	-	Simmon, 1979
Gene mutation assay			
V79 cells	NA	+	Wojciechowski et al., 1982

Cell mutation tk locus			
Mouse lymphoma L5178 Y cells, S9 +/-	16 – 1 780 µg/ml	+/-	Kirby, 1983a, b
UDS	0.1 – 1 000 µg/ml	-	Simmon, 1979
Human fibroblasts (WI-38), S9 +/-	0.1 – 1 000 µg/ml	+	Gentile et al., 1982
Human lung fibroblasts	0 – 100 µg/ml	-	SRI, 1979
Primary rat hepatocytes			
Alkaline comet assay			
HL	NA	+	Naravaneni & Jamil, 2005
	0.5 – 4.0 µM	+	Das et al., 2003
DNA fragmentation analysis			
Human skin fibroblasts	NA	+	Blevins et al., 1977b
SCE assay			
CHO cells, S9 +/-	12.5 – 312.5 µg/ml	-	Thilagar, 1983b
CHO cells, S9 +/-	12.5 – 2 500 µg/ml	+	Thilagar, 1983c
CHO-K1 cells	5 – 100 µg/ml	+	Gentile et al., 1982; Soloneski et al., 2008
CHO-W8 cells	0.04 – 0.32 µg/ml	+	Lin et al., 2007
HL	NA	+	Georgian et al., 1985
Chromosomal aberrations			
CHO cells, S9 +/-	50 – 2 500 µg/ml	-	Thilagar, 1983a
CHO-W8 cells	0.04 – 0.32 µg/ml	+	Lin et al., 2007
HL	NA	+/-	Naravaneni & Jamil, 2005
HL	100 – 300 µg/ml	+	Pilinskaia & Stepanova, 1984
HL[b]	NA	+	Das et al., 2003
Micronuclei assay			
CHO-K1 cells	10 – 100 µg/ml	+	Soloneski et al., 2008
HL, S9 +/-	0.008 µg/ml	+	Mladinic et al., 2009
Growth inhibition			
Escherichia coli	1 – 500 mg/ml	-	Simmon, 1979
Bacillus subtilis	1 – 500 mg/ml	-	Simmon, 1979
Alteration in CCP			
CHO-W8 cells	0.04 – 0.32 µg/ml	-	Lin et al., 2007
CHO-K1 cells	50 – 100 µg/ml	+	Soloneski et al., 2008
CHL cells	30 µM	-	Yoon et al., 2001
Brain tubulin assembly assay			
Porcine cells	100 – 2 000 µmol/l	+	Stehrer-Schmid & Wolf, 1995
Cell viability			
CHL cells	30 µM	-	Yoon et al., 2001
CHO-K1 cells	50 – 100 µg/ml	+	Soloneski et al., 2008

Apoptosis			
CHL cells	30 μM	-	Yoon et al., 2001
Mouse brain microvascular endothelial cells	3 - 30 μM	-	Jung et al., 2003
Rat cortical cells	500 μM	+	Kim et al., 2004
In vivo assays			
Reverse mutation			
Zea mays	NA	-	Gentile et al., 1982
Sex-linked recessive lethal test			
Drosophila melanogaster	0 - 10 ppm	-	DeGraff, 1983; Gee, 1983
Dominant-lethal mutagenicity			
Mice	0.025 - 0.5 mg/Kg/day	-	FMC, 1971
UDS			
Rat hepatocytes	5 - 10 ppm	-	Valencia, 1981; 1983
Alkaline comet assay			
Mouse peripheral lymphocytes	0.1 - 0.4 mg/Kg bw	-	Zhou et al., 2005
HL*	NA	+	Castillo-Cadena et al., 2006
SCE			
Mouse peripheral lymphocytes	NA	+	Gentile et al., 1982
Rat	NA	+	Aly, 1998
Chromosomal aberrations			
Allium cepa	20 - 80 ppm	+	Saxena et al., 2010
Allium sativum	20 - 80 ppm	+	Saxena et al., 2010
Drosophila melanogaster	NA	-	Woodruff et al., 1983
Mouse bone marrow cells	3.8 - 1.9 (for 4 days) mg/Kg bw	+	Chauhan et al., 2000
Mouse bone marrow cells	0.1 - 1.0 mg/Kg bw	-	Pilinskaia & Stepanova, 1984
Rat bone marrow cells	0.6 - 10 mg/Kg bw	-	Putman, 1983b, a
HL[b]	NA	+	Zeljezic et al., 2009
Micronuclei			
Mouse peripheral lymphocytes	0.1 - 0.4 mg/Kg bw	-	Zhou et al., 2005
Mouse bone marrow cells	5.7 - 1.9 (for 4 days)	-	Chauhan et al., 2000
Alteration in CCP			
Allium cepa	20 - 80 ppm	+	Saxena et al., 2010

On the other hand, Soloneski and co-workers (2008) reported a delay in the cell-cycle progression of Chinese hamster ovary cells after the insecticide treatment. Carbofuran was tested in vitro in the porcine brain tubulin assembly assay for detecting whether the chemical can be considered as a microtubule poison and an aneuploidy agent. A dose-dependent reduction in the degree of polymerization of tubulins was reported in porcine cells after in vitro treatment (StehrerSchmid & Wolf, 1995). Controversial results were reported for the cell viability assay in mammalian cells, e.g., Chinese hamster lung and ovary cells after the exposure (Yoon et al., 2001; Soloneski et al., 2008). Finally, whereas carbofuran-induced apoptosis has been reported in rat cortical cells (Kim et al., 2004), negative results have been also observed in mouse brain microvascular entothelial cells and Chinese lung fibroblasts (Yoon et al., 2001; Jung et al., 2003). Similar end-points for both genotoxicity and cytotoxicity were also applied in in vivo systems. Carbofuran has been reported as a non inducer agent of mutations in plants cells, at least in Zea mays (Gentile et al., 1982), in the Drosophila melanogaster sex-linked recessive lethal test (DeGraff, 1983; Gee, 1983), and in the mice dominant-lethal mutagenicity test (FMC, 1971). Negative results have been obtained for the induction of unscheduled DNA synthesis in primary rat hepatocytes (Valencia, 1981, 1983). Controversial observations have been reported for the induction of DNA single-strand breaks assayed by the alkaline comet assay. Positive results were reported in circulating erythrocytes from occupationally exposed workers (Castillo-Cadena et al., 2006) whereas no induction was observed in mouse peripheral lymphocytes exposed in vivo (Zhou et al., 2005). It should be noted that the former positive results could not be totally committed to carbofuran but to other pesticides, since the cohort of donors included in the study was exposed to a panel of other pesticides. Several reports were able to revealed that carbofuran increased the frequency of sister chromatid exchanges in mammalian cells from mouse and rats exposed in vivo (Gentile et al., 1982; Aly, 1998), and chromosomal aberrations in plants from Allium (Saxena et al., 2010), and mammals including occupationally exposed workers (Putman, 1983a, b; Pilinskaia & Stepanova, 1984; Chauhan et al., 2000; Zeljezic et al., 2009) but not in insects (Woodruff et al., 1983) as well as in rodent cells (Putman, 1983a, b; Pilinskaia & Stepanova, 1984). When the micronuclei induction end-point was employed in mouse, no induction was found either in bone marrow cells (Chauhan et al., 2000) or circulating lymphocytes (Zhou et al., 2005). Finally, alterations in the progression of the cell-cycle were reported to occur after carbofuran exposure in plants when the Allium cepa model was employed (Saxena et al., 2010).

PIRIMICARB. GENOTOXICITY AND CYTOTOXICITY PROFILES

Pirimicarb (2-dimethylamino-5,6-dimethylpyrimidin-4-yl dimethylcarbamate, CASRN: 23103-98-2) is a dimethylcarbamate insecticide member with both contact and systemic activity. Similar to carbofuran, pirimicarb is a derivative of carbamic acid being its chemical structure formula shown in Fig. 2.

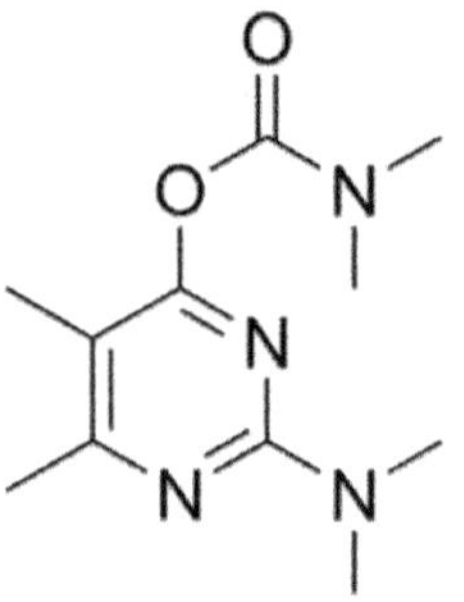

Figure 2. Chemical structure of pirimicarb. Source: INCHEM (www.inchem.org).

Based on its acute toxicity, pirimicarb has been classified as a moderately hazardous compound (class II) by WHO (http://www.who.int/ipcs/publications/pesticides hazard/en/) and slightly to moderately toxic (category II-III) by EPA (1974a). Among carbamate pesticides, pirimicarb is registered as a fast-acting selective aphicide mostly used in a broad range of crops, including cereals, sugar beet, potatoes, fruit, and vegetables, and is relatively non-toxic to beneficial predators, parasites, and bees (WHO-FAO, 2004, 2009). It acts by contact, translaminar, vapor, and systemic action. Its mode of action is inhibiting acetylcholinesterase activity (WHO-FAO, 2004, 2009).

Available information on the genotoxic and cytotoxic properties of pirimicarb is limited and inconsistent. Only few data are available in the literature (WHO-FAO, 2004, 2009). Genotoxicity and cytotoxicity studies have been conducted with this carbamate using several end-points on different cellular systems. A summary of the results reported so far is presented in Table 2.

Pirimicarb has been generally recognized as non-genotoxic in bacteria, yeast and fungi as well as in mammalian cells (EPA, 1974b). It has been reported to be non-mutagenic in Salmonella typhimurium when the Ames reversion mutagenicity test for the TA1535, TA1538, TA98, and TA100 strains after S9 metabolic activation has been used (Trueman, 1980; Callander, 1995). Furthermore, similar situation was observed in both Escherichia coli and Aspergillus nidulans when the reverse mutation assay or recessive lethal gene mutation test were respectively applied (Käfer et al., 1982; Callander, 1995).

Table 2. Evaluation of Pirimicarb-induced genotoxicity and cytotoxicity on different target systems. [a], expressed as reported by authors; UDS, unscheduled DNA synthesis; SCE, sister chromatid exchange; HL, human lymphocytes; CCP, cell-cycle proliferation; NA, data not available.

End-point/System	Concentration[a]	Results	References
In vitro assays			
Ames test			
Salmonella typhimurium, S9 +/-	2 500 µg/plate	-	Trueman, 1980
	5 000 µg/plate	-	Callander, 1995
Pol A reverse mutation			
Escherichia coli	5 000 µg/plate	-	Callander, 1995
Recessive lethal gene mutation			
Aspergillus nidulans	NA	-	Käfer et al., 1982
Cell mutation tk locus			
Mouse lymphoma L5178 Y cells	1 400 mg/ml - S9	-	Clay, 1996
	100 mg/ml + S9	+	Clay, 1996
Alkaline comet			
HL	50 - 500 µg/ml	+	Ündeger & Basaran, 2005
SCE assay			
CHO-K1 cells	100 - 200 µg/ml	+	Soloneski & Larramendy, 2010
Chromosomal aberrations			
CHO-K1 cells	10 - 300 µg/ml	+	Soloneski & Larramendy, 2010
HL, S9 +/-	500 µg/ml	-	Wildgoose et al., 1987
Alteration in CCP			
CHO-K1 cells	100 - 300 µg/ml	+	Soloneski & Larramendy, 2010
In vivo assays			
Eye mosaic system w/w+			
Drosophila melanogaster	NA	+	Aguirrezabalaga et al., 1994
Dominant lethal mutation			
Mice	20 mg/Kg bw	-	McGregor, 1974
UDS			
Rat liver cells	200 mg/Kg bw	-	Kennelly, 1990
Chromosomal aberrations			
Rat bone marrow cells	50/-100 mg/Kg bw	-	Anderson et al., 1980
HL*	NA	+	Pilinskaia, 1982
Micronuclei assay			
Cnesterodon decemmaculatus	50 - 157 mg/L	+	Vera Candioti et al., 2010b
Rhinella arenarum	80 - 250 mg/L	+	Vera Candioti et al., 2010a
Rat bone marrow cells	69.3 mg/Kg bw	-	Jones & Howard, 1989

Negative and positive results were obtained for the induction of mutagenicity in mouse lymphoma L5178Y cells regardless of the presence or absence of a rat liver metabolic activation system (Clay, 1996). Furthermore, the induction of DNA single strand breaks, estimated by the alkaline comet assay, was evaluated revealing positive results in human lymphocytes exposed in vitro to pirimicarb (Ündeger & Basaran, 2005). Similar positive results were found when the sister chromatid exchange assay was performed in Chinese hamster ovary cells (Soloneski & Larramendy, 2010). Although a significant increase in chromosomal aberrations has been reported in Chinese hamster ovary cells after pirimicarb exposure (Soloneski & Larramendy, 2010), Wildgoose and coworkers (1987) observed negative results in human lymphocytes with or without S9 metabolic activation. Finally, the induction of alterations in the cell-cycle progression in Chinese hamster ovary cells was reported to occur after in vitro exposure to pirimicarb (Soloneski & Larramendy, 2010).

In in vivo genotoxic and cytotoxic studies, pirimicarb was able to induce different types of lesions. It has been reported the ability of the insecticide to give positive results by using the eye mosaic system white/white+ (w/w+) somatic mutation and recombination test (SMART) when Drosophila melanogaster was employed as experimental model (Aguirrezabalaga et al., 1994). However, McGregor and co-workers (1974) reported negative results in mice when the dominant lethal mutation assay was performed. Similar negative results were found by Kenelly (1990) using the unscheduled DNA synthesis in rat liver cells. At the chromosomal level, pirimicarb did not induce chromosomal alterations in bone marrow cells of Wistar male rats after oral administration (Anderson et al., 1980). Contrarily, Pilinskaia (1982) observed a significant increase of chromosomal aberrations in the peripheral blood lymphocytes from occupational workers after pirimicarb exposure. Finally, when the micronuclei induction end-point was employed, positive results were reported in erythrocytes of the fish Cnesterodon decemmaculatus and Rhinella arenarum tadpoles by Vera Candioti and collaborators (Vera Candioti et al., 2010a, b). Lastly, when a mammal model was employed for the micronuclei detection, Jones and Howard (1989) found negative results in rat bone marrow cells.

COMPARISON OF THE GENOTOXICITY AND CYTOTOXICITY OF CARBOFURAN AND PIRIMICARB AND SOME ARGENTINEAN TECHNICAL FORMULATIONS

One of the goals of our research group is to compare the genotoxic and cytotoxic effects exerted by the pesticide active ingredients (Pestanal®, Riedel-de Haën, Germany) and their technical formulations commonly used in Argentina on vertebrate cells both in vitro and in vivo.

The evaluation was performed using end-points for genotoxicity [Sister Chromatid Exchange, Chromosome Aberration, and Micronuclei frequencies] and cytotoxicity [Mitotic Index, Cell Viability, Proliferative Rate Index, Erythroblasts/Erythrocytes Ratio, 3(4,5- Dimethylthiazol-2-yl)-2,5-diphenyltetrazolium bromide (MTT) and Neutral Red assays] (Soloneski et al., 2008; Soloneski & Larramendy, 2010; Vera Candioti et al., 2010a, b).

We comparatively evaluated the genotoxic and cytotoxic in vitro effects on mammalian Chinese hamster ovary cells induced by the pure insecticide carbofuran and its commercial formulation Furadan® (47% carbofuran, FMC Argentina S.A., Buenos Aires, Argentina). Similarly, the genotoxic and cytotoxic in vitro-in vivo effects induced by the pure insecticide pirimicarb and its commercial formulation Aficida® (50% pirimicarb, Syngenta Agro S.A., Buenos Aires, Argentina) on mammalian Chinese hamster ovary cells as well as circulating erythrocytes of the fish Cnesterodon decemmaculatus and the amphibian Rhinella arenarum tadpoles were also estimated.

A summary of the results obtained is presented in Fig. 3. The figure clearly reveals that all compounds assayed were able to inflict damage at chromosomal and cellular level regardless of the cellular system used as target.

We observed that carbofuran/Furadan® and pirimicarb/Aficida® caused SCEs on mammalian cells indicating that they have a clastogenic activity (Fig. 3A). It has been suggested that at the chromosomal level, the induction of SCEs is a reliable indicator for the screening of clastogens, since the bioassay is more sensitive than the analysis of clastogeninduced chromosomal aberrations (Palitti et al., 1982). The results also demonstrate the ability of pirimicarb/ Aficida® to induce DNA damage quali- and quantitative analyzed by the frequency of chromosomal aberrations (Fig. 3B). Furthermore, a putative clastogenic/aneugenic activity exerted in vitro by carbofuran/Furadan® and in vivo by pirimicarb/Aficida® was also demonstrated by the ability of the pesticide-induced micronuclei (Fig. 3C). The analysis of the proliferative replication (Fig. 3D) and the mitotic indexes (Fig. 3E) demonstrated that both carbofuran/Furadan® and pirimicarb/Aficida® were able to delay the cell-cycle progression as well as to exert a marked reduction of the cellular mitotic activity on mammalian cells in vitro. Besides, carbofuran/Furadan® and pirimicarb/Aficida® were able to induced a clear cellular cytotoxicity. This deleterious effect was estimated by a loss of lysosomal activity (indicated by a decrease in the uptake of neutral red), as well as alteration in energy metabolism (measured by mitochondrial succinic dehydrogenase activity in the MTT assay), as clearly revealed in insecticides-treated Chinese hamster ovary cells (Fig. 3F).

Overall, the results revealed, depending upon the endpoint employed, that the damage induced by the commercial formulations of both insecticides is, in general, greater than that produced by the pure pesticides (Fig. 3). Unfortunately, the identity of the components present in the excipient formulations was not made available by the manufacturers. These final remarks are in accord with previous observations not only reported by us but also by other research groups indicating the presence of xenobiotics within the composition of the commercial formulations with genotoxic and cytotoxic effects (David, 1982; Lin & Garry, 2000; Soloneski et al., 2001, 2002, 2003, 2008; González et al., 2007a, b, 2009; Elsik et al., 2008; Molinari et al., 2009; Soloneski & Larramendy, 2010). These observations highlight that, in agriculture, agrochemicals are generally not used as a single active ingredient but as part of a complex commercial formulations. Thus, both the workers as well as non-target organisms are exposed to the simultaneous action of the active ingredient and a variety of other chemical/s contained in the formulated product. Hence, risk assessment must also consider additional geno-cytotoxic effects caused by the excipient/s.

Finally, the results highlight that a whole knowledge of the toxic effect/s of the active ingredient of a pesticide is not enough in biomonitoring studies as well as that agrochemical/s toxic effect/s should be evaluated according to the commercial formulation available in market. Furthermore, the deleterious effect/s of the excipient/s present within the commercial formulation should be neither discarded nor underestimated. The importance of further studies on this type of pesticide in order to achieve a complete knowledge on its genetic toxicology seems to be, then, more than evident.

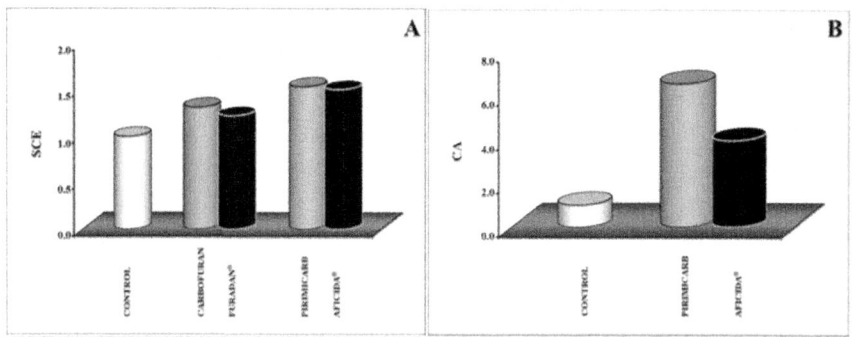

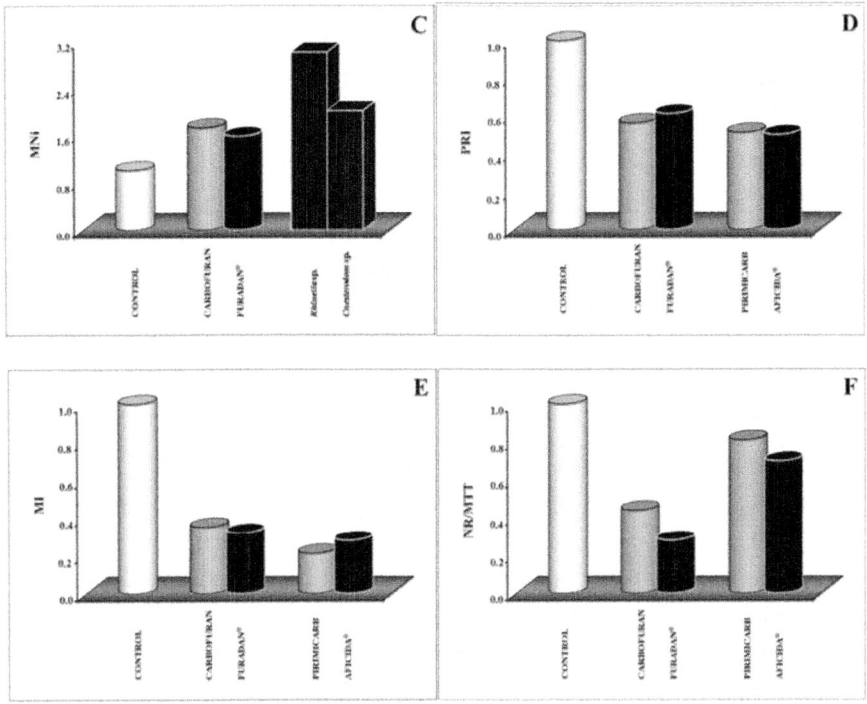

Figure 3. Comparative genotoxicity and cytotoxicity effects induced by carbofuran and pirimicarb pure herbicides Pestanal® (grey) and their technical formulations Furadan® and Aficida® (black) commonly used in Argentina on mammalian Chinese hamster ovary cells (cylinders) and in vivo piscine and amphibian erythrocytes (bars). Results are expressed as fold-time values over control data. Evaluation was performed using end-points for genotoxicity [Sister Chromatid Exchanges (A), Chromosome Aberrations (B), Micronuclei (C)] and cytotoxicity [Proliferative Rate Index (D), Mitotic Index (E), 3(4,5-Dimethylthiazol- 2-yl)-2,5-diphenyltetrazolium bromide (MTT) and Neutral Red (NR) (F)].

ACKNOWLEDGEMENTS

This study was supported by grants from the National University of La Plata (Grant Numbers 11/N564, 11/N619) and the National Council for Scientific and Technological Research (CONICET, PIP N° 0106) from Argentina.

REFERENCES

1. Aguirrezabalaga, I., Santamaría, I., Comendador, M.A., 1994. The w/w+ SMART is a useful tool for the evaluation of pesticides. Mutagenesis 9, 341-346.

2. Aly, M.S., 1998. Chromosomal damage induced in adult mice by carbofuran. Egyptian German Society of Zoology 26C, 1-10.
3. Anderson, D., Richardson, C.R., Howard, C.A., Bradbrook, C., Salt, M.J., 1980. Pirimicarb: a cytogenetic study in the rat. World Health Organization.
4. Blevins, R.D., Lee, M., Regan, J.D., 1977a. Mutagenicity screening of five methyl carbamate insecticides and their nitroso derivatives using mutants of Salmonella typhimurium LT2. Mutation Research 56, 1-6.
5. Blevins, R.D., Lijinksy, W., Regan, J.D., 1977b. Nitrosated methylcarbamate insecticides: effect on the DNA of human cells. Mutation Research 44, 1-7.
6. Bolognesi, C., 2003. Genotoxicity of pesticides: a review of human biomonitoring studies. Mutation Research 543, 251–272.
7. Bolognesi, C., Carrasquilla, G., Volpi, S., Solomon, K.R., Marshall, E.J., 2009. Biomonitoring of genotoxic risk in agricultural workers from five colombian regions: association to occupational exposure to glyphosate. Journal of Toxicological and Environmental Health. Part A 72, 986-997.
8. Bolognesi, C., Creus, A., Ostrosky-Wegman, P., Marcos, R., 2011. Micronuclei and pesticide exposure. Mutagenesis 26, 19-26.
9. Bolognesi, C., Parrini, M., Reggiardo, G., Merlo, F., Bonassi, S., 1993. Biomonitoring of workers exposed to pesticides. International Archives of Occupational and Environmental Health 65, S185-S187.
10. Callander, R.D., 1995. Pirimicarb: an evaluation of the mutagenic potential using S. typhimurium and E. coli. Central Toxicology Laboratory Report No. CTL/P/4798 GLP, Unpublished.
11. Canna-Michaelidou, S., Nicolaou, A.S., 1996. Evaluation of the genotoxicity potential (by Mutatox test) of ten pesticides found as water pollutants in Cyprus. The Science of the Total Environmental Research 193, 27-35.
12. Cantelli-Forti, G., Paolini, M., Hrelia, P., 1993. Multiple end point procedure to evaluate risk from pesticides. Environmental Health Perspectives 101, 15-20.
13. Castillo-Cadena, J., Tenorio-Vieyra, L.E., Quintana-Carabia, A.I., García-Fabila, M.M., Ramírez-San Juan, E., Madrigal-Bujaidar, E., 2006. Determination of DNA damage in floriculturists exposed to mixtures of pesticides. Journal of Biomedicine and Biotechnology 2006, 1-12.
14. Clark, H.A., Snedeker, S.M., 2005. Critical evaluation of the cancer risk of dibromochloropropane (DBCP). Journal of Environmental Science

and Health - Part C Environmental Carcinogenesis and Ecotoxicology Reviews 23, 215-260.
15. Clay, P., 1996. Pirimicarb: L5178Y TK+/- Mouse Lymphoma Mutation Assay. Central Toxicology Laboratory. Report No: CTL/P/5080 GLP, Unpublished.
16. Cole, D.C., Carpio, F., Julian, J., Léon, N., 1998. Assessment of peripheral nerve function in an Ecuadorian rural population exposed to pesticides. Journal of Toxicology and Environmental Health. Part A 55, 77-91.
17. Chauhan, L.K., Pant, N., Gupta, S.K., Srivastava, S.P., 2000. Induction of chromosome aberrations, micronucleus formation and sperm abnormalities in mouse following carbofuran exposure. Mutation Research 465, 123-129.
18. Das, A.C., Chakravarty, A., Sukul, P., Mukherjee, D., 2003. Influence and persistence of phorate and carbofuran insecticides on microorganisms in rice field. Chemosphere 53, 1033-1037.
19. David, D., 1982. Influence of technical and commercial decamethrin, a new synthetic pyrethroid, on the gonadic germ population in quail embryos. Archives d'Anatomie, d'Histologie et d'Embryologie Vol. 65, 99-110.
20. DeGraff, W.G., 1983. Mutagenicity evaluation of MC 10242 for the sex-linked recessive lethal test in Drosophila melanogaster. Study No. A 83-1060. Unpublished report prepared by Litton Bionetics, Inc. Kensington, MD, USA. Submitted to WHO by FMC Corp., Philadelphia, PA, USA.
21. Dorough, H.W., Casida, J.E., 1964. Insecticide metabolism, nature of certain carbamate metabolites of insecticide Sevin. Journal of Agricultural and Food Chemistry 12, 294- 304.
22. Elsik, C.M., Stridde, H.M., Tann, R.S., 2008. Glyphosate adjuvant formulation with glycerin. ASTM Special Technical Publication, pp. 53-58.
23. EPA, 1974a. Compendium of Registered Pesticides. US Government Printing Office, Washington, DC.
24. EPA, 1974b. Pesticide Fact Sheet: Pirimicarb. US Government Printing Office, Washington, DC.
25. EPA, 2004. ECOTOX: Ecotoxicology Database. USEPA/ORD/NHEERL, Mid-Continent Ecology Division. URL:http://www.epa.gov/ecotox/ecotox_home.htm.
26. EPA, 2006. Carbofuran I.R.E.D. Facts. U.S. Environmental Protection Agency, Office of Pesticide Programs, U.S. Government Printing Office: Washington, DC.

27. EPA, 2007. Revised N-methyl carbamate cumulative risk assessment. U.S. Environmental Protection Agency. Office of Pesticide Programs.
28. FMC, 1971. Dominant-lethal mutagenicity study [carbofuran]. NCT. 438.99. DPR Vol. 254- 029, Record 939879.
29. Gee, J., 1983. Drosophila sex-linked recessive lethal assay of 2,3 dihydro-2,2-dimethyl-7- benzofuranyI N-methyl carbamate (carbofuran, T-2047). University of Wisconsin. Lab. Project 103c. DPR Vol. 254-109, Record 47742.
30. Gentile, J.M., Gentile, G.J., Bultman, J., Sechriest, R., Wagner, E.D., Plewa, M.J., 1982. An evaluation of the genotoxic properties of insecticides following plant and animal activation. Mutation Research 101, 19-29.
31. Georgian, L.I., Moraru, I., Draghicescu, T., Tarnavschi, R., 1985. The effect of low concentrations of carbofuran, cholin salt of maleic hydrazine, propham and chlorpropham on sister-chromatid exchange (SCE) frequency in human lymphocytes in vitro. Mutation Research 147, 296-301.
32. González, N.V., Soloneski, S., Larramendy, M., 2009. Dicamba-induced genotoxicity oh Chinese hamster ovary (CHO) cells is prevented by vitamin E. Journal of Hazardous Materials 163, 337-343.
33. González, N.V., Soloneski, S., Larramendy, M.L., 2007a. The chlorophenoxy herbicide dicamba and its commercial formulation banvel induce genotoxicity in Chinese hamster ovary cells. Mutation Research 634, 60-68.
34. González, N.V., Soloneski, S., Larramendy, M.L., 2007b. Genotoxicity analysis of the phenoxy herbicide dicamba in mammalian cells in vitro. Toxicology In Vitro 20, 1481- 1487.
35. Gupta, R.C., 1994. Carbofuran toxicity. Journal of Toxicology and Environmental Health. Part A 43, 383-418.
36. Haworth, S.R., Lawlor, T.E., 1983. Salmonella/mammalian microsome plate incorporation mutagenicity assay (Ames test). Carbofuran technical. FMC Study No. A 83-868 (MBA Study No. 1921.501). Unpublished report prepared by Microbiological Associates, MD, USA. Submitted to WHO by FMC Corp., Philadelphia, PA, USA.
37. Hodgson, E., Levi, P.E., 1996. Pesticides: an important but underused model for the environmental health sciences. Environmental Health Perspectives 104, 97- 106.

38. Hour, T.C., Chen, L., Lin, J.K., 1998. Comparative investigation on the mutagenicities of organophosphate, phthalimide, pyrethroid and carbamate insecticides by the Ames and lactam tests. Mutagenesis 13, 157-166.
39. HSDB, 2011. Carbofuran. National Library of Medicine.
40. IARC, 1976. Some carbamates, thiocarbamates and carbazides. International Agency for Research on Cancer, Lyon.
41. IARC, 1987. Genetic and related effects: an updating of selected IARC monographs from volumes 1 to 42. International Agency for Research on Cancer, Lyon.
42. IARC, 2003. Monographs on the evaluation of carcinogenic risk to human. Vols. 5 - 53. International Agency for Research on Cancer, Lyon.
43. Jones, K., Howard, C.A., 1989. Pirimicarb (technical): an evaluation in the mouse micronucleus test. Unpublished report No. CTL/P/2641 from Central Toxicology Laboratory, Zeneca. Submitted to WHO by Syngenta Crop Protection AG. Conducte according to OECD 474 (1983). GLP compliant.
44. Jung, Y.S., Kim, C.S., Park, H.S., Sohn, S., Lee, B.H., Moon, C.K., Lee, S.H., Baik, E.J., Moon, C.H., 2003. N-nitroso carbofuran induces apoptosis in mouse brain microvascular endothelial cells (bEnd.3). Journal of Pharmacology Sciences 93, 489-495.
45. Käfer, E., Scott, B.R., Dorn, G.L., Stafford, R., 1982. Aspergillus nidulans: systems and results of tests for chemical induction of mitotic segregation and mutation. I. Diploid and duplication assay systems. A report of the U.S. EPA Gene-Tox Program. Mutation Research 98, 1-48.
46. Kamel, F., Boyes, W.K., Gladen, B.C., Rowland, A.S., Alavanja, M.C., Blair, A., Sandler, D.P., 2000. Retinal degeneration in licensed pesticide applicators. American Journal of Industrial Medicine 37, 618-628.
47. Karczmar, A., 1998. Anticholinesterases: dramatic aspects of their use and misuse. Neurochemistry International 32, 401-411.
48. Kennelly, J.C., 1990. Pirimicarb: Assessment for the induction of unscheduled DNA synthesis in rat hepatocytes in vivo. Unpublished report No. CTL/P/2824 from Central Toxicology Laboratory, Zeneca. Submitted to WHO by Syngenta Crop Protection AG. Conducte according to OECD 486 (1983). GLP compliant.
49. Kim, S.J., Kim, J.E., Ko, B.H., Moon, I.S., 2004. Carbofuran induces apoptosis of rat cortical neurons and down-regulates surface alpha7 subunit of acetylcholine receptors. Molecules and Cells 17, 242-247.

50. Kirby, P.E., 1983a. L5178Y TK+/- mouse lymphoma mutagenesis assay. FMC Study No. A 83-962 and A 83-988 (MBA Study No. T1982.701). Unpublished report prepared by Microbiological Associates, Bethesda, MD, USA. Submitted to WHO by FMC Corp., Philadelphia, PA, USA.

51. Kirby, P.E., 1983b. L5178Y TK+/- mouse lymphoma mutagenesis assay. FMC Study No. A 83-1064 (MBA Study No. T2124.201). Unpublished report prepared by Microbiological Associates, Bethesda, MD, USA. Submitted to WHO by FMC Corp., Philadelphia, PA, USA.

52. Lin, C.M., Wei, L.Y., Wang, T.C., 2007. The delayed genotoxic effect of N-nitroso N-propoxur insecticide in mammalian cells. Food and Chemical Toxicology 45, 928-934.

53. Lin, N., Garry, V.F., 2000. In vitro studies of cellular and molecular developmental toxicity of adjuvants, herbicides, and fungicides commonly used in Red River Valley, Minnesota. Journal of Toxicology and Environmental Health 60, 423-439.

54. McGregor, D.B., 1974. Dominant lethal study in mice of ICI PP062. Zeneca unpublished report No. CTL/C/256 from Inveresk Research International. Submitted to WHO by Syngenta Crop Protection AG. Conducte according to OECD 478 (1983). GLP compliant.

55. Metcalf, R., Fukuto, R., Collins, C., Borck, K., Abd El-Aziz, A., Munoz, R., Cassil, C., 1968. Metabolism of 2,2-dimethyl-2,3-dihydrobenzofuran-7-N-methylcarbamate (Furadan) in plants, insects, and mamals. Journal of Agricultural and Food Chemistry 16, 300-311.

56. Mladinic, M., Perkovic, P., Zeljezic, D., 2009. Characterization of chromatin instabilities induced by glyphosate, terbuthylazine and carbofuran using cytome FISH assay. Toxicology Letters 189, 130-137.

57. Molinari, G., Soloneski, S., Reigosa, M.A., Larramendy, M.L., 2009. In vitro genotoxic and citotoxic effects of ivermectin and its formulation ivomec® on Chinese hamster ovary (CHOK1) cells. Journal of Hazardous Materials 165, 1074-1082.

58. Moriya, M., Ohta, T., Watanabe, K., Miyazawa, T., Kato, K., Shirasu, Y., 1983. Further mutagenicity studies on pesticides in bacterial reversion assay systems. Mutation Research 116, 185-216.

59. Naravaneni, R., Jamil, K., 2005. Cytogenetic biomarkers of carbofuran toxicity utilizing human lymphocyte cultures in vitro. Drug and Chemical Toxicology 28, 359- 372.

60. Palitti, F., Tanzarella, C., Cozzi, R., Ricordy, R., Vitagliano, E., Fiore, M., 1982. Comparison of the frequencies of SCEs induced by chemical

mutagens in bone-marrow, spleen and spermatogonial cells of mice. Mutation Research 103, 191-105.

61. Pavanello, S., Clonfero, E., 2000. Biomarkers of genotoxic risk and metabolic polymorphisms. Indicatori biologici di rischio genotossico e polimorfismi metabolici 91, 431-469.

62. Peter, J.V., Cherian, A.M., 2000. Organic insecticides. Anaesthesia and Intensive Care 28, 11- 21.

63. Pilinskaia, M.A., 1982. Cytogenetic effect of the pesticide pirimor in a lymphocyte culture of human peripheral blood in vivo and in vitro. Tsitologiia Genetika 16, 38-42.

64. Pilinskaia, M.A., Stepanova, L.S., 1984. Effect of the biotransformation of the insecticide furadan on in vivo and in vitro manifestations of its cytogenetic activity. Tsitologiia Genetika 18, 17-20.

65. Putman, D.L., 1983a. Activity of FMC 10242 (T1982) in the in vivo cytogenetics assay in sprague-dawley rats. Unpublished report No: A83-972 prepared by Microbiological Associates, Bethesda, MD, USA. Submitted to WHO by FMC Corp., Philadelphia, PA, USA.

66. Putman, D.L., 1983b. Activity of FMC 10242 in the subchronic in vivo cytogenetics assay in male rats. Study No. A 83-1065 (MBA No. T2124.102). Unpublished report prepared by Microbiological Associates, Bethesda, MD, USA. Submitted to WHO by FMC Corp., Philadelphia, PA, USA USA.

67. Saxena, P.N., Gupta, S.K., Murthy, R.C., 2010. Carbofuran induced cytogenetic effects in root meristem cells of Allium cepa and Allium sativum: A spectroscopic approach for chromosome damage. Pesticide Biochemistry and Physiology 96, 93-100.

68. Saxena, S., Ashok, B.T., Musarrat, J., 1997. Mutagenic and genotoxic activities of four pesticides: captan, foltaf, phosphamidon and furadan. Biochemistry and Molecular Biology International 41, 1125-1136.

69. Shirasu, Y., 1975. Significance of mutagenicity testing on pesticides. Environmental and Quality Safety 4, 226-231.

70. Simmon, U.F., 1979. In vitro microbiological mutagenicity and unscheduled DNA synthesis studies of eighteen pesticides. Study No. 68-01-2458 (FMC Study No. A81-509, EPA 600/1-79-041). Report prepared by SRI International, Menlo Park, California, for Health Effects Research Laboratory Office of Research and Development. US Environmental Protection Agency, Research Triangle Park, NC, USA. Submitted to WHO by FMC Corp., Philadelphia, PA.

71. Soloneski, S., A, R.M., Larramendy, M.L., 2003. Effect of the dithiocarbamate pesticide zineb and its commercial formulation, the azzurro. V. Abnormalities induced in the spindle apparatus of transformed and non-transformed mammalian cell lines. Mutation Research 536, 121-129.
72. Soloneski, S., González, M., Piaggio, E., Apezteguía, M., Reigosa, M.A., Larramendy, M.L., 2001. Effect of dithiocarbamate pesticide zineb and its commercial formulation azzurro. I. Genotoxic evaluation on cultured human lymphocytes exposed in vitro. Mutagenesis 16, 487-493.
73. Soloneski, S., González, M., Piaggio, E., Reigosa, M.A., Larramendy, M.L., 2002. Effect of dithiocarbamate pesticide zineb and its commercial formulation azzurro. III. Genotoxic evaluation on Chinese hamster ovary (CHO) cells. Mutation Research 514, 201-212.
74. Soloneski, S., Larramendy, M.L., 2010. Sister chromatid exchanges and chromosomal aberrations in chinese hamster ovary (CHO-K1) cells treated with the insecticide pirimicarb. Journal of Hazardous Materials 174, 410-415.
75. Soloneski, S., Reigosa, M.A., Molinari, G., González, N.V., Larramendy, M.L., 2008. Genotoxic and cytotoxic effects of carbofuran and furadan® on Chinese hamster ovary (CHOK1) cells. Mutation Research 656, 68-73.
76. SRI, 1979. In vitro microbiological mutagenicity and unscheduled DNA synthesis studies of eighteen pesticides. Contract 68-0102458. DPR Vol. 254-110, Record 47751 (E. coli & B. subtilis), 47752 (S. cerevisiae), 47753 (WI-38).
77. Stehrer-Schmid, P., Wolf, H., 1995. Effects of benzofuran and seven benzofuran derivatives including four carbamate insecticides in the in vitro porcine brain tubulin assembly assay and description of a new approach for the evaluation of the test data. Mutation Research/Reviews in Genetic Toxicology 339, 61-72.
78. Thilagar, A., 1983a. Chromosome aberrations in Chinese hamster ovary (CHO) cells. FMC Study No. A 83-1096. Unpublished report prepared by Microbiological Associates, Bethesda, MD, USA. Submitted to WHO by FMC Corp., Philadelphia, PA, USA.
79. Thilagar, A., 1983b. Sister chromatid exchange assay in Chinese hamster ovary (CHO) cells. FMC Study No. A 83-1095 (MBA No.1982.334001). Unpublished report prepared by Microbiological Associates, Bethesda, MD, USA. Submitted to WHO by FMC Corp., Philadelphia, PA, USA.

80. Thilagar, A., 1983c. Sister chromatid exchange assay in Chinese hamster ovary (CHO) cells. FMC Study No. A 83-1097 (MBA No. 2124.334001). Unpublished report prepared by Microbiological Associates, Bethesda, MD, USA. Submitted to WHO by FMC Corp., Philadelphia, PA, USA.
81. Thilagar, A., 1983d. Unscheduled DNA synthesis in rat primary hepatocytes. FMC Study No. A 83-969 (MBA No. T1982.380) Unpublished report prepared by Microbiological Associates, Bethesda, MD, USA. Submitted to WHO by FMC Corp., Philadelphia, PA, USA.
82. Trueman, R.W., 1980. An examination of pirimicarb for potential mutagenicity using the Salmonella/microsome reverse mutation assay. Unpublished report No: CTL/P/540 form Central Toxicology Laboratory, Zeneca. Submitted to WHO by Syngenta Crop Protection AG. Conducted to a protocol that was consistent with the OECD guideline 471 (1983). GLP compliant.
83. Ündeger, Ü., Basaran, N., 2005. Effects of pesticides on human peripheral lymphocytes in vitro: induction of DNA damage. Archives in Toxicology 79, 169-176.
84. Valencia, R., 1981. Mutagenesis screening of pesticides using Drosophila. FMC Study No. A 83-1042 (EPA 600/1-81-017). Report prepared by Warf Institute Inc., Madison, WI, for Health Effects Research Laboratory. Office of Research and Development, US Environmental Protection Agency, Research Triangle Park, NC, USA. Submitted to WHO by FMC Corp., Philadelphia, PA, USA.
85. Valencia, R., 1983. Drosophila sex-linked recessive lethal assay of 2,3-dihydro-2,2-dimethyl-7- benzofuranyl- N-methyl carbamate (carbofuran). FMC Study No. A 83-1019 (MBA No. T1047.160). Unpublished report prepared by Microbiological Associates, Bethesda, MD and the University of Wisconsin, Zoology Department, USA. Submitted to WHO by FMC Corp., Philadelphia, PA, USA.
86. Vera Candioti, J., Natale, G.S., Soloneski, S., Ronco, A.E., Larramendy, M.L., 2010a. Sublethal and lethal effects on Rhinella arenarum (Anura, Bufonidae) tadpoles exerted by the pirimicarb-containing technical formulation insecticide Aficida. Chemosphere 78, 249-255.
87. Vera Candioti, J., Soloneski, S., Larramendy, M.L., 2010b. Genotoxic and cytotoxic effects of the formulated insecticide aficida on Cnesterodon decemmaculatus (Jenyns, 1842) (Pisces: Poeciliidae). Mutation Research 703, 180-186.
88. Waters, M.D., Sandhu, S.S., Simon, V.F., 1982. Study of pesticide genotoxicity. Basic Life Sciences 21, 275-326.

89. WHO-FAO, 2004. Pesticides residues in food-2004. FAO Plant Production and Protection paper World Health Organization and Food and Agriculture Organization of the United Nations, Rome, pp. 154-161.
90. WHO-FAO, 2009. Pesticides residues in food-2009. FAO Plant Production and Protection paper World Health Organization and Food and Agriculture Organization of the United Nations, Rome, pp. 1-426.
91. WHO, 1988. The WHO recommended classification of pesticides by hazard and guidelines to the classification 1988-1989. World Health Organization, Geneva.
92. WHO, 1990. Public health impacts of pesticides used in agriculture (WHO in collaboration with the United Nations Environment Programme, Geneva, 1990). World Health Organization, Geneva.
93. WHO, 2000-2002. International programme of chemical safety. World Health Organization, Geneva.
94. WHO, 2009. The WHO recommended classification of pesticides by hazard. World Health Organization, Geneva.
95. Wildgoose, J., Howard, C.A., Richardson, C.R., Randall, V., 1987. Pirimicarb: A cytogenetic study in human lymphocytes in vitro. Central Toxicology Laboratory. Report No. CTL/P/1655, GLP, Unpublished.
96. Wojciechowski, J.P., Kaur, P., Sabharwal, P.S., 1982. Induction of ouabain resistance in V-79 cells by four carbamate pesticides. Environmental Research 29, 48-53.
97. Woodruff, R.C., Phillips, J.P., Irwin, D., 1983. Pesticide-induced complete and partial chromosome loss in screens with repair-defective females of Drosophila melanogaster. Environmental Mutagenesis 5, 835-846.
98. Yoon, J.Y., Oh, S.H., Yoo, S.M., Lee, S.J., Lee, H.S., Choi, S.J., Moon, C.K., Lee, B.H., 2001. Nnitrosocarbofuran, but not carbofuran, induces apoptosis and cell cycle arrest in CHL cells. Toxicology 169, 153-161.
99. Zeljezic, D., Vrdoljak, A.L., Lucas, J.N., Lasan, R., Fucic, A., Kopjar, N., Katic, J., Mladinic, M., Radic, B., 2009. Effect of occupational exposure to multiple pesticides on translocation yield and chromosomal aberrations in lymphocytes of plant workers. Environmental Sciences Technololgy 43, 6370-6377.
100. Zhou, P., Liu, B., Lu, Y., 2005. DNA damaging effects of carbofuran and its main metabolites on mice by micronucleus test and single cell gel electrophoresis. Science in China. Series C, Life Sciences 48, 40-47.

Chapter 9

PESTICIDES AND AGRICULTURAL WORK ENVIRONMENTS IN ARGENTINA

M. Butinof[1], R. Fernández[2], M.J. Lantieri[1], M.I. Stimolo[3], M. Blanco[4], A.L. Machado[5], G. Franchini[1], M. Gieco[1], M. Portilla[1], M. Eandi[1], A. Sastre[5], and M.P. Diaz[1]

[1] Faculty of Medical Sciences, National University of Córdoba, Córdoba, Argentina
[2] Faculty of Medicine, Catholic University of Córdoba, Córdoba, Argentina
[3] Faculty of Economics, National University of Córdoba, Córdoba, Argentina
[4] Faculty of Agricultural Sciences, National University of Córdoba, Córdoba, Argentina
[5] Faculty of Psychology, National University of Córdoba, Córdoba, Argentina

INTRODUCTION

The use of chemical pesticides has brought benefits such as the increase of agricultural production, soil productivity and product quality, which is reflected in economic benefits, vector disease control and in general, in public health. However, given that only 10 percent of applied pesticides reach the target organism, a high percentage is deposited on non-target-areas (soil, water, sediments) and, as well as affecting public health, impacts non-target organisms such as wild life [1]. Also, the extended use of pesticides commonly results in residues in foods [2] generating continued human exposure by different pathways, which has led to widespread concern over the potentially adverse effects of these chemicals on human health.

Pesticides are an important aspect of agricultural practice in both developed and developing countries and, despite the many technological advances brought by the modern intensification of agriculture, the increased yields were achieved primarily through the use of fertilizers and pesticides [3].

Argentina is one of the major crop producers in Latin America, with the export of cereals and oilseeds being one of the principal axes of the national economy. The frontiers of farming have expanded greatly in the past 30 years, from 15 to the current 30 million hectares, with an increase of the area planted for grain production, particularly for soybeans, from 34,700 ha in the 1969/70 season to about 18 million ha in 2011/12 [4]. Today, Argentina is the world's leading producer of vegetable oils, the fourth largest producer and second largest exporter of sunflower oil, and the fourth producer and leading exporter of soybean oil. The country has one of the highest yields in the world in soybean, corn and wheat [5].

Argentina's extensive production of cereals and oilseeds for the international market coexists with intensive horticulture and family farming, with wide geographical distribution, mainly close to urban centers, and diversity of cultivated species, occupying an area of about 230,000 ha [6], giving an annual production of over 10,000,000 tons, primarily for domestic consumption.

Crop production has been accompanied by a steady increase in the use of agrochemicals; pesticide marketing has grown strongly, from 155 million pounds in 1995 up to 700 m.p. in 2012 [7]. In technology used for spraying pesticides, the country has a wide variety of equipment ranging from self-propelled sprayers, which also involve high technological complexity, with filtered air cabins, to activated charcoal filters, spray drag and power and manual backpacks used particularly in intensive farming. Each of these different technological environments is associated with different health and environmental risks.

The Province of Córdoba is in the central region of Argentina, with a total area of 165,321 km^2. Its location, as well as its political and physical characteristics, make this province a hub of articulation between different natural regions of the country. It has a population of 3,304,825 inhabitants, 88.7% of whom live in urban areas and 11.3% in rural areas [8]. Among the inhabitants of rural areas, 45.9% live in towns of less than 2,000 inhabitants, and the rest dispersed in the open countryside [9]. The northern and western areas are less populated and are the ones which concentrate most indicators of structural and cyclical poverty. The agricultural roots of this province mean that the settlements are mixed with agricultural developments, increasing the risk of non-occupational exposure of communities adjacent to cultivated fields.

The rural area in Cordoba devoted to extensive crops (soybean, maize, sorghum, peanut, wheat and sunflower), has expanded from 3,397,050 ha in 1994/95 to 7,300,000 ha in 2011/2012 [4].

The country's extensive agricultural model, based on glyphosate-resistant transgenic soybean farming, no-till and the intensive use of fertilizers and pesticides [10], is highly dependent on modern technologies [11]. In contrast, intensive crops such as fruits and vegetables are characterized by high demand of labor per unit of output. Typically, this is a small-scale activity usually performed by the peasant family production unit [12, 13], with all its members participating. The incidence of pesticide poisoning in these agricultural settings includes non-intentional child exposure, occupational exposure of young farm laborers, para-occupational exposure of the farm workers and their families and the adjacent community, and exposure to banned pesticides [14].

In Córdoba Province, exposure to different pesticides linked to agricultural production has long been recognized [10, 15-16], as well as the unavoidable soil contamination even decades after its application [17]. Our previous results of a population-based study in the province of terrestrial applicators of pesticides in extensive crops (n=880), emphasized that workers were highly exposed to pesticides, and we studied various determinants of this exposure, including the pesticides most frequently used or still in use by the applicators. We also reported the negative health consequences associated with their employment status. The weakness of compliance with the rules governing the activity was also highlighted as a factor that increases the health risk of agricultural workers and the general population [10].

The greenbelt surrounding the provincial capital is a zone of fruit and vegetable farming, providing fresh food to the local urban population. Its extension includes neighboring towns, forming a strongly integrated commercial and productive system. Almost 90% of the fruits and vegetables it provides are produced within the urban area [17]. Horticultural smallholders and farmworkers are often immigrant workers from neighboring countries [18-20]: according to the Ministry of Education [21], sixty percent of them are Bolivian citizens, which increases their risk of environmental and occupational illness and injury, as well as the health disparities typically associated with poverty [22].

This chapter offers a comparative analysis of two widely different agricultural settings (extensive and horticultural crops) and characterizes the pesticide applicator populations in each, including the health conditions associated with occupational pesticide use. We introduce two pesticide exposure assessment proposals, consisting of intensity and accumulated exposure indexes for both scenarios. The proposals include new results about the pesticide applicators of extensive crops, including an update of the differential characteristics of worker populations in homogeneous ecological areas of the province. We also introduce a new scenario consisting of horticultural smallholders and farmworkers, and describe their working conditions. The study and comparison of these different work settings allows us to tailor the exposure indexes developed in our previous publication [10] to the particular pesticide exposure of greenbelt situations, as well as to develop proposals for preventive measures for the reduction of human exposure and environmental impact, according to each scenario.

MATERIALS AND METHODS

Population Studies

We conducted a population-based study in Córdoba Province, Argentina, with two principal target populations: a) terrestrial applicators of pesticides of extensive crops; b) smallholders and farmworkers of the greenbelt of its capital city, Córdoba.

- In the first case, all the applicators attending the mandatory courses for obtaining the applicator license, provided by the Agriculture, Livestock and Food Ministry, were asked to participate in the survey, during the period 2007-2012. A self-administered questionnaire was used to obtain demographic data, pesticides and technologies used, crops sprayed, workers' lifestyle and family health information, as already described in a previous publication [10]. From 1479 completed questionnaires, a consistency analysis for several responses was carried out, with a sample size of 1327 for further analysis. We also performed a stratified analysis taking into account the Homogeneous Ecological Areas (HEAs) divisions of the province in order to describe differences between these.

- In the second case, 101 smallholders and farmworkers were contacted in Cordoba's wholesale fruit and vegetables market and in the greenbelt setting itself. The above questionnaire was adapted for this specific population, after an exploratory study through in-depth interviews during 2011. As described in the literature, the exploratory study shows that this is a difficult-to-reach population due to their migratory status and unstable working conditions.

Variables

Terrestrial Applicators of Pesticides of Extensive Crops

- Social and demographic variables: age (in years, as from birth date), education level (highest level of educational attainment in the formal system) and marital status (married or cohabiting and others).
- Technological and working practices variables: pesticide spray equipment (self-propelled crop sprayer with cab and activated charcoal filter; trailed crop sprayer with cab and activated charcoal filter), area worked (average hectares applied in the last year), seniority in the job

(years mixing/applying), written pesticide prescription signed by an agricultural engineer (yes/no).

- To assess the level of protection implemented by the terrestrial applicators, we adopt the proposal in [23], considering eight categories of personal protective equipment (PPE) used, alone or in combination: waterproof clothing, gas mask, chemical-resistant gloves, face shields or goggles, hat or helmet and other protective clothes (boots, apron, waterproof pants). The weighting of PPE elements is based on monitoring and measurement of occupational exposure during the task. A new measure called protection level was constructed [10]: unprotected (0% protection), partially protected (20 to 70% protection) and protected (90% protection).

  These variables were analyzed comparatively between homogeneous ecological areas (HEAs) of the province, according to soil and climatic characteristics, land use and production activities, as described in [10].

- Good agricultural practices: we considered two practices included in the local regulation aimed at reducing human risks and negative environmental impacts [24]: a) the triple washing of pesticide containers (yes/no). This practice consists in washing the empty container three times and draining for thirty seconds in upload position; and b) correct end use of pesticide containers (yes/no): properly cleaned containers must be transferred to an authorized registered storage center, to be destroyed in a pyrolytic oven; burial, burning, storage, sale or reuse are prohibited.

Smallholders and Farm Workers of the Greenbelt Surrounding the Capital City of Cordoba

To highlight the particularities of the horticultural work scenario and its worker population, new variables were incorporated into the analysis when necessary.

- Socio-demographic variables: age, education and marital status are described as mentioned above; origin (country and province of birth); household (members and their participation in horticultural work). Dwelling infrastructure and public services: running water installed (yes/no), bathroom installed (yes/no), domestic gas distribution network (yes/no); public service of urban solid waste collection (yes/no).

- Work practices, technology and other exposure variables: pesticides sprayed, use of PPE (as described above); crops grown in the last year (type of crop and annual average harvests); greenhouse for crop (yes/no); household distance to the nearest crop (meters); extension of the

productive unit in hectares: small: up to 10 ha; medium: between 11 to 40 ha; and large, more than 40 ha [25]: seniority in the job (years mixing/applying); pesticide spray equipment (self-propelled crop sprayer with cab and activated charcoal filter; trailed crop sprayer with cab and air intake activated charcoal filter or without air intake filter; trailed crop machine without cabin, manual and engine backpack).
- Good agricultural practices (as described above).

Health Worker Conditions in Both Agricultural Settings
- Symptoms: Perception of acute and sub-acute manifestations: Irritative symptoms (skin, nose and eye irritation, nausea or vomiting, chest discomfort); fatigue/tiredness; nervousness or depression; headache; excessive sweating. Occurrence of symptoms: Never/Rarely/Sometimes/Frequently;
- Medical consultations related to pesticide use effects: yes/no; and Hospitalization linked to tasks with pesticides: yes/no;
- Workers' risk perception of different pesticides: not dangerous/slightly dangerous/dangerous/highly dangerous.

Exposure Assessment

Based on proposed indexes of our previous work [10], the present study incorporates intensity level (ILE) and accumulated exposure (CEI) indexes into pesticide exposure, adapted to the smallholder and farmworker population of the greenbelt of Cordoba city, describing the principal differences among them. These indexes measure instantaneous exposure intensity and cumulative exposure taking into account the life years of worker exposure. To use these indexes in the horticultural worker population, we have carefully adapted the weighting score procedure to this particular context.

Statistical Analysis for Association

We used a modeling approach to check differences between ecological areas. Assuming counts or frequencies in each category of the variables as the outcome, we fitted Poisson and Gamma generalized models to estimate the parameters (effects). The latter was used since the empirical distributions of both indexes presented skewness. Association between two or three variables was studied through log-linear models in order to estimate the odds ratio as association measures.

RESULTS

Population of Extensive Crops

In a previous work [10] we identified different agricultural settings in the province, based on homogeneous ecological area (HEAs) divisions (Figure 1). Differences in basic characteristics of this population, such as their average age, instruction level and length of occupational exposure to pesticides allow us to hypothesize the existence of diverse risk scenarios in the province. In this chapter, an update of the characterization of workers was performed with an increased sample size, n=1327.

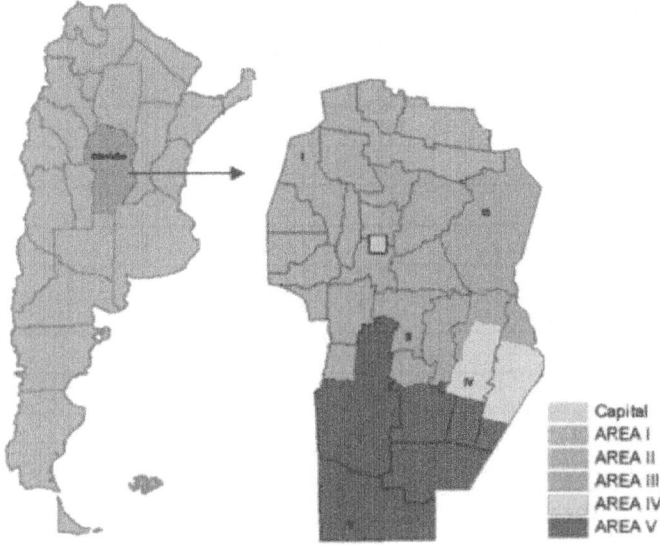

Figure 1. Homogeneous ecological areas (HEAs) of Córdoba province.

Significant differences among HEAs were found for age ($p<0.01$), education level ($p=0.03$) and marital status ($p<0.01$), as well as for seniority in the task ($p<0.05$), average/year of hectares sprayed ($p<0.03$), use of pesticides with written prescription signed by an agricultural engineer ($p<0.03$), and self-propelled crop sprayer with cab and activated charcoal filter ($p<0.03$) (Table 1). Protection level ($p<0.05$) also showed differences between HEAs I, II, III and HEA V ($p<0.05$), the latter having the fewest completely protected workers (31%). Only trailed crop sprayer with cab and activated charcoal filter results were similar in all the areas. It is important to highlight that HEA I showed the lowest percentage of applicators with complete secondary school level or higher (29.2%) and in subjects married or cohabiting (56.8%), but the most workers using complete protection (54%) and using pesticides with written prescription (58.1%) followed by HEA II (54.8%)

Table 1. Sociodemographic characteristics of pesticide applicators by Homogeneous Ecological Areas. Córdoba, Argentina. 2007- 2012.

	AREAS					
	I	II	III	IV	V	Total
n	41	641	230	156	259	1327
Age (years)						
Mean	32.3	35.8	34.9	37.6	34.8	35.6
Standard Deviation	8.6	11.6	9.9	10.9	11.9	11.3
14 – 24	16.2	16.5	16.1	12.8	21.2	16.9
25 – 24	48.6	36.0	38.1	30.2	32.8	35.4
35 – 44	27.0	24.9	28.7	30.2	25.2	26.3
> 45	8.1	22.6	17.0	26.8	20.8	21.4
Marital Status (%)1						
Married or cohabiting	56.8	66.8	61.7	78.9	58.6	65.5
Unmarried, separated, divorced or widower	43.2	33.2	38.3	21.1	41.4	34.5
Education (%)						
Incomplete Primary	2.4	11.1	13.5	7.1	10.8	10.7
Complete Primary	29.3	27.9	27.8	29.5	32.0	28.9
Incomplete Secondary	39.0	26.6	21.3	22.4	23.9	25.1
Complete Secondary, Technical or University studies	29.2	34.3	37.4	41.0	33.2	35.2

[i] - [1]Percentage considering the total of responses.

Pesticide with prescription signed by an agricultural engineer was used by only 33.7% of applicators in HEA V, which was different from the others ($p<0,03$); self-propelled crop sprayer with cab and activated charcoal filter was highest in HEA III (74,1%) and this was significantly different from HEAs II and V ($p<0,05$). No significant differences were found between areas in the use of the trailed crop sprayer with cab and activated charcoal filter, but this is a crop sprayer that is very little used in all the areas (Table 2).

Table 2. Protection Level, Area/year applied Seniority in the Job and Technology in the different Homogeneous Ecological Areas. Córdoba, Argentina, 2007-2012.

	AREAS					
	I	II	III	IV	V	Total
N	41	641	230	156	259	1327
Protection Level (%)1						
Unprotected	12.2	12.5	12.2	9.0	17	12.9
Partially Protected	34.1	48.8	46.1	55.8	52.1	49.4
Protected	53.7	38.7	41.7	35.3	30.9	37.8
Average area/year applied (ha)						
Mean	9717	5226	9923	6535	7182	6767
Years personally mixed/applied pesticides (%)						
≤ 1	26.3	11.2	22.4	10.3	17.0	14.6
2 – 5	36.8	31.6	34.5	27.7	38.3	33.1
6 – 10	18.4	23.5	23.8	23.9	21.3	23.0
11 – 20	10.5	21.4	13.5	23.9	15.8	18.9
21 - $\geq$ 30	2.6	11.7	5.4	14.2	6.3	9.5
Use pesticides with prescription signed by an agricultural engineer (%)						
Yes	58.3	54.8	46.4	53.8	33.7	49.9
Apply with Self-propelled Crop Sprayer with Cab and Activated Charcoal Filter (%)						
Yes	63.9	49.7	74.1	67.9	63.2	58.7
Apply with Trailed Crop Sprayer with Cab and Activated Charcoal Filter (%)						
Yes	2.9	8.4	4.9	6.2	7.9	7.3

[i] - [1]Percentage considering the total of responses.

Good agricultural practices were established to reduce the contamination that may be caused by empty pesticide containers and their geographical dispersion. Not all applicators carry out the triple washing of pesticide containers (89.9% do so), and only 10.5% are included in formally regulated programs to ensure the correct end use of empty pesticide containers; in many cases, empty containers of chemicals are burned, buried or reused.

Population of Smallholders and Farmworkers of the Green Belt around the Capital City of Córdoba

The green belt, in place since the founding of the city, has seen its landscape transformed over time through a steady process of land use change [26], extending to the nearby towns. Currently, the green belt is situated within an urban area with a sum of overlapping environmental hazards caused by agricultural activity and industrial activity (Figure 2).

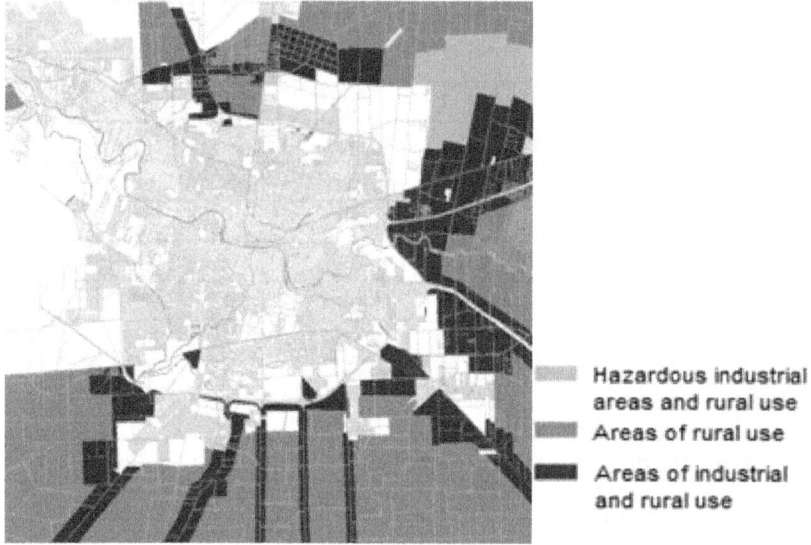

Figure 2. Land use map of the urban area of the city of Córdoba. Municipality of Córdoba, 2004 [17].

Our population consisted of male subjects, with only a single registered female. The mean age was 42.94 years (SD: 13.34), with 67% over 35 years (Table 3). 52% of subjects achieved low levels of education, with 24% who did not complete primary school and 28% who completed only this level. 71% were Argentine and 29% Bolivians. Of the Argentine farmers and workers, 13% were migrants from other provinces. One of the distinguishing characteristics of horticultural farms was their family origin, and this situation, with variations, was maintained over time [27]. 23% of respondents lived alone, while the remaining 77% lived with family members. Of these, 11% lived only with their partner, while 66% also lived with children and 14% with extended family members (older adults, uncles/aunts, cousins). In 31% of families, all took part in the horticultural work with different tasks and hourly loads (involving spouses, children and extended family members).

Among the job roles reported by the horticulturists are the owner, tenant, "mediero" and permanent or temporary employee, and combinations of the above. "Medieria" is a form of associative contract farming: the existence of a partner who provides land and part of the capital, while the other participant contributes labor and other inputs, sharing the product between them.

Part of the population of these small farmers and workers had unsatisfied basic needs, lacking such basic public services as a water network (23%) and a bathroom installed within the dwelling (13%). Precarious living conditions were associated with employment status and land tenure, with the "medieros" and employees having the highest chance of not satisfying these needs (p <0.048), as well as with nationality, to the detriment of the Bolivian-born small farmers (p <0.014, Table 3). An urban solid waste collection service was absent in 23% of cases and domestic gas network provision was lacking in large areas of the green belt (80%).

Table 3. Social and demographic characteristics of smallholders and farm workers of the Córdoba capital city green belt. 2012.

Sociodemographic characteristics	Number	Valid (%)1
Age (years)		
Mean	42.94	
Standard Deviation	13.34	
≤ 25	13	13
26 – 34	20	20
35 – 44	19	19
45 – 54	26	26
> 55	22	22
Education		
Incomplete Primary	24	24
Complete Primary	28	28
Incomplete Secondary	18	18
Complete Secondary, Technical or University studies	31	30
Marital Status		
Married or cohabiting	75	77
Unmarried, separated, divorced or widowed	22	23
Other members of the household working in crops		
Yes	33	31
Running water installed in the household		
Yes	75	77
Bathroom installed in the household		

Yes	85	87
Domestic gas distribution network		
Yes	20	20
Public service of urban solid waste collection		
Yes	66	67
Country of origin and internal migration		
Bolivia	29	29
Argentina	71	71
Born in Cordoba	62	87
Internal migrants	9	13

[i] - [1]Percentage considering the total of responses.

Table 4 shows that 58% of the productive units were classified as small in extension. Most of the smallholders and farm workers had long experience in the field, 61% with more than 15 years. 69% had their and their family's dwelling within the production unit where they work. 38% of the dwellings were located in close proximity to crops (less than 100 meters) and 50% within 500 m. The pesticide sprayer used by almost all the smallholders and farmworkers in the greenbelt was the backpack (85%), with the self-propelled crop sprayer with cab and activated charcoal filter reported by only one farmer. 13% of the productive units grew crops in greenhouses in the last year.

Table 4. Work practices and technology used by smallholders and farm workers of the Córdoba capital city greenbelt. 2012.

Work characteristics	Number	Valid (%)1
Extension of the productive unit (hectares)		
Small (≤ 10 ha)	57	58
Medium (11 to 40 ha)	34	33
Large (≥ 41 ha)	9	9
Area cultivated by worker (ha)		
≤ 10	70	71
11 - 20	11	11
21 - 40	9	9
≥ 41	7	6
Seniority in the horticultural work		
Average (years) 21.34 (SD: 14.58)		

≤ 5	15	17
6 - 10	13	14
11- 15	8	9
16 - 20	14	16
> 20	40	44
Dwelling distance to the nearest crop (meters)		
≤ 50	16	25
51 - 100	8	13
101 - 500	7	11
≥ 501	32	51
Greenhouse for crops in the productive unit		
Yes	13	13
Pesticide spray equipment		
Manual backpacks	77	77
Motor backpacks	7	7
Trailer crop sprayer without cab	28	31

[i] - [1]Percentage considering the total of responses.

The main vegetable crops cultivated in the green belt of Cordoba are leafy vegetables (Table 5): chard, lettuce, spinach, etc., with the particularity that they are grown throughout the year in a phased manner (Table 5). This means that on the farm at the same time there will be a patch prepared, a patch with the crop planted, another patch growing and another being harvested. These farms, located primarily in the northern greenbelt, are diversified with a large number of crops in many small lots. In these units, the farmer, tenant, and "mediero" work with their families or with hired laborers, carrying out the various farming tasks: transplanting, manual weed control, irrigation, pest control with manual (backpacks) sprays, harvesting, packing for market, loading and transport. The type of contract may be daily or for quantities.

In the southern area of the green belt, specialized farms have developed, devoted to potatoes as their main activity (22%) and rotation is incorporated into the production system with carrots (10%), wheat (9%) and soybeans (9%) variably according to the conditions of each crop year. These are large production units, with a greater degree of mechanization and automation. Most of the tasks are carried out with machinery, and pesticide application is performed with tractor-drawn and in some cases self-propelled machines. In these cases, labor is incorporated as needed, for example: chopping the seed potatoes and preparing them for planting, and harvesting at the manual collection stage. While potato harvester machines exist, they are not widespread in the green belt. The situation with carrots is similar.

Table 5. Principal crops grown in Córdoba capital city green belt, 2012.

Crops sprayed	Number	Valid (%)[1]	Average harvests per year
Chard *Beta vulgaris L. var. cicla*	71	75	3.75
Spinach *Spinacia oleracera L.*	70	69	3.28
Chicory *Cichorium intybus L.*	64	68	3.34
Scallion *Welsh onion Allium cepa L.*	68	67	2.79
Summer squash *Cucúrbita maxima*	66	65	2.46
Broccoli *Brassica oleracea L.*	65	64	2.93
Parsley *Petroselinum sativum Hoff*m.	62	61	2.98
White cabbage *Brassica oleracea*	61	60	3.22
Butterhead lettuce *Lactuca sativa L. var. Roman*a	59	58	3.47
Lettuce *Lactuca sativa L. var. crisp*a	58	57	3.83
Lettuce *Lactuca sativa L. var. capitat*a	57	56	3.61
Leek *Allium porrum L.*	56	55	2.13
Beet *Beta vulgaris L.*	56	55	3.59
Purple cabbage *Brassica oleracea L. var. capitat*a	55	54	3.11
Arugula *Eruca sativa L.*	55	54	4.08
Eggplant *Solanum melongena L.*	47	47	1.61
Cauliflower *Brassica oleracea var. botrytis L. subvar. Cauliflora*	41	41	2.40
Chinese cabbage *Brassica chinensis L.*	35	38	2.54
Radish *Raphanus sativus L.*	38	38	3.04

[i] - [1]Percentage considering the total of responses.

The most frequently used pesticides were herbicides (Table 6): glyphosate for 81% of the responses and metolachlor for 65%. In the group of insecticides, those most commonly handled were deltamethrin (72%), cypermethrin (65%), Imidacloprid (66%) and Chlorpyrifos (57%). The fungicides most frequently used were cabendazin (71%), mancozeb (63%), zineb (62%), and captan (50%).

Table 6. Most frequently used pesticides in the Córdoba capital city green belt, 2012.

Pesticides	(%)1
Insecticides	
Deltamethrin	72
Cypermethrin	65
Lambda-cyhalotrin	33
Cartap	40
Carbofuran	36
Carbaryl	34
Methiocarb	25
Chlorpyrifos	57
Dimethoate	50
Methamidophos	23
Imidacloprid	66
Endosulfan	46
Abamectine	35
Fungicides	
Azoxystrobin	47
Azoxystrobin + Ciproconazole	23
Carbendazim + Epoxiconazole	9
Mancozeb	63
Zineb	62
Maneb	15
Carbendazim	71
Captan	50
Chlorothanolil	38
Herbicides	
Glyphosate	81
Fluazifop p butil	46
Metolaclhor	65
2,4 D	19
Atrazine	12

Dicamba	14
Phenmediphan	27
Linuron	61
Metribuzin	31

[i] - [1]Percentage considering the total of responses.

The use (current or past) of banned pesticides was also surveyed: 33% reported having used Parathion, 16% Lindane, 10% Monocrotophos, 8% Methyl Bromide, 7% Malathion, 5% Aldicarb. Current use of Aldicarb was reported by two farmworkers; Monocrotophos and Aldrin were reported by a single case. Regarding good agricultural practices in this setting, 90% performed the triple washing of pesticide containers; but this was not accompanied by correct end use of the empty containers: 57% were stored, 17% were burned, 7% buried, and there were other misuses of contaminated containers.

Exposure Assessment

In a previous work [10], we proposed two indexes to describe pesticide exposure in applicators. The Intensity Level of pesticide Exposure (ILE) index measures instantaneous exposure intensity and the Cumulative Exposure Index (CEI) takes into account the average period of exposure, including the previous ILE information. Both indexes were constructed based on the Dosemeci proposal [23], carefully adapting the weighting procedure to our own context, and particularly to local professional opinion. The expressions of these measures are as follows:

$$ILE = (mix * PPE) + (\sum_{i=1}^{n} \frac{meth * PPE}{\# meth}) + (repair * PPE) + house\_dist$$

$$CEI = ILE + (\sum_{i=1}^{n} \log(1 + \frac{Ha/year}{55})$$

where *mix* represents a dichotomic response about mixing pesticides, *meth* the category of the method used with a certain *PPE*, *repair* the binary variable for which success is the positive response, *house_dist* the score indicating the applicator dwelling proximity to the nearest crop, and 55 the average of ha treated with a single load in the crop sprayer. These measures were denoted ILE_{EC} and CEI_{EC} for extensive crop worker's population. Lantieri et al. [10] calculated both measures for all subjects in the opening sample of terrestrial pesticide applicators of extensive crops (n=880) and using Bootstrap and Monte Carlo resampling methods, identified the most suitable theoretical stochastic distribution for each measure. In the present work, we assessed the two indexes once again but on a larger sample of applicators (n=1327) and stratifying by HEAs.

The ILE_{EC} and CEI_{EC} indexes were adapted to assess the specific exposure conditions of the population involved of farmworkers and smallholders in the green belt of Cordoba city. The methodology and definition criteria for the preliminary version of these two indexes were as described in [10]. These indexes are presented bellow (ILE_{GB} and CEI_{GB}):

$$IL\,E_{GB} = \left[(mix/load * syst) + \sum_{i=1}^{n} \frac{(meth\_apl)}{\#meth}\right] * PPE1 + (wash * PPE2) + (rep * PPE3) * hyg * spill$$

$$CEI_{GB} = IL\,E_{GB} * Duration * Frequency$$

where *mix/load* represents a dichotomic response about mixing or loading pesticide; *syst* the sprayer system (open or closed); *meth_apl* the method of performing pesticide application; *PPE1*, the score of use of Personal Protective Equipment for spraying crops, as described before; *wash* is also a dichotomic variable for washing the pesticide application equipment (backpack or machine); *PPE2*, the score of use of Personal Protective Equipment, as described before, for washing the machine and/or backpack; *Rep*, whether repairing application equipment; *PPE3*, the score of use of Personal Protective Equipment for repairing equipment; *hyg* hygiene mode after completing the task with pesticides; and *spill* the behavior during a pesticide spill on clothing, thus, whether the worker changes clothes immediately after the spill or not. The cumulative exposure index incorporates the intensity level of pesticide exposure, the *duration* (years) and *frequency* of exposures (number of days of applications per year).

Tables 7 and 8 show summary statistics for both the measures, constructed for exposure assessment in first population (extensive crops). As can be seen, mean values for both indexes were generally quite different from their medians, indicating empirical distributions different from the normal distribution. Significant differences between ecological areas were found for ILE_{EC} ($p=0.013$) and CEI_{EC} ($p=0.003$). For the former, areas I and III showed the lower and similar values ($p=0.201$) for exposure index, while area V had the highest average ($p<0.01$). As an intermediate group, there was no difference between areas II and IV ($p=0.203$), and these yielded higher values than those obtained in areas I and III ($p<0.001$).

Table 7. Summary statistics of Exposure Index distribution based on pesticide applicators of Extensive Crops (EI_{EC}) information, regarding to Homogeneous Ecological Area Classification. Córdoba, Argentina.

	AREAS					
	I	II	III	IV	V	Total
n	41	641	230	156	259	1327
Statistics	Exposure Index distribution					
Mean	2.04	3.02	2.37	2.76	3.59	2.92
Standard Deviation	2.14	2.36	2.29	2.19	2.57	2.40
Median	0.94	2.61	0.94	2.49	3.24	2.6
Standard Error	0.29	0.09	0.13	0.17	0.15	0.06
p25	0.61	0.86	0.72	0.84	0.89	0.82
p75	2.61	4.47	3.66	3.74	5.66	4.34
Minimum	0	0	0	0	0	0
Maximun	8.80	10.15	9.23	8.94	10.93	10.93

Table 8. Summary statistics of Cumulative Exposure Index distribution based on pesticide applicators of Extensive Crops (EI_{EC}) information, regarding to Homogeneous Ecological Areas Classification. Córdoba, Argentina.

	AREAS					
	I	II	III	IV	V	Total
N	41	641	230	156	259	1327
Statistics	Cumulative Exposure Index distribution					
Mean	23.28	44.43	42.34	62.13	59.97	48.02
Standard Deviation	56.46	67.88	66.84	78.25	86.38	72.76
Median	2.18	16.18	13.80	28.52	27.56	17.83
Standard Error	7.61	2.61	3.70	6.17	5.14	1.88
p25	0	0	0	4.31	2.58	0
p75	21.48	56.98	50.62	89.15	84.61	62.36
Minimum	0	0	0	0	0	0
Maximun	383.3	534.877	383.34	370.96	514.2	534.87

For the cumulative exposure index, the differences structure between areas was slightly different. Only areas II and IV were similar (p=0.270) showing intermediate values, while areas I and V yielded lower and higher averages (p<0.001) of the cumulative exposure measure. Figure 3 (first row) presents the box plots for both indexes for the five ecological areas.

When the log of applicator age was included as a covariate, the above results held. The estimate of regression coefficient (slope) for this covariate was equal to b=-0.40 (SE 0.15) and significant (p=0.044), showing that there is an inverse ratio between the exposure index and the log of age. Since the log is a mathematical monotone (increasing) function, this coefficient indicates that the younger workers have higher exposure. In contrast, the age pattern for the cumulative exposure index indicated a direct ratio: the coefficient estimate was 0.43 (SE 0.18), which means that, as expected, that older workers have higher values of cumulative exposure. Figure 3 (second row) illustrates this behavior.

Finally, personal protection was strongly associated with the differences between the areas for both indexes (p<0.001), indicating that in ecological areas with rural workers with lower cumulative exposure, the protection feature used was ideal (Figure 4). There was no association (p=0.695) between CEI_{EC} and the marital status of subjects.

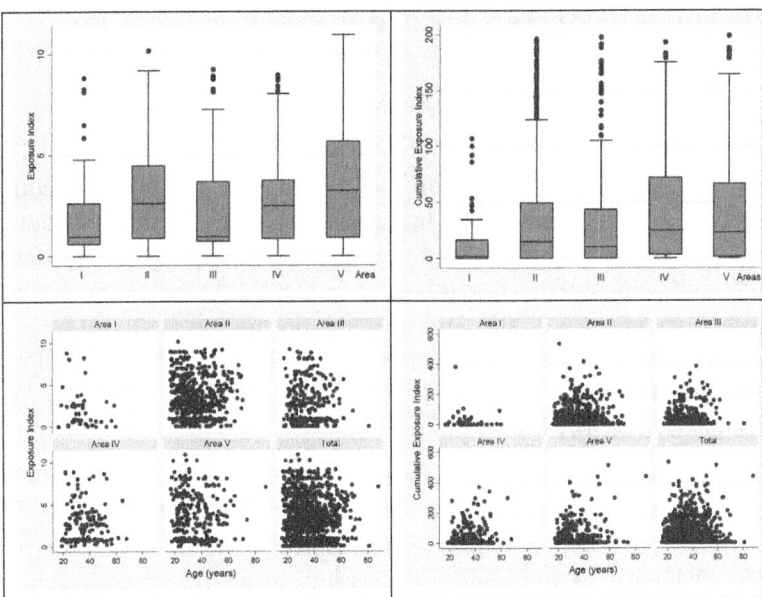

Figure 3. Box plots (above) of Exposure Index (EI_{EC}) and Cumulative Exposure Index (CEI_{EC}) and scatter plots (below) for these indexes *versus* age (years) of workers, for Ecological Areas.

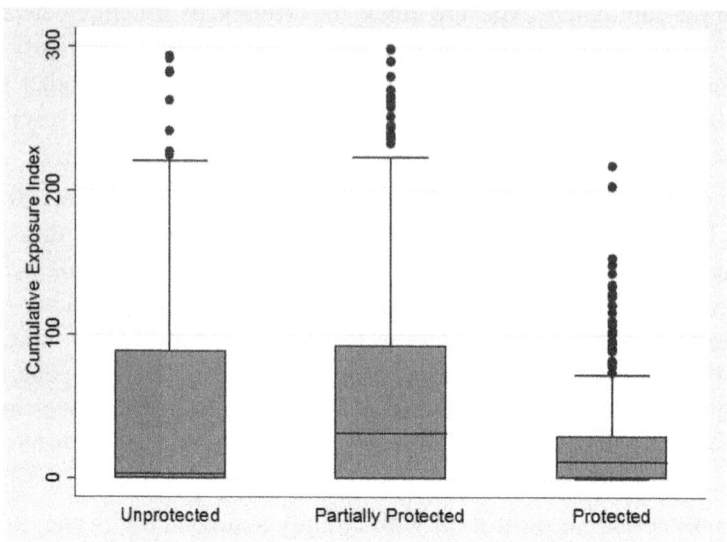

Figure 4. Box plot of Cumulative Exposure Index (CEI_{EC}) in terms of the personal protection category, Córdoba, Argentina.

Health Status of Workers Related to Pesticide Exposure

Extensive Crop Pesticide Applicator Population.

A previous study reported a high prevalence of symptoms: 47.4% with occasional or frequent irritative symptoms, 35.5% fatigue, 40.4% headache and 27.6% anxiety or depression [1]. Increased frequency of medical consultation and hospitalization was associated with the use of chlorpyrifos ($p<0.001$ and $p=0.05$) and endosulfan ($p<0.001$ and $p=0.021$) insecticides, exposure to multiple pesticides ($p<0.001$) and seniority in the job ($p<0.001$). Only 32% of workers were adequately protected. The proper use of Personal Protective Equipment (PPE) (OR: 0.45, SE. 1.56) and marital status (OR 0.16, SE. 1.62) were protective factors for hospitalization.

Within HEAs, there was a difference between Homogeneous Ecological Areas II and III in the probability of medical consultation at least once for reasons related to occupational exposure to pesticides ($p<0.02$), with agricultural workers of HEA III having more probability of medical consultation. In the other health-related variables, no statistical differences were found.

Smallholders and Farmworkers of the Green Belt Population

In this sensitive population, occasional or frequent manifestation of irritative symptoms affected 49.3%, fatigue 35.6%, headache 52.6%, nervousness or depression 30.6%, dizziness 13.7% and excessive sweating 16.7%, and 18% had had an accident with pesticides. The prevalence of medical consultation and hospitalization was lower than expected: 22.2% and 4% respectively (Table 9). No statistical association was found between these two variables and exposure to specific pesticides.

Table 9. Prevalence of symptoms, health assistance and accidents related to occupational exposure among smallholders and farmworkers of the Córdoba capital city green belt. 2012

Symptoms	Never / Rarely	Sometimes / Frequently	Number
Fatigue - tiredness	64.4	35.6	73
Nervousness ordepression	69.4	30.6	72
Headache	47.4	52.6	73
Irritative Symptoms	50.7	49.3	73
Dizziness or vertigo	86.3	13.7	73
Excessive sweating	83.3	16.7	72
Health assistance	Never	Once or more times	Number
Medical consultation	78.8	22.2	80
Hospitalization	96.0	4.0	75
Accident with pesticide			
Yes	82	18	95

[i] - [1]Percentage considering the total of responses.

Workers' Risk Perception of Different Agrochemicals

We studied the perceived threat level of pesticides used. In the extensive crop pesticide applicator population, there was a high perception of danger (85.76 - 98%) only for insecticides, with the highest perception of danger for organophosphate. Herbicides and fungicides were considered less hazardous (35.09% - 91.54% and 49.07 - 58.55% respectively). Glyphosate, the most widely used pesticide in crops (98% use in the past year), was considered hardly or not at all dangerous. The level of protection used did not vary according to the perception of risk.

Among workers and smallholders of intensive crops, the insecticide group was also seen as presenting the highest perception of risk: between 33.3% and 86% felt that they are dangerous or very dangerous, with organophosphates and organochlorines seen as the most dangerous. Fungicides and herbicides were perceived as less dangerous (29% - 38% and 33% - 65%).

DISCUSSION

This work presents an interesting update of our previous work on extensive crops of Córdoba province, stratified by the Homogeneous Ecological Areas (HEAs) [10]. It also includes, for first time in our country, a characterization of the horticultural smallholder and farmworker populations of the greenbelt of the provincial capital, both settings being recognized as vastly different in pesticide exposure determinants, based on professional judgment. The analysis of each agricultural scenario enabled groups with occupational exposure to pesticides to be identified in each particular labor context (extensive and horticultural crops), as well as the health conditions associated with occupational agrochemical use.

When evaluating the pesticide applicator population of extensive crops, we founded statistically significant differences between areas in age, education, marital status, seniority in the task, average/year of hectares sprayed, use of pesticides with prescription signed by an agricultural engineer, self-propelled crop sprayer with cab and activated charcoal filter, and protection level. Only trailed crop sprayer with cab and activated charcoal filter was similar in all the areas. Self-propelled and trailed crop sprayer combined showed an average 55.5% use in all areas, which means that a large percentage of workers used unsafe machinery, i.e., sprayer with no cab or cab without activated charcoal filter, and this was an important determinant of exposure and was more pronounced in HEA II ("Middle Agricultural and Livestock Area"), followed by HEA V ("South-eastern Agricultural and Livestock Area"). HEA I ("North-western Extensive Livestock Area") was traditionally characterized by grazing cattle but it is now a newly developed agricultural region, due to the nationwide agriculturization process. This area's applicators had the highest level of personal protection and of using pesticides with written prescription, followed by HEA II. Others areas with a historical agricultural tradition, such as HEA IV ("South-eastern Agricultural Area") and HEA V, did not have similar protective measures or a safe work environment; in fact, the highest rates of unprotected or partially protected applicators were found in these areas.

Our current results confirm previous works [10, 16] and MacFarlane's study [28] reporting no association between instruction level and personal protection. Indeed, HEA IV, with the highest percentage of applicators that had completed secondary school or higher, had only 35.3% of workers completely protected during the task. Likewise, we found no association between marital status and PPE use, as in the case of HEA IV, with the highest percentage of married or cohabiting subjects.

Based on two indexes proposed in previous work [10] for the assessment of pesticide exposure risk, the intensity and accumulated exposure indexes (ILE and CEI), the current assessment was performed in a larger sample of terrestrial pesticide applicators of extensive crops, stratifying by HEAs and showing significant differences among these for ILE_{EC} (p=0.013) and CEI_{EC} (p=0.003). The results reinforce the previous hypothesis of the emergence of different new risk scenarios in the province. As expected, HEA V yielded the highest averages for both indexes, followed by areas II and IV. It should be stressed that the differences between areas in both measures were strongly associated with the personal protection used (PPE).

As reported in a previous study, we continue to find a lack of enforcement of existing regulations (Law N° 9164) in all the agricultural settings of the province, with low use of pesticide prescriptions signed by an agricultural engineer, and poor implementation of good agricultural practices such as triple washing of pesticide containers and their correct disposal. Burning, burying or reusing agrochemical containers, a common practice in the study populations, add other risk factors for applicators, as well as abiotic and biotic environmental contamination.

As expected, in contrast with extensive crop settings, wide differences were found in exposure determinants in the greenbelt population of Cordoba city, between their social and demographical characteristics and compared with other agricultural scenarios of the province, as shown above and in previous works [10, 16]. Horticultural workers had a greater average age, long experience in the task, lower educational level, and a high proportion of Bolivian workers and national migrants. Part of the population had unsatisfied basic needs: 23% lacked a running water supply and 13% a bathroom in the dwelling. Precarious living conditions were associated with being a "mediero" (see below), or an employee and a migrant, particularly Bolivian. It is thus a heterogeneous and highly vulnerable population, which favors lax labor structures for their work, leading to scenarios in which a higher rate of occupational health risk is to be expected. Seniority in the job was associated with higher cumulative exposure to pesticides, in turn associated with various deleterious effects on health [29].

The heterogeneity of this population is also seen in the different job roles, employment status and land tenure conditions of the smallholders and farmworkers. The agrarian structure has become dominated by family farms, giving rise to processes of social differentiation, concentration of land and capital, and the emergence of a new social actor: the "mediero", a kind of sharecropper that almost monopolizes the supply of labor by having their family take part in the work. This has transformed the social organization of horticultural work and is extremely functional [30] in that the existence of "medieros" often hides the figure of an unregistered employee, with the advantage for the farmers of transferring some of the risk, while avoiding compliance with labor legislation, social security and occupational risk prevention [31]. It enables them to turn fixed labor costs into variable costs, distribute downward the fluctuations in prices and profitability that are typical of fresh vegetable production, obtain a more stable workforce, delegate responsibilities and reduce the need for control, among others.

The active participation of the family (31%), as in the greenbelt, and the short distance from the home to the cultivation sites (38% less than 100 m), as also reported by applicators in extensive crops, (almost half of them live within 500 m of the nearest crop), leads to non-occupational exposure of the worker after work and para-occupational exposure of the other family members. McCurdy et al., Chaio-Cheng et al., Clifford et al., Loewenherz et al, and Lu et al., [as cited by 32] reported studies suggesting a take-home pathway for pesticides. Applicators and farmworkers accumulate chemicals on their clothing and skin, and can carry these into their homes. The homes of agricultural workers have higher pesticide concentrations in house dust than other homes in the same agricultural community. Children living there have elevated urinary metabolites of organophosphorus pesticides. Regarding dwelling location, higher levels of pesticides were found in dust samples in farmers' dwellings and non-agricultural reference homes closer to orchards [33].

In the greenbelt, the staggered mode in which a diversity of crops are grown allows farmers to grow a large number of crops in small plots, leading to a higher frequency of pesticide application. There is thus a heavy burden of pesticides in both scenarios: in extensive crops, due to the extensive areas sprayed, and in horticultural crops, to the process of spraying throughout the year. This also implies significant environmental pollution, with approximately 47% of the product deposited in adjacent soils and waters or dispersed in the atmosphere [34], depending on climatic conditions such as rain and wind direction, geological features such as soil type and the presence of water currents, and other factors such as the formula and presentation of the product as well as the application technique. Other phenomena promoting environmental spread are photodegradation and volatilization, leaching and surface soil washing, both related to streams and rainfall [35].

Other modern phenomena aggravate the level of pollution and affect the dynamics of farming in the greenbelt. The advance of crops such as cereals and oilseeds, mainly soybeans, over horticultural production, causes the greenbelt to shift towards other neighboring districts [36]. Moreover, the increase of housing and of informal settlements in urban residential areas, coupled with inadequate planning and land management, further reduces and displaces horticultural production [37]. This is exacerbated by industrial development: the dominant industrial area (including dangerous industrial areas), increased from 8,000 ha (15.1%) to 12,000 (21%) between 2004 and 2012, while the predominantly rural area fell from 29% to 27.5% in the same period [38]. While about 40% of the area sown in the Capital Department is horticultural production [39], it is estimated that it has fallen from 11,000 hectares in 2004 [40] to an area of 5,500 hectares in 2012 [36]. Thus the greenbelt is now located in an urban area with a sum of overlapping environmental risks (caused by agriculture and pesticide pollution as well industrial pollution), making the Capital Department of Cordoba an area of high environmental risk [41].

The informality and precariousness of the situation endured by greenbelt workers is more complex than that of those in extensive crops, whose working conditions are more modern, regulated and safer. The wide diversity of greenbelt workers' tasks in contact with pesticides, the greater burden of insecticides resulting from the type of crops grown, and the application of risky technologies such as spraying with backpacks, also make this group of workers more vulnerable. The broad spectrum herbicide glyphosate, the most frequently used pesticide in this setting, is applied in the vicinity of the crops. Insecticides are also used several times during the crop cycle, as well as fungicides. The level of exposure and the likelihood of acute poisoning in these groups are thus substantially higher due to the continuous contact [34], which is for relatively short periods but is still intense and repetitive during the work day, causing toxic effects that vary depending on the type and amount of pesticide.

Work activity as a source of exposure to pesticides has been widely recognized in farm workers who mix, transport, carry, store or apply them [42]. The magnitude and severity of occupational pesticide exposure, its effects and consequences, cannot be measured exclusively by the classical indicators of mortality and morbidity. The apparent underreporting of cases of acute pesticide poisoning [43] hides the true extent of the problem in rural areas, where some authors report a deficit of up to 50% in reporting these events [44]. The adverse health effects reported in this study show a serious impact on exposed workers. The prevalence of acute and subacute symptoms reported in our study in both groups – extensive and intensive farming – with 47.4% and 49.3% irritative symptoms, 35.5% and 35.6% fatigue, 40.4% and

52.6% headache, 27.6% and 30.6% nervousness or depression, 35.6% and 22.2% rate of activity-related medical consultation, and 5.4% and 4% of hospitalization, respectively, show the high occupational exposure, and may be categorized as indirect indicators of the exposure level, unlike the recording of cases of pesticide poisoning. Argentina reported one of the highest indexes of agricultural accidents at work (94.8‰), with a mortality rate of 195 cases per million workers, only surpassed by the construction sector (229‰) [45]. The Province of Córdoba concentrates 88% of the labor sector in that area.

There are several factors involved in the occurrence of these high levels of accidents. The higher consumption of pesticides (kg/year), the toxicity and diversity of agrochemicals applied, the extent of the areas sprayed, the laxity of State monitoring, the prevailing weather conditions and, particularly, the everyday working conditions of applicators, are among the main variables that shape the patterns of occupational exposure to pesticides. This study provides evidence for this hypothesis and helps to analyze the risk. The association between the symptoms reported, as well as the increased hospitalizations and medical consultation among those exposed to certain insecticides, such as chlorpyrifos and endosulfan, as observed previously [46], provide evidence in this regard. Symptoms reported here, and the frequency of their occurrence, match other reports in Argentina and elsewhere showing a positive correlation between health effects and occupational exposure to pesticides [47-50].

Pesticide hazard perception can be associated with the occupational exposure risk prevention in agricultural settings. Our study found a low perception of hazard in relation to herbicides and fungicides and a higher perception to the group of insecticides in both populations, although the smallholders and farmworkers reported lower risk perceptions in all pesticide groups than terrestrial applicators of extensive crops. But it should be noted that the different risk perception reported in our study did not lead to variations in PPE. The hazard perception of insecticides may be explained by the acute toxicological data, and not by the volumes applied, the possibility of dispersal, environmental persistence and the likelihood of chronic health effects. Another explanation proposed for this behavior is that the pesticide use in agriculture is not perceived as risky for the environment due basically to trust in the improvement of product quality, in the technological innovation that has taken place in the last few years and in the work of official agencies responsible for approving pesticides [51].

The absence of the agrochemical prescription, as well as the lack of implementation of formally regulated programs to ensure the correct end use of empty pesticide containers in both agricultural settings studied, indicate the weakness of compliance with the provincial regulations in force [24].

The results of the two subject groups present a picture of highly vulnerable populations, which must be considered in risk assessment, and in particular in the implementation of prevention strategies. Comprehensive knowledge of the study population is a priority in designing and strengthening protective measures for improving the health and safety conditions of workers and their families. The presence of highly vulnerable groups, such as women of childbearing age and children at all stages of growth, must be taken into account in assessing the problem, including approach strategies [14].

We proposed an analytical approach to assess workers' exposure to pesticides that takes advantage of existing comprehensive information about pesticide uses as well as about the main working habits of subjects, which is of relatively simple application. The information from assessing the indexes includes some observations relative to the specific local exposure scenario [10] in which the different variables that influence or determine exposure have been weighted and combined. Even though this approach does not give accurate estimates of individual exposure but rather pragmatic information on the risks faced by the workers and, consequently, of the presence or absence of a need for preventive interventions, we believe that these measures provide a valuable monitoring tool in our context.

There are some limitations to this study. Because of the complexity described in labor relations in the greenbelt, there is some selection bias in this study population due to difficulties in accessing directly exposed workers. The laxity in the employment relationship, the informality with which employment contracts are made, the uncertainty regarding operating times and the undocumented status of many of these workers [18], are some of the reasons for this, as has also been reported by other authors [52]. Secondly, information on pesticide use and on PPE, as well as some work practices, was based on self-reporting in the interview questionnaires. Thus, errors in recall and reporting may have occurred. A preliminary validation study was conducted, though only for the population of extensive crop workers (n=60), using a short version of questionnaire. Results (not shown here) indicated that the match between the volunteer farmers' questionnaire responses on both occasions was acceptable. Finally, the potential for differential exposure misclassification as reported by terrestrial applicators has been recognized in the present study by proposing the assessment of specific indexes describing the exposure. However, these measures weight, substantially, the use and the amount of pesticides applied in their usual work. Data from the National Cancer Institute studies found little evidence for differential recall of pesticides by farmers [53]. Since applicators are heavily involved in all aspects of pesticide manipulation/operation and this is practically their single occupation, they have a good memory for all the pesticides used. Further research will be carried out explore this in our populations.

CONCLUSION

The evidence presented describes a problem whose complexity is difficult to cover through the usual approaches. Exposure to pesticides in workers responsible for applying these is high. A variety of economic and socio-cultural factors affect exposure and only through a proper evaluation can its true dimension be identified and quantified. The assessment and monitoring of these populations allows us to obtain information about the risk factors associated with occupational exposure and the consequent health damage.

Recognizing the complexity of the processes underlying the vulnerability of these populations to pesticide exposure is a first step to significant change in preventive health. Adopting a comprehensive view of the different aspects of the problem will favor the reception of preventive proposals and their chances of application. The exposure reported here seriously conspires against this activity's desired goal of sustainability, creating serious health and environmental risks with costs that are underestimated in the balance of these operating models. From an economic perspective, action to reduce the risks of exposure and adverse effects of the use of pesticides and to contribute to maintaining and improving public health and the quality of life, supports economic development in all sectors of the country, especially in production. Workers and their families improve their quality of life and their family's economy and social security. Companies do not incur high costs of care for acute and chronic intoxication, disability and compensation. Employers benefit from a real decrease in absenteeism and staff turnover, and the country has a more dynamic and competitive work force. Consequently, such action is a factor that strengthens the development of the country.

ACKNOWLEDGEMENTS

Grants: This study was supported by National Agency of Technological and Scientific Promotion, through the Fund for Scientific and Technological Research (FONCYT) grant PICT 2008-1814 and Secretary of Science and Technology, University of Córdoba grant SECyT-UNC - 162/12. We are grateful to the Agriculture, Livestock and Food Ministry of Cordoba Province.

We would also like to thank all of the workers who agreed to participate in this study.

REFERENCES

1. Ortiz-Hernández ML., Sánchez-Salinas E., Olvera-Velona A., Folch-Mallol JL. Pesticides in the Environment: Impacts and its Biodegradation as a Strategy for Residues Treatment. Rijeka: In Tech; 2011. http://www.intechopen.com/books/pesticides-formulations-effects-fate (accessed 13 June 2013).
2. Osman KA. Pesticides and Human Health. Rijeka: In Tech; 2011. http://www.intechopen.com/books/pesticides-in-the-modern-world-effects-of-pesticides-exposure (accessed 8 March 2013).
3. Schaaf AA. Uso de pesticidas y toxicidad: relevamiento en la zona agrícola de San Vicente, Santa Fe, Argentina. Revista Mexicana Ciencias Agrícolas 2013;4(2): 323-331.
4. Sistema Integrado de Información Agropecuaria IIA. Estimaciones Agrícolas. Datos de la Dirección de Información Agrícola y Forestal. Ministerio de Agricultura, Ganadería y Pesca 2013. http://www.siia.gov.ar/series (accessed 8 June 2013).
5. Viglizzo EF., Jobbágy E. Expansión de la frontera agropecuaria en argentina y su impacto ecológico-ambiental. Buenos Aires: Ediciones INTA; 2006.
6. Censo Nacional Agropecuario. Dirección General de Estadísticas y Censos. Provincia de Córdoba. http://web2.cba.gov.ar/actual_web/estadisticas/censo_agropecuario/index.htm (accessed 8 June 2013).
7. Cámara de Sanidad Agropecuaria y Fertilizantes (CASAFE). Estadísticas. Datos del Mercado Argentino de Fitosanitarios. Crece el volumen de uso de fitosanitarios, pero cae su toxicidad. http://www.casafe.org/biblioteca/estadisticas/ (accessed 19 August 2013).
8. Dirección General de Estadísticas y Censos. Gobierno de la Provincia de Córdoba. Censo Nacional de Población, hogares y viviendas 2010. http://web2.cba.gov.ar/actual_web /estadisticas/censo2010/index.htm (accessed 13 August 2013).
9. Alvarez FM., Harrington ME., Maccagno M.A., Maciá MR., Ribotta BS., Peláez E. Características sociodemográficas de la población de la Provincia de Córdoba. Ensayos Demográficos y Sociales. Córdoba: Centro de Estudios de la Población y el Desarrollo (CEPyD); 2004. http://www.cepyd.org.ar/pdfs/Cordobeses-Fasciculo1.pdf (accessed 25 August 2013).
10. Lantieri MJ., Butinof M., Fernández R., Stimolo MI., Blanco M., Díaz MP. Work Practices, Exposure Assessment and Geographical Analysis of

Pesticide Applicators in Argentina. Rijeka: In Tech; 2011. http://www.intechopen.com/books/pesticides-formulations-effects-fate (accessed 13 June 2013).

11. Lara S. Nuevas experiencias productivas y nuevas formas de organización flexible del trabajo en la agricultura mexicana. México: Procuraduría Agraria, Juan Pablo Editor; 1998.

12. Paulino E, De Almeida R. Terra e território a questão camponesa no capitalismo. Sao Paolo: Editora Expressão Popular; 2010.

13. FAO. Buenas Prácticas Agrícolas para la Agricultura Familiar. Cadena de las principales hortalizas de hojas en Argentina. http://www.fcagr.unr.edu.ar/PHR/9%20BPA%20para%20Hortalizas%20de%20Hoja%202010.pdf (accessed 5 August 2013).

14. Ministerio de Salud. Secretaría de Ambiente y Desarrollo Sustentable. La problemática de los agroquímicos y sus envases, su incidencia en la salud de los trabajadores, la población expuesta por el ambiente. Buenos Aires: Organización Panamericana de la Salud (OPS) y Asociación Argentina de Médicos por el Medio Ambiente (AAMMA). http://www.ambiente.gov.ar/archivos/web/UniDA/File/LIBRO%20Agroquimicos.pdf (accessed 3 July 2013).

15. García Fernandez J., Casabella C., Marzi AA., Astolfi E., Roses O., Donnewald H., Villamil E. Organochlorinated pesticides in the Argentine Antarctic sector and Atlantic coastline waters. Geogr Med 1979;9: 28-37.

16. Lantieri MJ., Meyer Paz R., Butinof M., Fernández RA., Stimolo MI., Díaz MP Exposición a plaguicidas en agroaplicadores terrestres de la provincia de Córdoba: Factores condicionantes. Agriscientia 2009;26(2): 43-54.

17. Municipalidad de Córdoba. Córdoba, una ciudad en cifras. Guía Estadística de la Ciudad de Córdoba. http://www.cordoba.gov.ar/cordobaciudad/principal2/docs/informacionestrategica/sie/GuiaEstadistica2004.pdf (accessed 8 August 2013).

18. Machado AL., Ruiz MV., Sastre MA., Butinof M., Blanco M., Lantieri MJ., Fernández R., Stimolo MI., Franchini G., Díaz MP. Exposición a plaguicidas, cuidado de la salud y subjetividad. Revista Kairos 2012; 30. http://www.revistakairos.org/k30-archivos/Machado.pdf

19. Benencia R., Gazzotti A. Migración limítrofe y empleo: precisiones e interrogantes. Estudios Migratorios Latinoamericanos 1999;31: 513-609.

20. Benencia R., Quaranta G. Reestructuración y contratos de mediería en la región pampeana argentina. Revista Europea de Estudios Latinoamericanos y del Caribe 2013;74: 65-83.

21. Ministerio de Educación. La horticultura en Argentina. Buenos Aires: Instituto Nacional de Educación Tecnológica. http://www.inet.edu.ar/actividades/foros/Horticultura.doc (accessed 29 August 2013).
22. Arcury AT., Quandt SA., Russell GB. Pesticide Safety among Farmworkers: Perceived Risk and Perceived Control as Factors Reflecting Environmental Justice. Environ Health Perspect 2002;110(2): 233–240.
23. Dosemeci M., Alavanja MCR., Rowland AS., Mage D., Zahm SH., Rothman N., Lubin JH., Hoppin JA., Sandler DP., Blair A. A quantitative approach for estimating exposure to pesticides in the Agricultural Health Study. Ann Occup Hyg 2002;46(2): 245-260.
24. Ley Provincial 9164: Productos químicos o biológicos de uso Agropecuario. Legislatura de la Provincia de Córdoba 2004. http://web2.cba.gov.ar/web/leyes.nsf/85a69a561f9ea43d03257234006a8594/1139770d4dcfd55603257250005a60fc?OpenDocument (accessed 15 June 2013).
25. Tártara E., Apezteguía J., Roberi A., Bocco M., Adib O. Características de los sistemas frutihortícolas bajo riego del Cinturón Verde de Córdoba. Córdoba: Universidad Nacional de Córdoba; Secretaría de Agricultura, Ganadería y Recursos Renovables; Municipalidad de Córdoba; 1998.
26. Dirección de urbanismo. Municipalidad de Córdoba. Córdoba en su situación actual: bases para un diagnóstico. http://www.cordoba.gov.ar/cordobaciudad/principal2/DOCS/PLANEAMIENTO/Cordoba%20en%20su%20situacion%20actual%20TEXTO%202000.PDF (accessed 1 August 2013).
27. Instituto Nacional de Educación Tecnológica (INET). La horticultura en Argentina, 2010. http://www.inet.edu.ar/actividades/foros/Horticultura.doc (accessed 27 July 2013).
28. Macfarlane E., Chapman A., Benke B., Meaklim J., Sim M., McNeil J. Training and other predictors of personal protective equipment use in Australian grain farmers using pesticides. Occup Environ Med 2008;65(2): 141-146.
29. Butinof M., Fernandez RA., Stimolo MI., Lantieri MJ., Blanco M., Machado AL., Franchini G., Díaz Mp. Agricultural applicators health profile in Córdoba Province, Argentina. Cadernos de Saude Publica. In revision.
30. Benencia R. La horticultura bonaerense: lógicas productivas y cambios en el mercado de trabajo. Desarrollo Económico 1994;34(133): 53-73.
31. García M., Lemmi S. Política legislativa y trabajo en la horticultura del Área Metropolitana de Buenos Aires (Argentina): Orígenes y

continuidades de la precarización laboral en la horticultura. Secuencia 2011;79: 89-112.

32. Thompson B, Coronado GD, Grossman JE, et al. Pesticide Take-Home Pathway among Children of Agricultural Workers: Study Design, Methods, and Baseline Findings. J Occup Environ Med 2003;45: 42–53.

33. Quandt SA., Hernández-Valero MA., Grzywacz JG., Hovey JD., Gonzales M., Arcury TA. Workplace, Household, and Personal Predictors of Pesticide Exposure for Farmworkers Environ. Health Perspect 2006;114: 943–952.

34. Organización Mundial de la Salud (OMS), Organización Panamericana de la Salud (OPS), Centro Panamericano de Ecología Humana y Salud. Serie Vigilancia, 9. Plaguicidas organoclorados. México: OMS/OPS; 1990.

35. Organización Mundial de la Salud (OMS), Organización Panamericana de la Salud (OPS), División Salud y Ambiente. Plaguicidas y Salud en las Américas. Washington: OMS/OPS; 1993.

36. Secretaria de comercio interior. Corporación del mercado central de Buenos Aires. La producción de hortalizas en Argentina. http://www.mercadoc entral.gob.ar/ziptecnicas/la_produccion_de_hortalizas_en_argentina.pdf. (accessed 21 July 2013).

37. Legeren AA. Parcelamiento rural en la sustentabilidad de una ciudad posmoderna, Córdoba, Argentina, 2000. Observatorio Geográfico América Latina. http://observatoriogeograficoamericalatina.org.mx/egal8/Geografiasocioeconomica/Geografiarural/18.pdf (accessed 10 August 2013).

38. Municipalidad de Córdoba. Córdoba, una ciudad en cifras. Guía Estadística de la Ciudad de Córdoba año 2012. http://www.cordoba.gov.ar/cordobaciudad/principal2/default.asp?ir=26_16 (accessed 8 August 2013).

39. Dirección General de Estadísticas y Censos. Gobierno de la Provincia de Córdoba. Censo Nacional Agropecuario año 2008. web2.cba.gov.ar /actual_web/ estadísticas /censo_agropecuario/index.htm (accessed 6 August 2013).

40. Moya G. Análisis para la gestión integrada de los recursos hídricos de los ríos Suquía y Xanáes, provincia de Córdoba. MSc thesis. Universidad Nacional de Córdoba; 2004.

41. Defensoría del Pueblo de la Nación. Atlas de Riesgo Ambiental de la Niñez de Argentina. Buenos Aires: PNUD, UNICEF, OPS, OIT; 2009.

http://defensoresymedios.org.ar/wp-content/uploads/2010/04/Atlas.pdf (accessed 27 July 2013).

42. López CL. Exposición a plaguicidas organofosforados. Perspectivas en Salud Pública. México: Instituto Nacional de Salud Pública, 1993.

43. Henao S, Nieto O (s.f.). Curso de autoinstrucción en diagnóstico, tratamiento y prevención de intoxicaciones agudas causadas por plaguicidas. División de Salud y Ambiente de la Organización Panamericana de la Salud (HEP/OPS), Centro Panamericano de Ingeniería Sanitaria y Ciencias del Ambiente (CEPIS/OPS) y proyecto PLAGSALUD de la OPS/OMS. 2008; 129. http://www.cepis.org.pe/tutorial2/e/unidad1/index.html (accessed 8 Jun 2011).

44. Altamirano JE., Franco M., Bovi Mitre G. Modelo epidemiológico para el diagnóstico de intoxicación aguda por plaguicidas. Jujuy, Argentina. Rev Toxicol 2004;21: 98-102.

45. Ministerio de Salud de la Nación. Boletín Epidemiológico Anual 2009. http://www.msal.gov.ar/htm/site/sala_situacion/PANELES/boletines/BEPANUAL_2009.pdf (accessed 13 Aug 2013).

46. Alavanja MC., Sandler DP., McDonnell CJ., et al. Factors associated with self-reported, pesticide-related visits to health care providers in the agricultural health study. Environ Health Perspect 1998 Jul; 106(7): 415-20.

47. Matos EL., Loria DJ., Albiano N., et al. Efectos de los plaguicidas en trabajadores de cultivos intensivos. Rev Med Inst Mex Seguro Soc 2008;46(2): 145-152.

48. Mourand TA. Adverse impact of insecticides on the health of Palestian farm workers in the Gaza Strip. Int J Occup Environ Health 2005;11: 144-149.

49. Palacios-Nava ME., Moreno-Tetlacuilo LMA.. Diferencias en la salud de jornaleras y jornaleros agrícolas migrantes en Sinaloa, México. Salud Pública Mex 2004;46: 286-293.

50. Cortés-Genchi P., Villegas-Arrizón A., Aguilar-Madrid G., et al. Síntomas ocasionados por plaguicidas en trabajadores agrícolas. Rev Med Inst Mex Seguro Soc 2008;46(2): 145-152.

51. Ramirez O. Percepción del riesgo del sector agroindustrial frente al uso agrícola de plaguicidas: la soja transgénica en la Pampa Argentina. Ambiente y desarrollo 2010;14(26): 35-62.

52. Arcury TA., Quandt SA., Barr DB., Hoppin JA., McCauley L., Grzywacz JG., Robson MG. Farmworker Exposure to Pesticides: Methodologic

Issues for the Collection of Comparable Data Environ Health Perspect 2006;114: 923–928.
53. Blair A., Zahm SH. Patterns of pesticide use among farmers: implications for epidemiologic research. Epidemiology 1993;4: 55-6233.

Chapter 10

PHOTOSYNTHETIC RESPONSE OF TWO RICE FIELD CYANOBACTERIA TO PESTICIDES

Binata Nayak[1], Shantanu Bhattacharyya[1], and Jayanta K. Sahu[2]

[1]School of Life Sciences, Sambalpur University Jyoti Vihar, Burla, Odisha, India
[2]Trust Fund College, Bargarh, Odisha, India

INTRODUCTION

It was 2.4 billion (Ga) years ago that oxygen accumulated atmosphere began in our planet and cyanobacteria (earlier known as Bluegreen algae) are inhabitant of almost 3.5 billions years ago. This oxygenic atmosphere lead to the evolution of life on the earth. The exponential growth rate and long life span of human beings now a days creating a population bomb which is going to affect the environmental stability. In present day scenario of population explosion, it is essential to increase food production to meet the food demands and to maintain the socioeconomic status of the people in all the developing countries including India. In the year 2050, India will reach to the highest population (1.22 billion in 2012), within this globe total population of world will be approx 9.1 billion in 2050 (Carvalho, 2006). The immediate response to increase food production in limited agricultural land areas is possible by intensive use of agrochemicals. Agrochemicals include two large groups of compounds: chemical fertilizers and pesticides. The use of chemical fertilizers tremendously increased worldwide since 1960s and was largely responsible for the "green revolution", i.e. the massive increase in production obtained from the same surface of land with the help of mineral fertilizers and intensive irrigation. The revolution was assisted also with the introduction of more productive varieties of crops.

The use of pesticides, including insecticides, fungicides, herbicides, rodenticides, etc., to protect crops from pests, allowed to significantly reduce the losses and to improve the yield of crops. The application of different agrochemicals is region specific. In the tropical regions, where insect pests and plant diseases are more frequent, pesticides are generally applied in massive amounts, both in small farms as well as in cash crop. It has been reported that especially the organochlorine and organophosphorus pesticide residues, are found in soils, atmosphere and in the aquatic environment in relatively high concentrations (Carvalho et al., 1997). Pesticides are poisons, intentionally dispersed in the environment to control pests but they also act upon other species causing serious side effects on non-target species and destabilise the ecosystem. Cyanobacteria the natural nitrogen engineer of the soil are also adversely affected by indiscriminate use of pesticides.

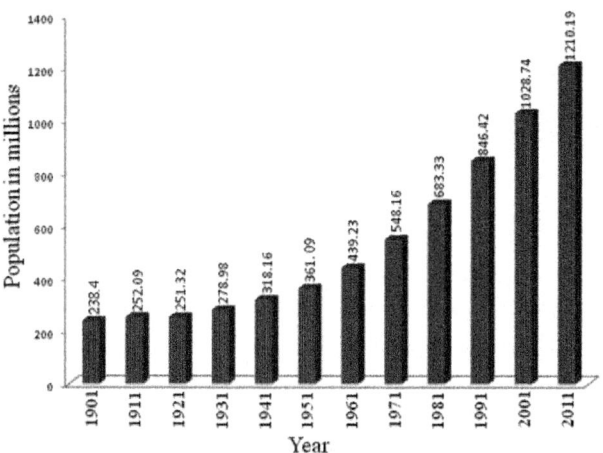

Figure 1. Increasing rate of population

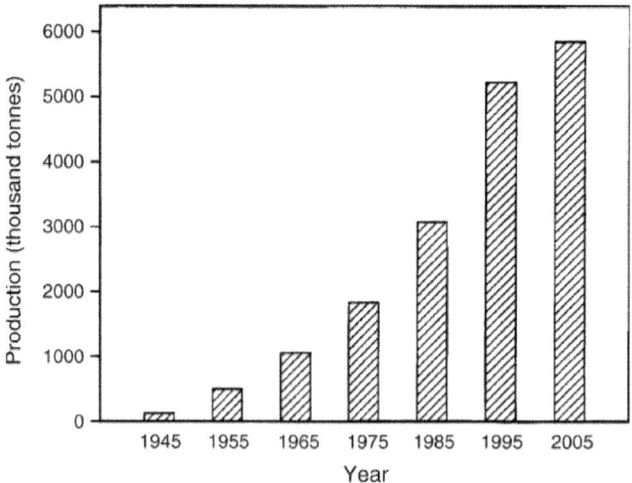

Figure 2. World production of formulated pesticides. Data for year 2005 is estimated (sourse Agrochemical service, 2000).

Table 1. Physicochemical characteristics of Monocrotophos and Endosulfan

Name	Chemical name	Molecular Weight	CAS registry Number	Formulla	Structure	Activity	Solubility at 25°C in water (mg/L)
Endosulfan	6,7,8,9,10-hexachloro-1,5,5a,6,9,9a-hexahydro-6,9-methano-2,4,3-benzadioxathiepin 3-oxide	406.96	115-29-7	$C_9H_6O_3SCl_6$		Insecticidal	36%
Monocrotophos	dimethyl (E)-1-methyl-2-(methylcarbamoyl)vinyl phosphate	223.2	2157-98-4	$C_7H_{14}NO_5P$		Insecticidal	35%

Cyanobacteria are the most diversified ecologically, most successful and evolutionarily most important group of photosynthetic prokaryotes

(Peschek et al., 1994) and maintain the homeostasis of nitrogen budget of the rice agroecosystem by photobiological nitrogen fixation in a specialized cell called heterocyst (Fay et al., 1968) at almost zero cost (Mishra & Pabbi 2004). Diversity and evolutionary information of cyanobacteria are available in the internet and for images one can search for "cyanobacteria, images" using Google. Most paddy soils have a natural population of cyanobacteria as they grow and multiply at the simple expense of water, light and air (Fay 1983). Soil nitrogen is the main source of nitrogen for crop growth and rice plant consume 50% of soil nitrogen (Fernandez-Valiante et al., 2000). Several reports are available on the adverse effect of agrochemicals on cyanobacteria (Marsac& Houmard, 1993; Das & Adhikary, 1996; Kapoor & Arora, 2000a, 2000b; Shikha & Singh, 2004; Xia, 2005; Kim & Lee, 2006). Although a lot of work has been done on the effect of pesticide in general, no attempt has been made on the effect of pesticide in locally growing cyanobacteria of Western Odisha, India. Farmers of this region use Monocrotophos and Endosulfan on a large scale in rice fields as both the pesticide have broad spectrum activity and they control the attack of insects; the physicochemical properties of both pesticides are given in Table 1.

BGA biofertilizer are added to rice fields to increase the fertility of soil and to minimize dependence on chemical fertilizer. The aim of the present study was to investigate whether *Anabaena sp.* and *Nostoc sp.*, the locally isolated rice field cyanobacteria can be recommended to use as biofertilizer by tolerating the deleterious effect of Monocrotophos and Endosulfan. The chapter presents experimental results to illustrate the effects of Monocrotophos and Endosulfan in time and concentration dependent manner on growth, pigments and photosynthesis of these two alga.

MATERIAL AND METHODS

Two species of heterocystous cyanobacteria belonging to genera *Anabaena* and *Nostoc* isolated from rice field. Selection of these two genus was based on their relatively better growth rate and wider occurance. Two commercial grade pesticide i.e.Monocrotophos (organophosphate 36%SL) and Endosulfan (organochlorine 35%EC) were used in the investigation. Fresh stock solutions of these pesticides were prepared in double distilled water and added to the culture medium to obtain the desired concentration. pH of all the medium was adjusted to 7.4 prior to sterilisation. Experiments were conducted in 15×150mm Borosil test tube containing 10 ml of nitrogen free BG11 medium (Rippka et al.,1979) and by inoculating equal amount of homogenized culture suspension (absorbance of the inoculum of each organism from their exponential growth phase at 760nm was 0.4 always). The medium contained various concentration of Monocrotophos (20,50,100 &150ppm) and Endosulfan (1,3,5,10 and 15 ppm),

Growth was measured by light scattering technique by taking the absorbance at 760nm, *Chl a* pigment of the cyanobacterial cells were extracted with 80% chilled acetone. Absorbance of the acetone extract was recorded at 660 nm and the amount of *Chl a* was determined using extinction coefficient ofMackinney, 1941.

Algal suspension was homogenized in a glass hand-homogenizer for 5 minutes and then centrifuged at 3500 rpm for 10 minutes. After centriguation, the pellet containing algal cells was resuspended in 50 mM tris-HCl buffer, pH 7.8 containing 175 mM NaCl. Room temperature *Chl a* fluorescence emission of algal suspension was measured as per Panda (1999) in a spectrofluorimeter (Hitachi, model, 650-40, Japan). For all scanning, a slit width of 10 nm was used. The whole cell algal suspension equivalent to 10 g of Chl in a total volume of 3 ml containing 50 mM Tris-HCl buffer and 175 mM NaCl (pH 7.8) was excited at 450 nm and emission was recorded at 685 nm for PS II and 735 nm for PS I emission.

The same algal suspension was also used to measure the excitation emission. During scanning, the emission was monitored at 685 nm and a slit width of 10 nm was maintained. Excitation emission was recorded at 439 nm, 471 nm and 485 nm. The excitation energy transfer from Car to Chl was measured by exciting the algal suspension at 475 nm and 600 nm. The emission was recorded at 685 nm for PS II and 735 nm for PS I. Efficiency of the energy transfer was assessed by calculating the ratio of excitation at 475 nm to 600 nm as described by Gruszescki et al., 1991.

Fluorescence polarization was measured by exciting the algal suspension at 620 nm and polarization was recorded at 685 nm. Polarization (P) was calculated as per the following formula of Swain et al., 1990.

$$P = \frac{I_{vv} - \dfrac{I_{vh} \times I_{hv}}{I_{hh}}}{I_{vv} + \dfrac{I_{vh} \times I_{hv}}{I_{hh}}}$$

where,

I = intensity of fluorescence

v = vertical geometry of the polarizer

h = horizontal geometry of the analyzer

The 2,6-dichlorophenol indophenol (DCPIP) photoreduction was measured spectrophotometrically as described by Swain et al., 1990 with modification. 3 ml of reaction mixture contained whole cell algal suspension (equivalent to 10 g Chl), 50 mM Tris-HCl buffer (pH 7.8) and 175 mM NaCl. This reaction mixture was illuminated for 30 seconds with saturating white light (7×10^4 ergs cm^{-2} sec^{-1}) coming from a projector lamp. The incident radiation beam was passed through a water filter to minimize infrared radiation. The photoreduction of the dye was measured at 600 nm. The reduction of the dye is expressed as moles DCPIP reduced/mg Chl/hr.

Photosynthetic efficiency of algal suspension in terms of chlorophyll fluorescence was measured at room temperature using a Plant Efficiency Analyzer (Handy PEA, Hansatech Instruments, Norfolk, UK). The Fv/Fm of algal suspension was measured by the Handy PEA after 20 minutes dark-adaption.

RESULTS AND DISCUSSION

The nitrogen-fixing cyanobacteria represent as one of the prominent component of microbial population in wetland soils, especially in rice fields. They significantly contribute to soil fertility as a natural biofertilizer (Kumar & Kumar, 1998). Some cyanobacterial strains that thrive and grow in rice fields release small quantities of the major fertilizing product ammonia and small polypeptides during active growth whereas most of the other fixed products become available mainly through autolysis and decomposition (Hammouda, 1999). Therefore, cyanobacteria are considered as a vital component of the rice agroecosystem. However, excessive use of pesticides has a detrimental effect on the growth of these beneficial microorganisms, soil fertility and ultimately on the crop productivity. The effect of pesticides on the population of nitrogen fixing organisms varies with characteristics of the species and chemical nature of the pesticide.

Changes in the Growth Pattern

Growth response of two different species of heterocystous cyanobacteria namely *Anabaena* sp. and *Nostoc* sp. to different concentrations of insecticide Monocrotophos is shown in *Fig.3*. Experiments showed that *Anabaena* sp. tolerated up to 100 ppm whereas *Nostoc* sp. tolerated upto 150 ppm of the insecticide where as for Endosulfan its limit was 5ppm & 15 ppm respectively. Growth curves indicate that both the algae showed lag phase up to 3rd day of incubation followed by rapid growth up to 12th day in case of control and treated (20 ppm for *Anabaena* sp. and 50 ppm for *Nostoc* sp.) samples. The present study of growth pattern of *Anabaena* sp. and *Nostoc* sp suggest that the tolerance capacity of *Nostoc* is more compared to *Anabaena* for both the pesticides. Both the alga also tolerate higher doses of pesticides Monocrotophos compared to Endosulfan. Endosulfan has more inhibitory effect on growth of both the BGA. These findings support the observation of Das and Adhikary, 1996 that organophosphate insecticide is less toxic than organocholorine. Several authors (Rath & Adhikary, 1994; Goyal et al., 1994; Das &Adhikary, 1996; Anand & Subramanian 1997; Kaur &Ahluwalia, 1997;Kapoor & Arora, 1998; 2000 a, 2000b; Xia, 2005; Chen et al.,2007; Kumar et al.,2008; Bhattacharyya et al., 2011) have reported inhibitory effect of various pesticides on the growth of cyanobacteria. The inhibition of growth in different concentrations of the pesticides is due to alteration in synthesis of nucleic acids, amino acids and proteins (Kumar et al., 2011) as well as due to impairment in photosynthetic activity (Lal &Saxena, 1980) of the BGA.

Changes in Chlorophyl *a* Contents

Almost all oxygenic photosynthesizer, with the exception of *Acaryochloris* a cyanobacterium, use chlorophyll a (BjÖrn et al., 2009). The amount of Chl content in the photosynthetic unit of cyanobacterial cell indicates its growth and physiological status. *Fig.4* depicts the kinetics of *Chl a*accumulation and loss in *Anabaena* sp. and *Nostoc* sp. treated with different concentrations of Monocrotophos *(Fig.4B) and* Endosulfan *(Fig.4A)* in the BG11 medium over 15 days of incubation along with the control. The kinetics pattern was closely similar to that of growth kinetics with minor variations for both the treated and control samples. Except in *Anabaena* (control sample), the *Chl a* content reached its maximum level on the 10th day of incubation followed by decline both in control and treated samples. The rate of pigment loss in treated sample was more than that of control. As indicated from the levels, the pigment synthesis was very less in *Anabaena* with 100 ppm Monocrotophos, 5ppm Endosulfan and *Nostoc* with 150 ppm Monocrotophos, 15 ppm Endosulfan treatment. In these two concentration of pesticides the pigment levels were almost same with the initial level of the pigment in both the samples through out the experimental period of 15 days. The *Chl a*content was maximum on 10th day though growth rate was maximum at 12th day of incubation. The pigment content declined after 10th day of incubation in all treated samples and control of *Nostoc* sp. On the other hand, except in the control samples of *Anabaena* sp. pigment level was highest on 10th day then followed by a sharp decline.

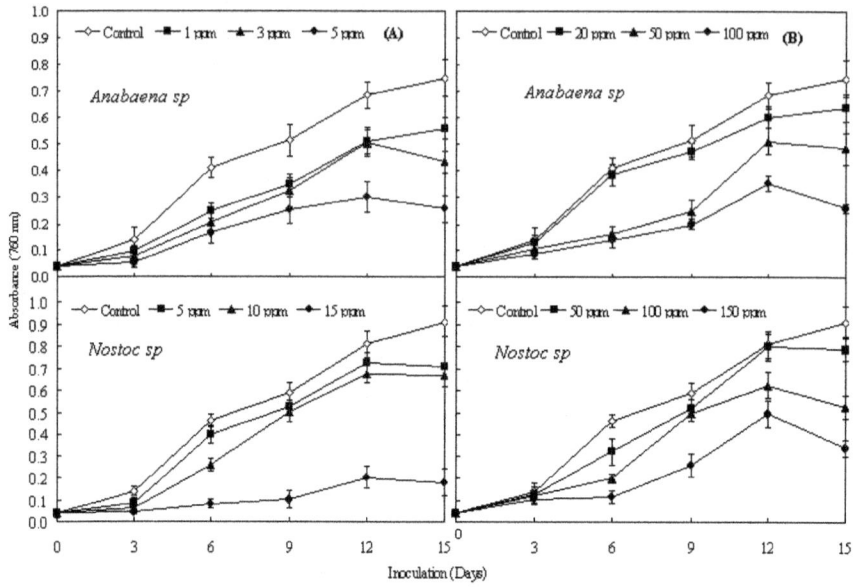

Figure 3. Effect of different concentration of Endosulfan (A) &Monocrotophos (B) on growth of *Anabaena* and *Nostoc* cultured under laboratory condition.

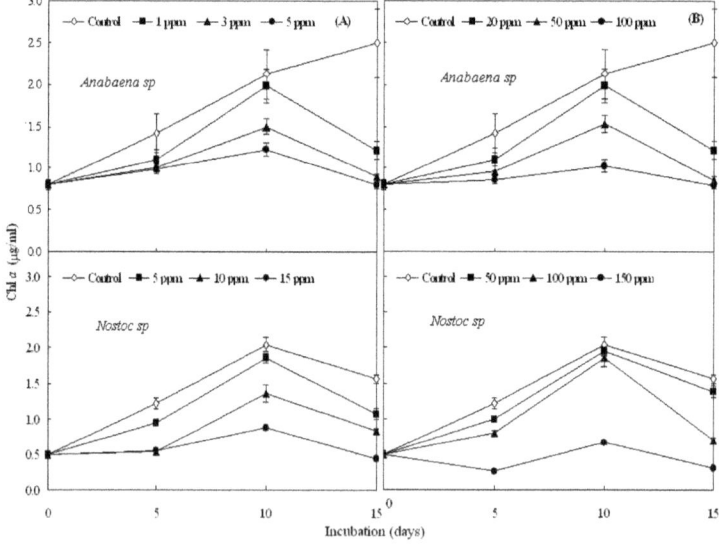

Figure 4. Effect of different concentration of Endosulfan(A) &Monocrotophos (B) on *Chl a* of *Anabaena* and *Nostoc* cultured under laboratory condition.

The data indicates that *Chl a* accumulation and loss in the present study is also time and concentration dependent manner *(Fig 4)*. The pesticides are known to interfere with the synthesis of *Chl a* pigment by inhibiting the formation of porphyrin rings (Moreland, 1980; Lal and Saxena, 1980). In the present work, the low level of Chl in pesticide treated samples supports the observations of Das & Adhikary, 1996, Megharaj et al., 2011, Kumar et al., 2008, Battah et al., 2001; Sikha & Singh, 2004 and Xia, 2005.

Changes in Electron Transport Activity

The DCPIP photoreduction reflects the photochemical potential of PS II. The activities also reflect the coupling between light absorption and photochemical reaction of the thylakoid membrane. In the present study the rate of dye reduction in control and pesticides treated samples of both the alga resemble with the kinetics of *Chl a* accumulation and loss. The rate of dye reduction in treated samples is low compared to the control (Table 2 and 3). This could be due to loss of pigments and protein content of the organism under the pesticide treated conditions due to stress-induced formation of ROS (Behera & Choudhury, 2001; Hideg & Vass, 1996) and possible changes in the thylakoid microenvironment. The degradation of D1 protein under stress condition may be another reason (Long & Humphries, 1994). These observations are similar to the findings of Shikha and Singh,2004,Bhattacharyya et al., 2011.

Table 2. Electron transport efficiency of PS II in terms of DCPIP photoreduction of *Anabaena sp* grown under different concentrations of Monocrotophos and Endosulfan in laboratory condition. (±SD)

μ mol of DCPIP Reduced (*Anabaena sp*)

Treatment	Dose (ppm)	0 day	5 day	10 day	15 day
Control		20±1.2	101±3.6	175±3.5	241±5.5
Monocrotophos	20	20±1.2	94±2.7	180±6.2	152±2.9
	50	20±1.2	83±1.1	140±2.8	132±3.6
	100	20±1.2	65±5.3	95±3.3	85±2.6

Treatment	Dose (ppm)	0 day	5 day	10 day	15 day
Control		20±1.2	101±1.8	175±5.9	241±8.5
Endosulfan	1	20±1.2	91±22	165±8	135±6
	2	20±1.2	80±0.8	120±3	110±7
	3	20±1.2	65±1.6	95±5	75±3

Table 3. Electron transport efficiency of PS II in terms of DCPIP photoreduction of *Nostoc Sp* grown under different concentrations of Monocrotophos and Endosulfan in laboratory condition. (±SD)

μ mol of DCPIP Reduced (*Nostoc Sp*)

Treatment	Dose (ppm)	0 day	5 day	10 day	15 day
Control		29±2.6	145.8±4.2	190±8	142±6.5
Monocrotophos	50	29±2.6	94±6	126±6	84±3.2
	100	29±2.6	55.5±4	96±7	49.6±3.6
	150	29±2.6	25.2±2.6	45.8±5.4	30±6

Treatment	Dose (ppm)	0 day	5 day	10 day	15 day
Control		29±2.6	145.8±7	190±11	142±8.2
Endosulfan	5	29±2.6	100.8±3.8	116±7	54.6±5
	10	29±2.6	84±3.3	102±6	50±6
	15	29±2.6	82±3.4	94±8	32.4±4

Measurement of Fluorescence Characteristics

Analyses of fluorescence emission, excitation emission, fluorescence polarization and excitation energy transfer of thylakoids provide information about the structural organization and the microenvironment of thylakoid membrane. The analyses also give information about the degree of coupling of different pigment complexes. Information about the coupling of light absorption and photochemical reactions could also be obtained by monitoring fluorescence characteristics of whole cells of the cyanobacteria. Therefore, to determine the structural and functional status of the thylakoid, fluorescence excitation, emission and polarization measurements are very much important. Campbell et al.,1998 have opined that fluorescence analysis is an integral part of the studies of photosynthesis in BGA. Shikha & Singh, 2004 have extensively used fluorescence studies to monitor photosynthetic status of *A. doliolum* treated with herbicide glyophosphate.

Cyanobacteria fluorescence characteristics are distinct from those of plants due to their specific structural and functional properties (Campbell et al., 1998). These include significant fluorescence emission from the light harvesting phycobiliproteins, large and rapid changes in fluorescence yield (state transitions) which depend on metabolic and environmental conditions as well as flexible and overlapping respiratory and photosynthetic electron transport chains. In cyanobacteria, the photosynthetic system is tightly linked to other principal metabolic pathways and is itself a major metabolic sink for iron, nitrogen and carbon skeletons. Therefore, Chl fluorescence signals can provide rapid, real-time information on both photosynthesis and overall acclimation status of cyanobacteria.

Fluorescence Emission

The fluorescence characteristics of test organismsboth in control and insecticides treated samples are shown in Table4 and 5. There is gradual increase in fluorescence intensity at F685 and F735 in all conditions over 15 days of incubation under laboratory condition except on 5th day of control. The ratios of F685 to F735 increased from 5th day till the end of experiment both in control and treated samples. The ratios were also more in treated samples than in the control. The ratios gradually increased as the concentration of the insecticide and treatment period increased in both the organisms.

Table 4. Chlorophyll a fluorescence emission of *Anabaena* grown in control and different concentrations of Monocrotophos and Endodahan in laboratory conditions.

Fluorescence Intensity (Arbitrary unit)													
Treatment	Dose (ppm)	F685				F735				F685/F735			
		0 day	5 day	10 day	15 day	0 day	5 day	10 day	15 day	0 day	5 day	10 day	15 day
Control	0	91.0	88.0	137.0	195.0	26.5	25.0	30.0	36.0	3.4	3.52	4.50	5.40
Monocrotophos	20	91.0	100.0	167.0	202.0	26.5	28.0	34.0	40.0	3.4	3.57	4.90	5.05
	50	91.0	122.0	178.0	240.0	26.5	32.0	36.0	42.0	3.4	3.81	4.94	5.71
	100	91.0	139.0	195.0	270.0	26.5	36.0	39.0	46.0	3.4	3.86	5.00	5.84
Endosulfan	1	91.0	110.0	183.0	299.0	26.5	27.0	37.0	54.0	3.4	4.07	4.90	5.53
	3	91.0	122.0	193.0	314.0	26.5	29.0	38.0	55.0	3.4	4.20	5.07	5.70
	5	91.0	152.0	218.0	338.0	26.5	35.0	41.0	56.0	3.4	4.34	5.30	6.03

Table 5. Chlorophyll *a* fluorescence emission of *Nostoc* grown in control and different concentrations of Monocrotophos and Endodahan in laboratory conditions.

Fluorescence Intensity (Arbitrary unit)													
Treatment	Dose (ppm)	F685				F735				F685/F735			
		0 day	5 day	10 day	15 day	0 day	5 day	10 day	15 day	0 day	5 day	10 day	15 day
Control		97.5	92.0	146.0	204.0	21.6	20.0	24.6	34.0	4.5	4.6	5.9	6.0
Monocrotophos	50	97.5	108.0	177.0	220.0	21.6	22.1	29.0	36.0	4.5	4.9	6.1	6.1
	100	97.5	134.0	188.0	257.0	21.6	25.2	30.3	39.1	4.5	5.3	6.2	6.6
	150	97.5	147.0	203.0	284.0	21.6	26.3	31.7	42.3	4.5	5.6	6.4	6.7
Endosulfan	5	97.5	102.0	151.0	280.0	21.6	21.8	25.0	44.0	4.5	4.7	6.0	6.4
	10	97.5	114.0	164.0	294.0	21.6	21.6	24.8	42.5	4.5	5.3	6.6	6.9
	15	97.5	148.0	198.0	328.0	21.6	26.6	28.4	45.8	4.5	5.6	6.6	7.1

Chlorophyll *a* fluorescence emission spectra of whole algal cells measured at room temperature exhibit usually two emission maxima, the first at 685 (F685) nm and the second at 735 (F735) nm. F685 is considered as the emission from PS II and F735 from PS I (Papageorgiou, 1975). A gradual increase in fluorescence intensity (Table 4 and 5) at 685 nm (F685) and 735 nm (F735) is observed over 15 days of incubation of both the alga in control as well as insecticides treated samples. Increase in fluorescence intensity particularly during developmental stage has been ascribed due to improved organization of light harvesting (antenna) complexes of the thylakoid which results in trapping of more solar energy. However, if proportional increase in the photochemical activity will not take place, then the absorbed energy is emitted as fluorescence (Krause & Weis, 1991; Krieger et al., 1992). On the other hand, uncoupling of the photosynthetic pigments and RC during natural ageing or under stress conditions may also lead to increase in the fluorescence intensity. Continuous increase in the fluorescence intensity in the control sample could be due to higher trapping of solar energy as the algal cell improves their thylakoid organization during the culture. However, without proportional increase in photochemical activities, the excitation energy is emitted as fluorescence. On the other hand increase in the fluorescence intensity in the insecticides treated samples is much higher than the control. This suggests that the pesticides have induced uncoupling of light harvesting system and electron transport resulting emission of excitation energy as fluorescence. Higher susceptibility of PS II compared to PS I to different stress such as water stress (Deo and Biswal, 1998), light stress (Behera et al., 2002), oxidation stress (Behera and Choudhury, 2001) etc. have been reported in different plant systems. The gradual increase in the ratio of F685 and F735, when the concentration of pesticides increase suggests that PS II is more affected by the treatments.

Fluorescence Excitation

Table-6 depicts the effect of different concentrations of pesticides on the ratio of peak heights of fluorescence excitation emission of *Anabaena* sp. The ratios of 471 nm to 439 nm and 485 nm to 439 nm increased throughout the 15 days of incubation in all concentrations of Monocrotophos and Endosulfan used along with the control sample. However, compared to the control, the ratio declined with insecticide treatment as well as with the increase in concentration of both pesticides.

Table 6. Effect of different concentrations of Monocrotophos and Endosulfan on ratio of peak heights of fluorescence excitation of *Anabaena* grown in laboratory conditions.

Treatment	Dose (ppm)	Ratio of Peak Heights(a.u)							
		E471/E439				E485/E439			
		0 day	5 day	10 day	15 day	0 day	5 day	10 day	15 day
Control		0.489	0.988	1.213	1.500	0.625	1.195	1.325	1.632
Monocroto-phos	20	0.489	0.940	1.208	1.478	0.625	1.010	1.217	1.417
	50	0.489	0.891	1.112	1.155	0.625	0.921	1.200	1.253
	100	0.489	0.695	0.789	0.918	0.625	0.900	1.182	1.208
Endosulfan	1	0.489	0.825	1.094	1.132	0.625	1.093	1.131	1.348
	3	0.489	0.821	1.087	1.021	0.625	0.956	1.121	1.187
	5	0.489	0.624	0.721	0.802	0.625	0.795	0.860	0.934

The ratio of peak heights of fluorescence excitation emission of *Nostoc* sp. is represented in Table-7 both in control and pesticides treated (50, 100 and 150 ppm of Monocrotophos and 5, 10 and 15 ppm of Endosulfan) samples over 15 days of incubation under laboratory condition. Similar trend of increase in the ratio of 471 nm to 439 nm was also noted in Monocrotophos and Endosulfan treated samples as well as in control over 15 days of incubation. However, the increase was less in insecticide treated samples and more so when the concentration of the insecticide was more. On the other hand, except in control, no definite increasing or decreasing trend in the ratio of 485 nm to 439 nm was noted in the treated samples.

Table 7. Effect of different concentrations of Monocrotophos and Endosulfan on ratio of peak heights of fluorescence excitation of *Nostoc* grown in laboratory conditions.

Treatment	Dose (ppm)	Ratio of Peak Heights(a.u)							
		E471/E439				E485/E439			
		0 day	5 day	10 day	15 day	0 day	5 day	10 day	15 day
Control	0	0.627	1.055	1.368	1.600	1.304	1.505	1.602	1.678
Monocro-tophos	50	0.627	1.040	1.150	1.320	1.304	1.202	1.310	1.408
	100	0.627	0.932	0.983	1.152	1.304	1.084	1.093	1.101
	150	0.627	0.729	0.765	0.997	1.304	0.935	1.012	1.087
Endosul-fan	5	0.627	0.757	0.925	1.162	1.304	1.454	0.835	1.303
	10	0.627	0.747	0.854	1.051	1.304	1.359	0.519	1.131
	15	0.627	0.629	0.765	0.908	1.304	0.909	0.429	0.933

The study of fluorescence excitation characteristics of chloroplast is used to explain the spatial arrangement and coupling of different pigment molecules in the thylakoid membrane (Behera & Choudhury, 1997). The changes in the relative peak values of fluorescence excitation at 471 (E471) nm and 485 (E485) nm with reference to peak at 439 (E439) nm reflects the alterations in pigment protein complexes in the thylakoid domain during development of the organism. Table 6 and 7 indicate the changes in the ratio of peak heights in control and with different concentration of Monocrotophos and Endosulfan. The decrease in the ratio of 471 nm to 439 nm is attributed to gradual decrease in coupling between Chl and Car with the increase of the duration of incubation period with the insecticides.

Efficiency of Energy Transfer

The simple and direct proof of excitation energy transfer from Car and phycobilisomes (PBS) to Chl comes from the contribution of the light absorbed by the Car and PBS in Chl *a* fluorescence. At shorter wavelength, only Chl, Car and PBS and at longer wavelength only Chl is responsible for the absorption. There is a significant decline in the capacity of energy transfer in PS II for all concentrations of Monocrotophos and Endosulfan in both the alga compared to the control (Table 8 and 9). The temporal kinetics of energy transfer follows similar pattern like the kinetics of DCPIP photoreduction (Table 2and 3) and photosynthetic efficiency of PS II during the 15 days of incubation. The decrease in the *Chl a* contents in the insecticide treated samples (*Fig. 4*) may be correlated to certain conformational changes in the pigment protein complex in the photosystem in turns affecting the excitation energy transfer (Gruszescki et al., 1991). The energy transfer from Car to Chl is increasingly hampered (Table 8 and 9) as the concentration of the pesticides increased.

Though the values are different, the kinetics of energy transfer in PS I is similar to that of PS II for both*Anabaena* sp. and *Nostoc* sp. in control and treated samples. However, compared to PS II, PS I is less susceptible to the insecticide treatment in both the alga. Smaller changes in excitation energy transfer in PS I suggest that PS I is less effected even under stress condition. Relatively less susceptibility of PS I compared to PS II to various stress conditions has been shown earlier by various authors (Choudhury & Choe, 1996; Deo & Biswal, 1998; Campbell et al., 1998; Behera et al., 2002)

Table 8. Efficiency of excitation energy transfer from carotenoids to chlorophyll of PS II and PS I of *Anabaena* grown under control and different concentrations of Monocrotophos and Endosulfan in laboratory conditions.

Treatment	Dose (ppm)	Efficiency of excitation energy transfer (a.u)							
		PS II				PS I			
		0 day	5 day	10 day	15 day	0 day	5 day	10 day	15 day
Control	0	1.48 (100)	1.95 (131)	2.15 (145)	2.10 (142)	1.22 (100)	1.64 (134)	1.83 (150)	1.80 (147)
Monocrotophos	20	1.48 (100)	1.87 (126)	2.00 (135)	1.92 (130)	1.22 (100)	1.59 (130)	1.74 (143)	1.69 (138)
	50	1.48 (100)	1.75 (118)	1.93 (130)	1.80 (122)	1.22 (100)	1.52 (125)	1.67 (137)	1.58 (130)
	100	1.48 (100)	1.67 (112)	1.70 (114)	1.62 (109)	1.22 (100)	1.41 (116)	1.52 (125)	1.46 (120)
Endosulfan	1	1.48 (100)	1.82 (123)	1.98 (134)	1.88 (127)	1.22 (100)	1.58 (129)	1.70 (149)	1.67 (137)
	3	1.48 (100)	1.70 (115)	1.87 (126)	1.82 (123)	1.22 (100)	1.54 (126)	1.64 (134)	1.56 (128)
	5	1.48 (100)	1.55 (105)	1.64 (111)	1.60 (108)	1.22 (100)	1.46 (120)	1.54 (126)	1.50 (122)

Table 9. Efficiency of excitation energy transfer from carotenoids to chlorophyll of PS II and PS I of *Nostoc* grown under control and different concentrations of Monocrotophos and Endosulfan in laboratory conditions.

Treatment	Dose (ppm)	Efficiency of excitation energy transfer (a.u)							
		PS II				PS I			
		0 day	5 day	10 day	15 day	0 day	5 day	10 day	15 day
Control	0	0.558 (100)	0.714 (128)	0.822 (147)	0.812 (145)	0.469 (100)	0.615 (131)	0.724 (154)	0.703 (150)
Monocrotophos	20	0.558 (100)	0.708 (127)	0.794 (142)	0.760 (136)	0.469 (100)	0.600 (128)	0.613 (146)	0.656 (140)
	50	0.558 (100)	0.675 (121)	0.730 (131)	0.712 (127)	0.469 (100)	0.572 (122)	0.637 (136)	0.609 (130)
	100	0.558 (100)	0.655 (113)	0.647 (116)	0.608 (109)	0.469 (100)	0.539 (115)	0.562 (120)	0.529 (113)
Endosulfan	5	0.558 (100)	0.704 (126)	0.783 (140)	0.753 (135)	0.469 (100)	0.605 (129)	0.680 (145)	0.656 (140)
	10	0.558 (100)	0.671 (120)	0.725 (130)	0.704 (126)	0.469 (100)	0.586 (125)	0.635 (136)	0.609 (130)
	15	0.558 (100)	0.619 (111)	0.636 (114)	0.638 (110)	0.469 (100)	0.558 (119)	0.572 (122)	0.548 (117)

Fluorescence Polarization

Changes in fluorescence polarization of the two algal species under control and insecticides treated conditions give further information on the status of pigment-protein complexes with reference to their microenvironment in thylakoid membrane. The increase in polarization during the initial stage of incubation in all treated samples compared to the control (*Fig.5*) could be due to the poor coupling between the pigment protein complex and RC (Behera & Choudhury, 1997). On the other hand, increase in polarization in the later phase of growth (10-15 days of incubation) could be due to disorganization of pigment protein complexes and RC, leading to a decrease in quantum migration. Alternatively, peroxidation of lipid during later stage may induce gel phase of the thylakoid membranes restricting the mobility of Chl dipole (Panda & Biswal, 1989 and 1990). This may cause an increase in the polarization value at 100 and 150 ppm of Monocrotophos and 5 and 15 ppm of Endosulfan treatment to the *Anabaena* sp and *Nostoc* sp. respectively. Significant high levels of polarization suggests a greater disorganization of thylakoid membrane due to high lipid peroxidation (Kumar et al., 2008).

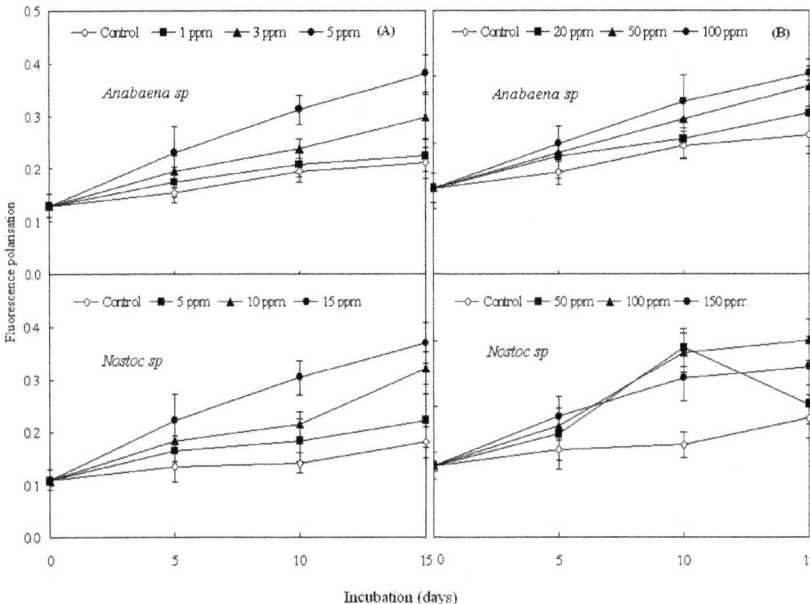

Figure 5. Effect of different concentration of Endosulfan(A) &Monocrotophos (B)on fluorescence polarization of *Nostoc* and *Anabaena* cultured under laboratory condition.

Photosynthetic Efficiency

Photosynthetic efficiency of PS II can be measured by monitoring the ratio Fv/Fm. It is known that photoinhibition occurs when the rate of excitation energy captured exceeds the rate of consumption in photosynthetic reactions (Osmond, 1981; Powles, 1984). Photoinhibition in terms of Fv/Fm has been found both in higher plants (Panda et al., 2006; Rodrigues et al., 2007) as well as in algae (Ying and Hader, 2002; Xia, 2005; Chen et al., 2007). The primary site of photoinhibition is the reaction centre (D1 protein) of PS II (Demming and Bjorkman, 1987; Jordan, 1996). Photoinhibition is manifested as a decrease in oxygen evolution (Krause, 1988) and photochemical efficiency (Falk and Samuelsson, 1992). The data on the measurement of Fv/Fm during the laboratory incubation of the different samples (Table 10 and 11) show similar kinetics like that off the photosynthetic pigment and protein accumulation and loss and DCPIP photoreduction of both the BGA in control and treated (pesticide) samples. As the concentration of pesticides increased, photosynthetic efficiency decreased. This shows that both the species of cyanobacteria are sensitive to higher concentration of pesticides. Pesticides, particularly at higher concentration may (directly or indirectly) cause damage to D1 protein of PS II leading to photoinhibition. The decrease in Fv/Fm ratio in pesticide treated sample could also be due to decrease in Phycocyanin, Phycoerythrin and Allophycocyanin content, which results in a decrease of light energy absorption by phycobilisomes and reduction of photochemical efficiency (Fv/Fm) of PS II. Similar observations have been reported by Xia, 2005 in *N. sphaeroids*. The present finding is in confirmatory to the observation of Xia, 2005.

Table 10. Photosynthetic efficiency (in term of F_v/F_m) of PS II of *Nostoc* grown under different concentrations of Monocrotophos and Endosulfan in laboratory condition.

PS II efficiency										
Nostoc										
Treatment	Dose (ppm)	0 day	5 day	10 day	15 day	Treatment	Dose (ppm)	0 day	5 day	10 day	15 day
Control		0.370 (100)	0.524 (142)	0.562 (152)	0.536 (145)	Control		0.370 (100)	0.524 (142)	0.562 (152)	0.536 (145)
Monocrotophos	50	0.370 (100)	0.500 (135)	0.515 (139)	0.473 (128)	Endosulfan	5	0.370 (100)	0.458 (124)	0.520 (141)	0.467 (126)
	100	0.370 (100)	0.431 (116)	0.500 (135)	0.429 (116)		10	0.370 (100)	0.387 (105)	0.455 (123)	0.375 (101)
	150	0.370 (100)	0.375 (101)	0.404 (109)	0.387 (105)		15	0.370 (100)	0.313 (85)	0.333 (90)	0.316 (85)

Table 11. Photosynthetic efficiency (in term of F_v/F_m) of PS II of *Anabaena* grown under different concentrations of Monocrotophos and Endosulfan in laboratory condition.

Anabaena										
Treatment	Dose (ppm)	0 day	5 day	10 day	15 day	Treatment	Dose (ppm)	0 day	5 day	10 day	15 day
Control		0.350 (100)	0.520 (149)	0.556 (159)	0.515 (147)	Control		0.350 (100)	0.520 (149)	0.556 (159)	0.515 (147)
Monocrotophos	20	0.350 (100)	0.482 (138)	0.500 (143)	0.468 (142)	Endosulfan	1	0.350 (100)	0.404 (115)	0.511 (146)	0.455 (130)
	50	0.350 (100)	0.442 (126)	0.482 (138)	0.429 (123)		3	0.350 (100)	0.316 (90.3)	0.419 (119.7)	0.327 (93.42)
	100	0.350 (100)	0.375 (107)	0.419 (120)	0.351 (100.3)		5	0.350 (100)	0.313 (89.42)	0.375 (107)	0.308 (88)

CONCLUSION

The insecticide Monocrotophos and Endosulfan affected photosynthetic function which may have inhibited the growth of both the cyanobacteria by affecting the production of photosynthetic pigments in the antenna complex, electron transfer, and photosynthetic efficiency of PS-II. Both the cyanobacteria responded differently to both the pesticides with time and concentration dependent manner. Endosulfan has more inhibitory effect than Monocrotophos. Tolerance capacity of *Nostoc sp.* is more than *Anabaena sp.*. Our data indicate that use of Endosulfun may pose a risk to diazotropic cyanobacterium, and consequently to the nitrogen economy of the soil. It is currently understood from the extensive studies conducted so far on impacts of many pollutants on cyanobacteria and microalgae that evaluation with a wide taxonomic range in different ecosystem is necessary to arrive at a generalization on the nontarget effects of pollutant (Ramakrishnan et al., 2010) However, the effect of pesticides on the population of nitrogen-fixing cyanobacteria in rice fields also depends on pesticide concentrations, moreover, toxicity is affected not only by the types of pesticide, but also by the taxonomic groups and species. Since Endosulfan has more deleterious effect on natural engineer of the rice field (BGA) it's use be limited to maintain the stability of paddy ecosystem.

ACKNOWLEDGEMENT

Authors are thankful to HOD school of life Sciences, Sambalpur University for giving all types of facilities.

REFERENCES

1. Agrochemical service,2000World Mackenize Consultants Limited. Edinborg, UK
2. N. Anand, T. D. Subramanian, 1997Effect of certain pesticides on the physiology of Nostoc calcicola. Phykos, 361520
3. M. G. Battah, E. F. Shabana, I. A. Kobbia, H. M. Eladel, 2001Differential effects of thiobencarb toxicity on the growth and photosynthesis of Anabaena variabilis with changes in phosphate level. Ecotoxicol. Environ. Saf., 49235239
4. L. M. Behera, N. K. Choudhury, 1997Changes in chlorophyll fluorescence characteristics of chloroplasts from intact pumpkin cotyledous, caused by organ excision and kinetin treatment. Photosynthetica, 34161168
5. R. K. Behera, N. K. Choudhury, 2001Photosynthetic characteristics of chloroplasts of primary wheat leaves grown under different irradiances. Photosynthetica, 391115
6. R. K. Behera, P. C. Mishra, N. K. Choudhury, 2002High irradiance and water stress induce alterations in pigment composition and chloroplast activities of primary wheat leaves.J. Plant Physiol., 159967973

7. S. Bhattacharyya, B. Nayak, N. K. Choudhury, 2011Response of Diazotropic cyanobacterium Nostoc carneum under pesticide and UV-B stress. Chemosphere 84131135

8. L. O. Bjorn, G. C. Papageorgiou, R. E. Balnkenship, Govindjee, 2009A view Point Why Chl a ? photosynth Res. 998598

9. D. Campbell, H. Vaughan, K. C. Adrian, G. Petter, O. Gunnal, 1998Chlorophyll fluorescence analysis of cyanobacterial photosynthesis and acclimation.Microbiol. Mol. Biol. Rev., 62667683

10. Carvalho F P.2006Agriculture, Pesticides, Food Security, and Food Safety, Encironmental science and policy, Elsevier,9685692

11. F. P. Carvalho, S. W. Fowler, J. Villeneuve, P. , M. Horvat, 1997Pesticide residues in the marine environment and analytical quality assurance of the results. In: Proceedings of an International FAO/IAEA Symposium on the Environmental Behaviour of Crop Protection Chemicals. IAEA, Vienna, 3557

12. Z. Chen, P. Juneau, B. Qia, 2007Effect of three pesticides on the growth, photosynthesis and phtoinhibition of the edible cyanobacterium Nostoc (Ge-Xian-Mi). Aqu. Toxicol., 81256265

13. N. K. Choudhury, H. T. Choe, 1996Photoprotective effect of kinetin on pigment content and photochemical activities of wheat chloroplasts aging in vitro.Biol. Plant., 386169

14. M. K. Das, S. P. Adhikary, 1996Toxicity of three pesticides on several rice field cyanobacteria. Trop. Agric. (Trinidad), 73155157

15. B. Demming-Adams, O. Bjorkman, 1987Comparison of the effect of excessive light on chlorophyll fluorescence (77k) and photon yield of O2 evolution in levels of higher plants. Planta., 171171184

16. P. M. Deo, B. Biswal, 1998Water stress induced alterations in chloroplast activity and nitrogen metabolism during development of clusterbean cotyledons in low lightintensity. In: Garab G. (Ed.), Photosynthesis: Mechanism and Effects, Vol.IV, Kluwer Academic Publishers, Dordrecht, Netherland, 25412544

17. S. Falk, G. Samuelsson, 1992Recovery of photosynthesis and photosystem II fluorescence in Chlamydomonas reinhardtii after exposure to three levels of high light. Physiol. Plant., 856168

18. P. Fay, W. D. P. Stewart, A. E. Walsby, G. E. Fogg, 1968Is The Heterocyst Is The Site Of Nitrogen Fixation In Blue Green Algae? Nature 220810812

19. P. Fay, 1983The Blue greens (Cyanophyta-Cyanobacteria) First Published in Great Britain,188

20. E. A. Fernández-Valiente, U. A. Quesada, F. Leganés, R. Careres, 2000Contribution of N Fixing Cyanobacteria to Rice Production: Availability of Nitrogen from N-Labelled Cyanobacteria and Ammonium Sulphate to Rice.Plant Soil, 211107112
21. D. Goyal, P. Roychoudhury, B. D. Kaushik, 1994Effect of two new herbicides on the growth and nitrogen fixation in Anabaena and Tolypothisx. Acta Bot. Indica, 192528
22. W. I. Gruszescki, K. Veeranjaneyulu, B. Zelent, R. M. Leblane, 1991Energy transfer process during senescence: Fluorescence and phtoacoustic studies of intact pea leaves. Biochim. Biophys. Acta, 1056173180
23. O. Hammouda, 1999Response of the paddy field cyanobacterium Anabaena doliolum to carbofuran. Ecotoxicol. Environ. Saf., 44215219
24. E. Hideg, I. Vass, 1996UV-B induced free radical production in plant leaves and isolated thylakoid membranes. Plant Sci., 115251260
25. B. R. Jordan, 1996The effects of ultraviolet-B radiation on plants: A molecular perspective. Adv. Biol. Res., 2297162
26. K. Kapoor, L. Arrora, 1998Ecophysiological observations on the influence of agrochemicals on growth and metabolism of Anabaena doliolum Bhardwaja. Ecol. Env. Cons., 41317
27. K. Kapoor, L. Arrora, 2000aComparative studies on the effect of herbicides on nitrogen fixing cyanobacteria Cylindrospermum majukutz. Indian J. Env. Sci., 48996
28. K. Kapoor, L. Arrora, 2000bInfluence of some pesticides on cyanobacterium in vitro conditons. Indian J. Env. Ecoplanning, 3219226
29. M. S. D. Kaur, A. S. Ahluwalia, 1997Response of diazotrophic cyanobacteria to Butachlor. Phykos, 3693101
30. Kim, J.D. and Lee C.G.2006Differential response of two freshwater cyanobacteria, Anabaena variabilis and Nostoc commume to sulfonylurea herbicide bensulfuron-methyl. J. Micorbiol. Biotechnol., 165256
31. G. H. Krause, 1988Photoinhibition of photosynthesis- An evaluation of damaging and protective mechanisms. Physiol. Plant, 74566574
32. G. H. Krause, E. Weis, 1991Chlorophyll fluorescence and photosynthesis: The basics. Annu. Rev. Plant Physiol, Plant Mol. Biol., 42313349
33. A. Krieger, I. Moyo, E. Weis, 1992Energy dependent quenching of chlorophyll a fluorescence: Effect of pH on stationary fluorescence and picosecond relaxation kinetics in thylakoid membranes and PS II preparation. Biochim. Biophys. Acta, 1102167176

34. A. Kumar, H. D. Kumar, 1998Nitrogen fixation by blue-green algae. In: S.P. Sen (Ed.), Proceeding of Plant Physiological Research, 1st International Congress of Plant Physiology, Society for Plant Physiology and Biochemistry, New Delhi, India, 85103
35. S. Kumar, K. Habib, T. Fatma, 2008Endosulfun induced biochemical changes in nitrogen-fixing cyanobacteria. The Science of the Total Environment, (ELSEVIER). 403130138
36. N. J. I. Kumar, A. Bora, M. K. Amb, R. N. Kumar, 2011An evaluation of esticide Stress Induces Proteins in Three Cyanobacterial Species-Anabaena fertilissima, Aulosira fertilissima and Westillopsis prolifica Using SDS-PAGE. Advances in Environmental Biology 54739745
37. S. Lal, D. M. Saxena, 1980Cytological and biochemical effects of pesticides on microorganisms. Residue Rev., 734986
38. S. P. Long, S. Humphries, 1994Photoinhibition of photosynthesis in nature. Annu. Rev. Plant Physiol. Plant Mol. Biol., 45633662
39. G. Mackinney, 1941Absorption of light by chlorophyll solution.J. Biol. Chem., 140315322
40. N. T. D. Marsac, J. Houmard, 1993Adaptation of cyanobacteria to environmental stimuli: New step towards molecular mechanism.FEMS Microbiol. Rev., 104119190
41. M. Megharaj, K. Venkateswarlu, R. Naidu, 2011Effects of carbaryl and 1-Napthol on soil Population of Cyanobacteria and Microalgae and select cultures of diazotropic cyanobacteria Bull Environ Contam Toxicol 87324329
42. U. Mishra, S. Pabbi, 2004Cyanobacteria: a potential biofertilizer for rice. Resonance, 610
43. D. E. Moreland, 1980Mechanism of action of herbicides.Annu. Rev. Plant Physiol., 31597638
44. C. B. Osmond, 1981Photorespiration and Photoinhibition. Biochim. Biophys. Acta, 6397798
45. D. Panda, D. N. Rao, S. G. Sharma, R. J. Strasser, R. K. Sarkar, 2006Submergence effects on rice genotypes during seedling stage: Probing of submergence driven changes of phtosystem 2 by chlorophyll a fluorescence induction O-J-I-P transients. Phtosynthetica, 4417
46. M. Panda, 1999Growth response of blue-green algae to paper mill waste water: Its possible role in pollution abatement, Ph.D. Thesis, Sambalpur University, Sambalpur, India.

47. S. Panda, U. C. Biswal, 1989Aging induced changes in thylakoid membrane organization and photoinhibition of pigments. Photosynthetica, 23507516
48. S. Panda, U. C. Biswal, 1990Effect of magnesium and calcium ions on photoinduced lipid peroxidation and thylakoid breakdown of cell-free chloroplast. Indian J. Biochim. Biophys., 27159163
49. G. Papageorgiou, 1975Chlorophyll fluorescence: An intrinsic probe of photosynthesis. In: Govindjee (Ed.), Bioenergetics of Photosynthesis, Academic Press, New York, 319371
50. G. A. Peschek, C. Obinger, D. M. Sherman, L. A. Sherman, 1994Immunocytochemical localization of the cytochrome-c xidase in a cyanobacterium, Synechococcus PCC7942 (Anacystis nidulans). Biochim. Biophys. Acta 1187369372
51. S. B. Powles, 1984Photoinhibition of photosynthesis induced by visible light. Annu. Rev. Plant Physiol., 351519
52. B. Ramakrishnan, M. Megharaj, K. Venkateswarlu, R. Naidu, N. Scthunathan, 2010The impact of Environmental pollutants on microalgae and cyanobacteria. Crit. Rev Environ Sci Technol 40699821
53. B. Rath, S. P. Adhikary, 1994Relative tolerance of several nitrogen fixing cyanobacteria to commercial grade furadan. Ind. J. Expt. Biol., 32213215
54. R. Rippka, J. Deruelles, J. B. Waterbuy, M. Herdman, R. Y. Stanier, 1979Genetic assignments, strain histories and properties of pure cultures of cyanobacteria. J. Gen. Microbiol., 111161
55. V. Rodrigues, R. Bhandari, J. P. Khurana, P. K. Sharma, 2007Movement of chloroplasts in mesophyll cells of Garcinia indica in response to UV-B radiation. Current Science, 9216101613
56. Shikha and Singh D.P.2004Influence of glyphosphate on photosynthetic properties of wild type and mutant strains of cyanobacterium Anabaena doliolum. Current Science, 86571576
57. N. K. Swain, N. K. Choudhury, M. K. Raval, U. C. Biswal, 1990Differential changes in fluorescence characteristics of photosystem II rich grana fraction during ageing in light and dark. Photosynthetica, 24135142
58. J. Xia, 2005Response of growth, photosynthesis and photoinhibition of the edible cyanobacterium N. sphaeroids colonies to thiobencarb herbicide. Chemosphere, 59561566

59. H. Ying-Yu, D. P. Hader, 2002UV-B induced formation of ROS and oxidative damage of the cyanobacterium Anabaena sp.: Protective effects of ascorbic acid and N-acetyl-C-cystein. J. Photochem. Photobiol., 66115124

CITATION

CHAPTER 1

Davor Zeljezic and Marin Mladinic (2011). Novel Approaches in Genetic Toxicology of Pesticides by Applying Fluorescent in Situ Hybridization Technique, Pesticides - The Impacts of Pesticides Exposure, Prof. Margarita Stoytcheva (Ed.), ISBN: 978-953-307-531-0, InTech, DOI: 10.5772/13426.

CHAPTER 2

Neiva Knaak, Diouneia Lisiane Berlitz and Lidia Mariana Fiuza (2012). Toxicology of the Bioinsecticides Used in Agricultural Food Production, Histopathology - Reviews and Recent Advances, Dr. Enrique Poblet (Ed.), ISBN: 978-953-51-0866-5, InTech, DOI: 10.5772/52070.

CHAPTER 3

Danieli Benedetti, Fernanda Rabaioli Da Silva, Kátia Kvitko, Simone Pereira Fernandes and Juliana da Silva (2014). Genotoxicity Induced by Ocupational Exposure to Pesticides, Pesticides - Toxic Aspects, Dr. Sonia Soloneski (Ed.), ISBN: 978-953-51-1217-4, InTech, DOI: 10.5772/57319.

CHAPTER 4

A. Sassolas, B. Prieto-Simón and J. Marty, "Biosensors for Pesticide Detection: New Trends," American Journal of Analytical Chemistry, Vol. 3 No. 3, 2012, pp. 210-232. doi: 10.4236/ajac.2012.33030.

CHAPTER 5

Sorin Avram, Simona Funar-Timofei, Ana Borota, Sridhar Rao Chennamaneni, Anil Kumar Manchala, and Sorel Muresan (2014). Quantitative estimation of pesticidelikeness for agrochemical discovery. Journal of Cheminformatics 2014 6:42. doi:10.1186/s13321-014-0042-6

CHAPTER 6

U. Acharya, P. Subedi and K. Walsh, "Evaluation of a Dry Extract System Involving NIR Spectroscopy (DESIR) for Rapid Assessment of Pesticide Contamination of Fruit Surfaces," American Journal of Analytical Chemistry, Vol. 3 No. 8, 2012, pp. 524-533. doi: 10.4236/ajac.2012.38070.

CHAPTER 7

Rachid Rouabhi (2010). Introduction and Toxicology of Fungicides, Fungicides, Odile Carisse (Ed.), ISBN: 978-953-307-266-1, InTech, DOI: 10.5772/12967.

CHAPTER 8

Sonia Soloneski and Marcelo L. Larramendy (2012). Genetic Toxicological Profile of Carbofuran and Pirimicarb Carbamic Insecticides, Insecticides - Pest Engineering, Dr. Farzana Perveen (Ed.), ISBN: 978-953-307-895-3, InTech, DOI: 10.5772/30137.

CHAPTER 9

M. Butinof, R. Fernández, M.J. Lantieri, M.I. Stimolo, M. Blanco, A.L. Machado, G. Franchini, M. Gieco, M. Portilla, M. Eandi, A. Sastre, and M.P. Diaz (2014). Pesticides and Agricultural Work Environments in Argentina, Pesticides - Toxic Aspects, Dr. Sonia Soloneski (Ed.), ISBN: 978-953-51-1217-4, InTech, DOI: 10.5772/57178.

CHAPTER 10

Binata Nayak, Shantanu Bhattacharyya and Jayanta Sahu (2012). Photosynthetic Response of Two Rice Field Cyanobacteria to Pesticides, Pesticides - Advances in Chemical and Botanical Pesticides, Dr. R.P. Soundararajan (Ed.), ISBN: 978-953-51-0680-7, InTech, DOI: 10.5772/46232.

INDEX

A

Acetylcholinesterase 87, 107
Acetylthiocholine 87
Agricultural Production 231, 233
Anticarsiagemmatalis 37
Aptamers 86, 103, 123, 124
Artificial Neural Networks (Anns) 86
Azoxystrobin 183, 184

B

Bacillus Thuringiensis 35, 38, 48, 51, 52, 53, 54, 55
Base Excision Repair (Ber) 68
Biofungicide 182, 200, 205
Biorecognition Element 85, 86
Biotinylated Proteins 42

C

Carbamates 58, 208, 209, 225
Carbofuran 209, 210, 211, 212, 215, 219, 221, 222, 223, 224, 225, 226, 228, 229, 230
Central Nervous System (Cns) 63
Chloradane 1
Cholinesterase 87, 89, 93, 105
Chromatographic Methods 86
Chromosomal Aberrations (Ca) 66
Cultivated Species 232
Cumulative Exposure Index (Cei) 246
Cyanobacteria 265, 268, 270, 271, 275, 281, 282, 284, 285, 286, 287
Cyanobacterial Cells 269
Cyanogenic Glycosides 36

D

Dichlorovos 90
Dithiocarbamate 163
Dithiocarbamates 1, 12, 13, 58

E

Ecoguard 182
Electrochemical Transducers 86
Environmental Protection Agency (Epa) 2, 4
Enzymatic Detection 87
Exposure 289

F

Fluorochrome 42, 43
Food Production 35
Fungicides 181, 182, 184, 185, 186, 190, 191, 194, 196, 198, 200, 204, 206

G

Gas Chromatography 162
Green Revolution 265
Green Tobacco Sickness\" (Gts) 69

H

H-Bond Acceptors (Hba) 135
Heterocystous Cyanobacteria 268, 270
Hexachlorocyclohexane 1
Homogeneous Ecological Areas (Heas) 234, 252
Hours After Treatment (Hat) 45

I

Integrated Pest Management (Ipm 35, 44
International Agency For Research On Cancer (Iarc) 4

L

Liquid Chromatography 162
Logistics V

M

Methiocarb 209
Methomyl 209
Molecular Imprinted Polymers (Mips) 86
Molecular Weight (Mw) 135, 150
Mouse Lymphoma Assay (Mla) 7

N

Near Infrared Spectroscopy 162, 163, 165, 169, 170, 171, 173
N-Methyl Carbamate 209, 210, 224, 229

O

Organochlorine 266, 268
Organochlorines 58, 208
Organophosphorous 1, 19, 27, 58, 83
Organophosphorus Pesticide Residues 266
Oxamyl 209

P

Pendimethalin 3
Personal Protective Equipment (Ppe) 235
Pesticides 57, 58, 59, 65, 66, 75, 77, 79, 80, 83
Phenoxyacetates 58
Pirimicarb 209, 215, 216, 218, 219, 221, 228, 229
Polarization 269, 280
Product Quality 231, 256
Propoxur 209, 226
Public Health 231, 258
Pyraclostrobin 183
Pyrethrium 1

Q

Quantitative Estimate Of Drug-Likeness (Qed) 135
Quantitative Estimates Of Fungicide- (Qef) 133
Quantitative Estimates Of Herbicide- (Qeh) 133

Quantitative Estimates Of Insecticide- (Qei) 133
Quantitative Estimates Of Pesticide-Likeness (Qep) 133, 150
Quartz Crystal Microbalance (Qcm) 88

R

Reactive Oxygen Species (Ros) 59
Receiver Operating Curve (Roc) 151
Rice Agroecosystem 268, 270
Rotatable Bonds (Rb) 135, 150

S

Sister Chromatid Exchange (Sce) 67
Spodoptera Frugiperda 37, 46, 47, 52, 53, 54
Supply Chain V
Support Vector Machine (Svm) 162
Synthetic Pyrethroids 58

T

Technological Environments 232
Thiocarbamates 58, 65
Thiodicarb 209
Trichoderma Harzianum 182
Trifloxistrobin 183